12
Springer Series on Fluorescence

Methods and Applications

Series Editor: O.S. Wolfbeis

For further volumes:
http://www.springer.com/series/4243

Springer Series on Fluorescence

Series Editor: O.S. Wolfbeis

Recently Published and Forthcoming Volumes

Fluorescent Proteins II
Application of Fluorescent Protein Technology
Volume Editor: G. Jung
Vol. 12, 2012

Fluorescent Proteins I
From Understanding to Design
Volume Editor: G. Jung
Vol. 11, 2012

Advanced Fluorescence Reporters in Chemistry and Biology III
Applications in Sensing and Imaging
Volume Editor: A.P. Demchenko
Vol. 10, 2011

Advanced Fluorescence Reporters in Chemistry and Biology II
Molecular Constructions, Polymers and Nanoparticles
Volume Editor: A.P. Demchenko
Vol. 9, 2010

Advanced Fluorescence Reporters in Chemistry and Biology I
Fundamentals and Molecular Design
Volume Editor: A.P. Demchenko
Vol. 8, 2010

Lanthanide Luminescence
Photophysical, Analytical and Biological Aspects
Volume Editors: P. Hänninen and H. Härmä
Vol. 7, 2011

Standardization and Quality Assurance in Fluorescence Measurements II
Bioanalytical and Biomedical Applications
Volume Editor: Resch-Genger, U.
Vol. 6, 2008

Standardization and Quality Assurance in Fluorescence Measurements I
Techniques
Volume Editor: U. Resch-Genger
Vol. 5, 2008

Fluorescence of Supermolecules, Polymeres, and Nanosystems
Volume Editor: M.N. Berberan-Santos
Vol. 4, 2007

Fluorescence Spectroscopy in Biology
Volume Editor: M. Hof
Vol. 3, 2004

Fluorescence Spectroscopy, Imaging and Probes
Volume Editor: R. Kraayenhof
Vol. 2, 2002

New Trends in Fluorescence Spectroscopy
Volume Editor: B. Valeur
Vol. 1, 2001

Series Editor

Prof. Dr. Otto S. Wolfbeis

Institute of Analytical Chemistry
Chemo- and Biosensors
University of Regensburg
93040 Regensburg
Germany
otto.wolfbeis@chemie.uni-regensburg.de

Aims and Scope

Fluorescence spectroscopy, fluorescence imaging and fluorescent probes are indispensible tools in numerous fields of modern medicine and science, including molecular biology, biophysics, biochemistry, clinical diagnosis and analytical and environmental chemistry. Applications stretch from spectroscopy and sensor technology to microscopy and imaging, to single molecule detection, to the development of novel fluorescent probes, and to proteomics and genomics. The *Springer Series on Fluorescence* aims at publishing state-of-the-art articles that can serve as invaluable tools for both practitioners and researchers being active in this highly interdisciplinary field. The carefully edited collection of papers in each volume will give continuous inspiration for new research and will point to exciting new trends.

Preface

A plethora of reviews, popular science books, and scientific textbooks have been written on the significance of fluorescent proteins in the life sciences. More than 30,000 references can be found in bibliographic databases which refer to at least one among the members of this protein family (see Fig. 1). Most of these narrate on how fluorescent proteins may be used to label gene products, how they may be visualized in cellular compartments by fluorescence microscopy, or how they may be expressed in individual cells, thus provoking novel findings in ontogenesis. In most of the experiments described, fluorescent proteins are being exploited as miniaturized light bulbs, the length scale is that of microns, and the time scale is that of seconds or longer. There is no doubt that fluorescent protein technology has revolutionized life sciences in that proteins have become universal and standard tools in molecular biology laboratories.

A minor fraction of roughly 5% of all publications deals with the *nanoscopic* properties of fluorescent proteins (FPs) acting as light bulbs. Early achievements include the crystallographic analysis of their molecular structure [1, 2], the discovery of excited-state proton transfer in the naturally occurring FP [3, 4], and the erratic light emission of individual members of FPs [5, 6]. Especially the last experiments, along with low temperature studies [7, 8], have revealed that FPs exhibit a tremendous heterogeneity in terms of structure and dynamics.

It is therefore not astonishing that FPs have had a large impact on other areas of biophysical research, e.g., in studies on protein folding [9–11]. However, the irregular emission of light by FPs also has impacted experiments in the life sciences: most operators of fluorescent protein technology, whom I was talking to, were concerned about weird experimental features like rapid initial fading in time-lapse microscopy, sometimes with sudden fluorescence recovery, or changing FRET-ratios upon continuous illumination. Such annoying findings can be traced back to the wealth of light-driven processes in the proteins, and I am quite sure that more surprises of that kind have been experienced by others. It should be emphasized here that such "strange" photodynamics have initiated seminal studies on protein diffusion and high-resolution microscopy [12–14].

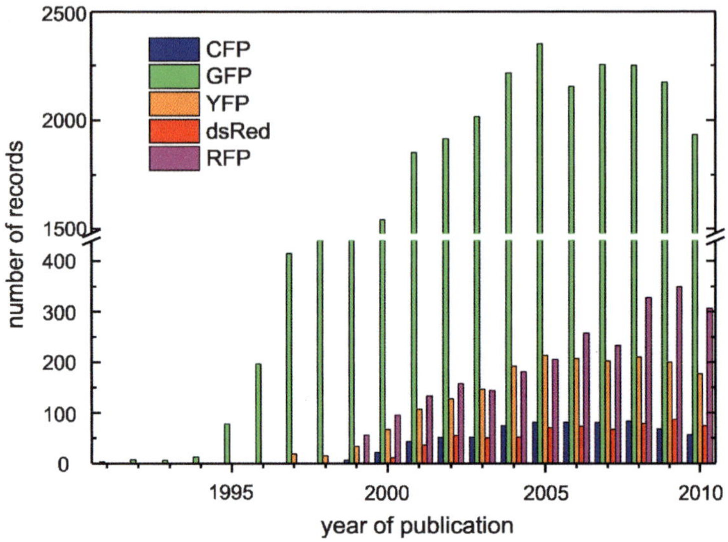

Fig. 1 Number of references related to fluorescent proteins (database: web-of-science). The number of articles dealing with Green Fluorescent Protein has reached saturation now at a level of typically 2,000 articles per year for almost a decade. Those on the Red Fluorescent Proteins are still increasing

Volumes 11 and 12 of the *Springer Series on Fluorescence* deal with various aspects of fluorescent proteins. The first volume (*Fluorescent Proteins I*) is devoted to the molecular, i.e., mainly optical, properties of fluorescent proteins. In the first part, the primary processes leading to fluorescence are discussed: excitation, relaxation, and other processes in the excited state and in emission. Fluorescence proteins are treated as "ordinary" fluorophores, and one article is highlighting our opportunities to circumvent the synthetic limitations given by nature. The second part focuses on the mechanisms that make the difference to conventional fluorophores: isomerization, protonation, as well as reversible and irreversible photochemical reactions. The knowledge on how these processes are affected by the surrounding of the FP allows for tailoring it with respect to spectacular applications, applications that are not conceivable with "ordinary" fluorophores.

In the second volume (*Fluorescent Proteins II*), the key aspect is on applications. Its first part is giving an overview on how many unconventional photophysical properties latently exist in naturally occurring and how double-resonance experiments enable the information to be extracted from microscopy data in an unprecedented way. More on high-resolution microscopy will be found in forthcoming volumes of this series. Quantitation, a central objective of analysis, is the comprehensive caption of the articles in the next part. We may state, justifiably, that researchers have reliable tools at hand to quantify some of the most abundant ions after more than a decade of development. Other physiological parameters of overwhelming importance like the transmembrane potential still need to experience this development. The last part reports on three examples of utmost biological relevance and how ultrasensitivity in bioanalysis, i.e., single-molecule technology, is merged with FP technology. This combination has resulted in an understanding of processes on a molecular level and in detection limits that were not even thought of some 15 years ago.

A preface is also always the occasion to deeply acknowledge the support by others. First of all, I have to thank my family who tolerated my commitment to this experience. I also would like to express my thanks to my coworkers, to my colleagues, and to the representatives of Saarland University for their understanding. In times of growing competitiveness in many academic areas, it is not self-evident to dedicate a substantial amount of time to such a book project. For the same reason, I especially appreciate the immense work of all authors of these two volumes who are all passionate, but busy scientists and who (more or less) voluntarily spared no pains to complete their manuscripts in a wonderful and highly professional way. By now, it also may be appropriate to apologize for my e-mail bombardments!

Saarbrücken, Germany Gregor Jung

References

1. Ormö M et al (1996) Science 273:1392–1395
2. Yang F et al (1996) Nat Biotechnol 14:1246–1251
3. Chattoraj M et al (1996) Proc Natl Acad Sci USA 93:8362–8267
4. Lossau H et al (1996) Chem Phys 213:1–16
5. Dickson R et al (1997) Nature 388:355–358
6. Pierce D et al (1997) Nature 388:338
7. Creemers T et al (1999) Nat Struct Biol 6:557–560
8. Seebacher C et al (1999) J Phys Chem B 103:7728–7732
9. Craggs T (2009) Chem Soc Rev 38:2865–2875
10. Hsu S et al (2009) Chem Soc Rev 38:2951–2965
11. Mickler M et al (2007) Proc Natl Acad Sci USA 104:20268–20273
12. Yokoe E, Meyer T (1996) Nat Biotech 14:1252–1256
13. Patterson G, Lippincott-Schwartz J (2002) Science 297:1873–1877
14. Betzig E et al (2006) Science 313:1642–1645

Contents

Part I
Fluorescence Microscopy Beyond Imaging

Fluorescent Proteins: Nature's Colorful Gifts for Live Cell Imaging

Jörg Wiedenmann, Cecilia D'Angelo, and G. Ulrich Nienhaus

Abstract Fluorescence of marine organisms has fascinated researchers since the early twentieth century. The successful application of the green fluorescent protein (GFP) from the jellyfish *Aequorea victoria* in 1994 as genetically encoded marker resulted in a massive increase in interest for naturally fluorescent proteins. Methods are now established that allow the fast isolation of new genes encoding GFP-like proteins from marine creatures, resulting in an impressive array of glowing proteins with different biochemical and optical properties. Protein engineering has been applied to render natural variants into advanced optical tools for live cell imaging, promoting studies of protein localization and movement, gene activity, sensing of intra- and extracellular condition, and tracking of whole cells and organisms. Finally, photoactivatable proteins were discovered that enable pulse-chase experiments and live cell imaging of proteins with a resolution beyond the diffraction barrier of optical microscopy. Phylogenetic sequence analyses revealed interesting details about the molecular evolution of these proteins including the convergent evolution of colors. Marine organisms, especially corals, still harbor a huge number of GFP-like pigments, the majority of which are yet to be studied. Consequently, further important discoveries of useful marker proteins can be expected in the future.

J. Wiedenmann (✉)
National Oceanography Centre Southampton, University of Southampton, Southampton SO14 3ZH, UK
e-mail: joerg.wiedenmann@noc.soton.ac.uk

C. D'Angelo
Institute of General Zoology and Endocrinology, University of Ulm, 89081 Ulm, Germany

G.U. Nienhaus
Institute of Applied Physics and Center for Functional Nanostructures, Karlsruhe Institute of Technology (KIT), Wolfgang-Gaede-Str. 1, 76131 Karlsruhe, Germany
and
Department of Physics, University of Illinois at Urbana-Champaign, 1110 West Green Street, Urbana, IL 61801, USA

G. Jung (ed.), *Fluorescent Proteins II*, Springer Ser Fluoresc (2012) 12: 3–34,
DOI 10.1007/4243_2011_21
Published online: 28 April 2011

Keywords Anthozoa · Color morph · Coral · dsRed · EosFP · Expression · Fluorescent protein · Function · GFP-like protein · Green fluorescent protein · IrisFP · Kaede · Light · Live cell imaging · mCherry · Mechnism · mRuby · Photoprotection · Red fluorescent protein · Reef corals · Regulation

Contents

1 Introduction

The functional expression of the green fluorescent protein (GFP) in a nematode worm launched the era of live cell imaging and opened up new horizons for biomedical research [1]. Initially discovered during studies of jellyfish bioluminescence by Osamu Shimomura, the unusual biochemical properties soon made GFP from *Aequorea victoria* (avGFP), an indispensable tool for cell biology [2]. The protein can be expressed in its functional form in virtually any type of cell, facilitating the use of GFP as a genetically encoded marker of gene activity or for tracking of proteins in living cells [2]. The outstanding impact of GFP technology on life sciences research was recognized by the award of the Nobel Prize in Chemistry 2008 to Osamu Shimomura, Martin Chalfie, and Roger Tsien for the "discovery and development of the green fluorescent protein, GFP". Multicolor labeling was enabled by the generation of blue and yellow variants, and GFP-based sensor systems were developed that report changes in both intracellular and extracellular conditions. The discovery of GFP-like proteins in non-bioluminescent sea anemones and related organisms gave access to a great variety of homologous proteins with novel optical properties [3–6]. The gene hunt in the oceans resulted in novel variants including red fluorescent and photoactivatable proteins [7–9]. The evolution of the diverse spectroscopic properties of GFP-like proteins was analyzed using molecular biology and bioinformatics tools [10]. The phylogenetic tree of GFP-like proteins shows a clustering of optical features in certain taxonomic groups. This knowledge can be exploited in targeted searches for novel lead structures. Here, we outline the history of the development of the fluorescent marker protein technology, introducing marine organisms as source of novel marker proteins that enable fascinating live cell imaging applications.

2 Natural Sources of Fluorescent Proteins

2.1 History of Fluorescent Protein Research

The striking phenomenon of cnidarian bioluminescence in the marine realm was first described by Pliny the Elder (first century AD) and by Claudius Aeliani (second century AD) [11]. About eighteen centuries passed until the GFP of the hydromedusae *A. victoria* (avGFP) emanated as a "by-product" from studies of bioluminescence [12].

In 1925, Harvey observed the appearance of bluish fluorescence in previously nonfluorescent, light-emitting tissue upon stimulation of luminescence of the ctenophore *Mnemiopsis* [13]. The fluorescence was probably emitted by a substance similar to the blue fluorescent protein that appears as an intermediate in the luminescence reaction of the photoprotein aequorin isolated from *A. victoria* [12, 14, 15]. A yellow-green fluorescence was observed in the luminescent tissue of *Aequorea* and *Halistaura* [16]. During isolation and characterization of the photoprotein aequorin, Shimomura et al. identified the green fluorescent pigment of *A. victoria* as a protein [12].

FPs were also described for the hydroid *Obelia*, the hydromedusae *Aequorea*, and the pennatullacean *Renilla* [17–19]. In vitro, the bioluminescence reaction of these species produces bluish light with broad emission spectra, with maxima between 460 and 486 nm [18, 19]. However, the in vivo luminescence showed narrow peaks with maximal emission at 508 nm, matching the fluorescence emission spectrum of GFPs. The absence of the blue emission during the in vivo luminescence reaction indicates a non-radiative energy transfer between the light-generating proteins (luciferases, aequorin) and GFP [18]. GFPs or tissue fluorescence peaking around 508 nm was identified in numerous bioluminescent hydromedusae, hydropolyps, and pennatularians [11, 18–26]. GFPs with shorter emission wavelengths were found in *Halistaura* (497 nm) and *Phialidium* (498 nm) [26, 27]. In these bioluminescent cnidarians, GFPs were exclusively found in the photogenic cells [11, 24, 28].

The imidazolone structure of GFP was proposed by Shimomura in 1979 and later confirmed by Cody et al. [29]. During these early years of GFP research, details about the biochemical and optical properties also became available [27].

Finally, Prasher and coworkers determined the amino acid sequence of avGFP in 1992 [30]. The application potential of avGFP was fully realized when Chalfie and coworkers achieved the functional expression in the nematode worm *Caenorhabditis elegans* [1]. Their utility as genetically encoded marker is enabled by the autocatalytic formation of the chromophore in the presence of molecular oxygen [31–33].

The possibility to produce avGFP in unlimited quantities in recombinant systems stimulated further research on the biochemical and photophysical properties [2]. The molecular structure of avGFP was resolved by X-ray crystallography [34, 35], which enabled rational approaches to molecular engineering of the protein. Finally, the possibility to alter the amino acid sequence by mutagenesis techniques opened the opportunity to customize GFP for imaging applications. These studies yielded, for instance, blue- and yellow-shifted emitters useful for multicolor labeling [2].

Today, it is well established that marine cnidarians host a variety of GFP-like protein pigments. However, already before the recent systematic studies, these pigments attracted sporadic interest from researchers. In 1927, UV-induced green fluorescence was demonstrated for a sea anemone from a rock pool in Great Britain [36]. Kawaguti noted in 1944 that green pigments of scleractinian corals in Palao exhibited green fluorescence [37].

Red fluorescence from a sea anemone was first observed by Marden [38] during a dive in the Red Sea. At a depth of 20 m, where the red components of the down-welling light are readily attenuated from the spectrum, a sea anemone appeared in bright red. He explained the phenomenon by the presence of red fluorescent pigments excited by blue-green light. Wobber [39] documented the red fluorescence of *Corynactics californicus* by photographing the animals under natural light at a depth of 40 m. A note on the fluorescence of the corals *Montastrea cavernosa* and *Mussa angulosa* was published by Read [40]. Orange and red fluorescence could be induced by exciting the corallimorpharian *C. californicus*, the coral *Balanophyllia elegans*, and a tube anemone *Cerianthus* sp. with ultraviolet, blue or green light [41]. Using UV light for excitation, Catala described the fluorescence of numerous corals [42–45]. Species belonging to 16 genera displayed fluorescence. For example, representatives of the genus *Flabellum* collected at a depth of 35–40 m showed intensive green fluorescence in the fleshy parts, whereas *Trachyphyllia* emitted orange fluorescence. In some specimens, he also observed a change of fluorescence color from green to pink upon prolonged or frequent irradiation with UV light [42]. UV-induced fluorescence was reported also for various corals, corallimorpharians, and actiniarians under irradiation [46].

A chromatophore system containing fluorescent pigments was found in the entodermal layer of the coral *Leptoseris fragilis* [47–50].

Mazel [51–55] provided photographic documentation and spectral characterization of fluorescent pigments of corals, corallimorpharians, and sea anemones from the Caribbean Sea. The pigments could be arranged in four major classes, with emission maxima around 486, 515, 575, and 685 nm. The red emission peaking at 685 nm could be attributed to chlorophyll of the symbiotic algae [54, 55]. Salih et al. [56] found fluorescent morphs among 124 species of 56 genera of Great Barrier Reef corals.

Despite the urgent need for red fluorescent marker proteins, the red-shifted emitters found in non-bioluminescent cnidarians were not considered as potential candidates for two reasons. (1) It was assumed that these pigments represent flavin-like compounds or phycobiliproteins. Both pigment types are products of complex biosynthesis pathways and therefore not suitable as genetically encoded markers. (2) In those days, GFPs were found only as secondary emitters in bioluminescent organisms and, consequently, their existence in non-bioluminescent cnidarians was ruled out.

The GFP-like protein nature of green and red fluorescent and the nonfluorescent pink pigments was realized by Wiedenmann in 1997 [4] and confirmed by the cloning of several FPs with emission colors from cyan to red by Matz et al. [3] and Wiedenmann et al. [5, 6].

The following years yielded numerous natural FPs with novel spectral properties [57–73]. Protein engineering rendered them in even more useful tools. Milestones of the discovery and engineering of fluorescent proteins are outlined in Fig. 1.

2.2 *Marine Organisms as Sources of GFP-Like Properties*

2.2.1 Distribution Among Animal Phyla

As yet, GFP-like proteins have only been isolated from marine organisms (Fig. 2). Most of them belong to the phylum cnidaria [10]. However, green fluorescent homologs were also isolated from the taxa crustacea [65], ctenophora [74], and chordata [75]. The wide distribution suggests that, in principle, any metazoan organism can harbor GFP-like proteins. However, the taxon anthozoa proved to be the most rewarding source for innovative fluorescent marker proteins such as red fluorescent and photoactivatable proteins [7, 9].

2.2.2 Color Morphs

The existence of several morphs with striking color differences is common among many species of reef corals and sea anemones [76–79]. Already in the nineteenth century, numerous color morphs of *Anemonia sulcata* (=viridis) were described [80–82]. Five distinct color morphs of this species can be distinguished based on the presence of four GFP-like proteins in the tentacles [77] (Fig. 3). Also the morphs of the reef coral *M. cavernosa* owe their colors to differing tissue concentrations of cyan, green, and red fluorescent proteins [61, 79]. Interestingly, color morphs of both *A. sulcata* and *M. cavernosa* express the whole collection of pigments characteristic for each species. The color differences result from transcript levels that differ relative to each other among the morphs [61, 83]. This implies that novel GFP-like proteins can be discovered that display colors different from the color of the animal under study. Our studies of the sea anemone *Calliactis parasitica* revealed that red fluorescent proteins can even be cloned from animals that appear to be nonfluorescent ([67]; Gamber and Wiedenmann, unpublished).

2.2.3 Distribution in the Organism

GFP-like proteins can contribute up to 14% to the total soluble cellular proteins in the expressing tissue of some corals [79, 83]. In contrast to bioluminescent cnidarians where the expression of GFPs seems to be restricted to the photogenic tissue,

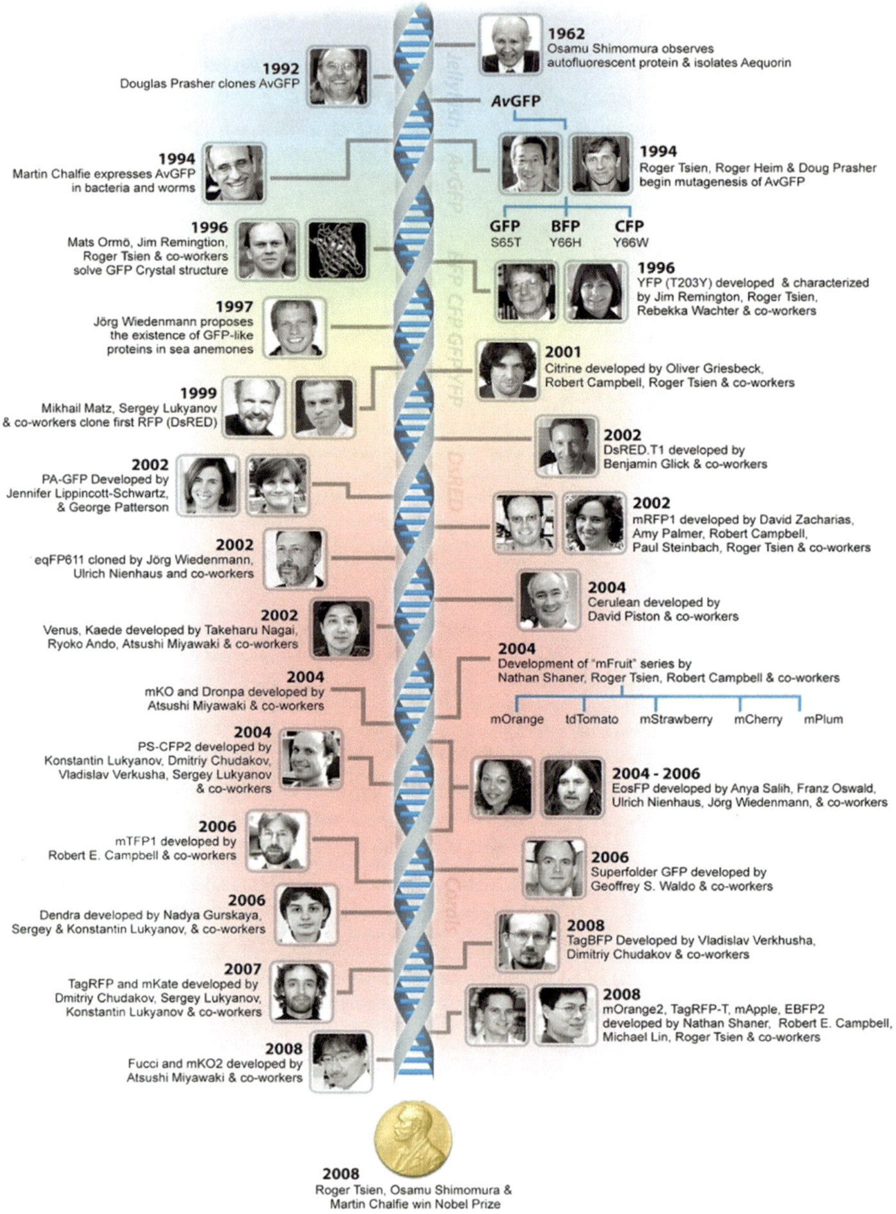

Fig. 1 Timeline of discovery and engineering of fluorescent proteins from cnidarians. This modified version of an image from reference [8] is a kind gift of Michael W. Davidson, Florida State University

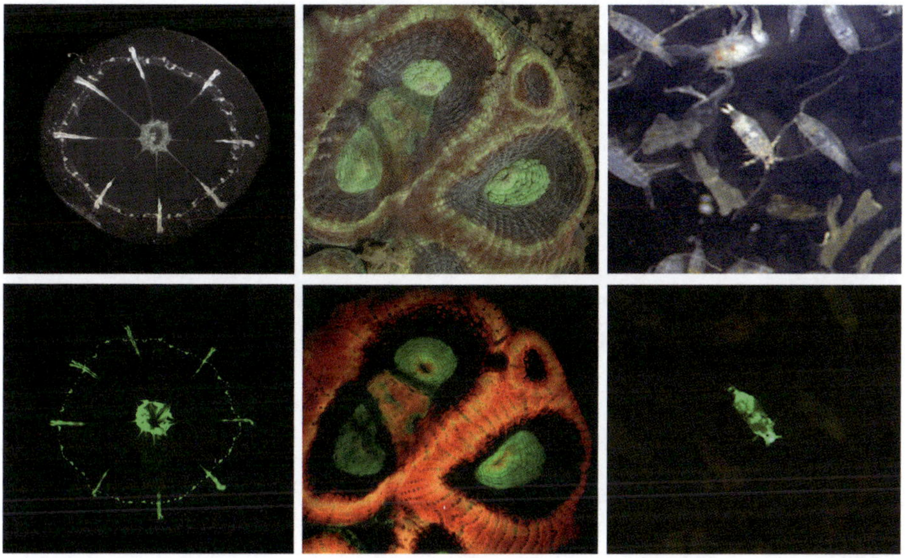

Fig. 2 Marine organisms as sources of fluorescent proteins. The *upper row* shows daylight photographies of a hydrozoan jellyfish (*left*), a faviid coral (*center*) from the Great Barrier Reef and plankton sample (mainly copepods) from the English Channel (*right*). Fluorescence images show the distribution of green and red fluorescent proteins. Only the calanoid copepod in the center of the image contains GFPs

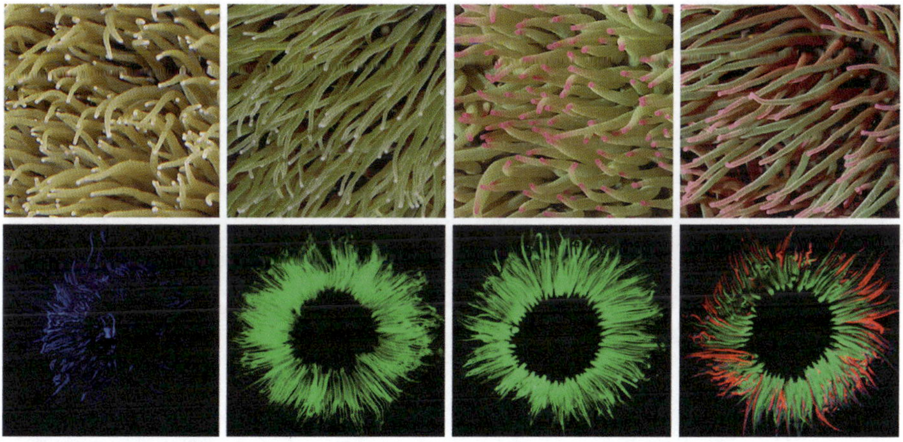

Fig. 3 Color morphs of *Anemonia sulcata* (=*viridis*) are defined by the tentacle content of green fluorescent proteins (asF499; asFP522), a red fluorescent protein (asFP595), and nonfluorescent, pink chromoproteins (asCP562). From *left* to *right*: var. rustica (no detectable content); var. viridis (GFPs); var. smaragdina (GFPs, asCP562); var. rufescence (GFPs, asFP595, asCP562) (not shown: var. vulgaris with asCP562). Animals were photographed under daylight (*upper row*) and UV (366 nm) light (*lower row*)

no uniform picture of the distribution of GFP-like proteins exists in non-bioluminescent anthozoans [11, 24, 28, 84].

Sea anemones and reef corals, in particular, show differing expression patterns of cyan, green, and red fluorescent proteins and nonfluorescent chromoproteins [77, 78, 83, 84]. Nevertheless, a few notable tendencies do exist. In many corals and sea anemones, the GFP-like proteins are found in highest abundance in the light-exposed part of the animals, such as the upper side of a colony or in the tentacles of polyps [77, 85]. Specifically, nonfluorescent chromoproteins are often localized in the tips of tentacles or branches and growth margins of coral colonies [77, 85]. Large differences in the distribution of GFP-like proteins in the tissue of scleractinian corals have also been documented at the microscopic scale.

Peloux [86] demonstrated that fluorescence was emitted from pigment granules localized in the entoderm in *Goniopora lobata*. In contrast, in *Euphyllia picteti*, the pigment granules were found in elongated cells of the ectoderm [86]. The orange-red fluorescent pigment of the solitary coral *Cynarina lacrymalis* appeared to be distributed in a diffuse, nongranular manner in the ectoderm. Kawaguti observed a membrane surrounding the pigment granules of *Lobophyllia robusta* as well as a close proximity of smaller granules with the endoplasmatic reticulum [87]. In the ectodermal cells of *Isophyllia sinuosa*, pigment granules are often surrounded by two membranes [88].

Salih and coworkers found that fluorescent granules were localized mainly above the symbiotic algae in corals from high light habitats, whereas in specimens from low light habitats, they were dispersed among or under the zooxanthellae layer [56]. Both, in *Lobophyllia hemprichii* and *M. cavernosa*, fluorescent pigments can engulf the vacuoles containing the symbiotic algae [79] (Fig. 4).

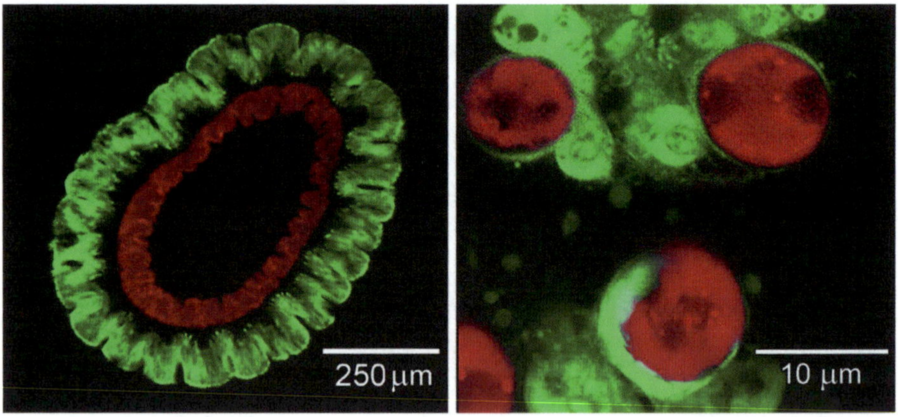

Fig. 4 Microscopic images of the distribution of GFPs in the tissue of anthozoans. The tentacle cross-section of the sea anemone *Anemonia sulcata* (var. smaragdina) (*left*) shows a dominant localization of GFPs in the ectodermal cells. The red fluorescence of the gastrodermal cells is caused by the chlorophyll of the zooxanthellae. In the reef coral *Montastrea cavernosa* (*right*), GFPs are also found in gastrodermal cells surrounding the red fluorescent zooxanthellae. Images modified from references [79, 89]. Copyright Elsevier 2007 and 2007 John Wiley & Sons Ltd

The green and red fluorescent pigments of the sea anemone *A. sulcata* are evenly distributed in the ectodermal cells of the tentacles [89] (Fig. 4).

2.2.4 Regulation of GFP-Like Protein Expression in Cnidarians

In many reef building corals, both fluorescent and nonfluorescent GFP-like proteins are most abundant in light-exposed parts of the colonies. For numerous species, this can be explained by a strong response of FP-encoding genes to the intensity of light experienced by the corals [85]. The blue region of the spectrum proved to be most effective in stimulating the expression of FPs on the transcriptional level. Interestingly, different groups can be distinguished based on their response to changes in the light climate (Fig. 5). Members of the low threshold group are upregulated already at low light intensities; however, under higher light levels, the tissue concentration of these proteins reaches a plateau or even decreases. In contrast, the amount of FPs belonging to the high threshold groups is negligible under low light, but increases nearly proportional to the amount of incident photons [85]. Representatives of a third group express FPs at high levels independent of the light intensity in the habitat [79, 83]. A striking example of light-independent expression of GFPs are tube anemones from the deep sea that are brightly fluorescent in the virtual absence of sunlight [90] (Fig. 5).

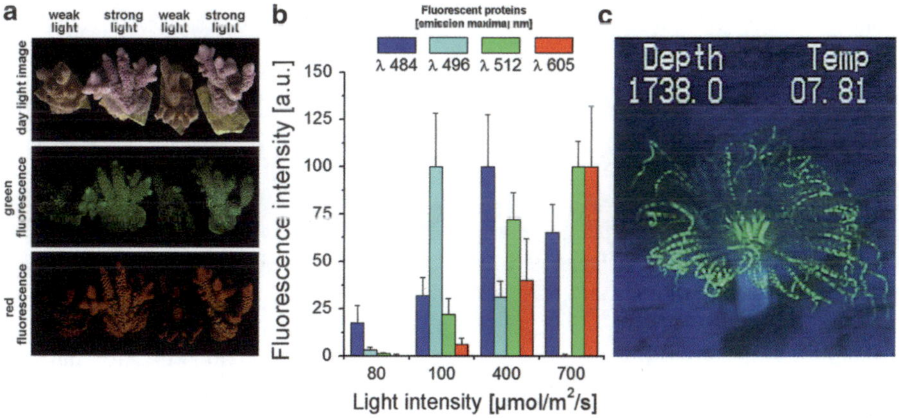

Fig. 5 Differential expression of fluorescent proteins in *Acropora millepora* in response to the light intensity (**a**). Cyan fluorescent proteins (amFP584, amFP496) are downregulated at high light intensities. Tissue concentrations of green and red fluorescent proteins (amFP512; amFP605) increase nearly proportional with the amount of incident light (**b**). The graph shows averages of five replicate measurements. Error bars represent standard deviation. (**c**) Fluorescence image of a tube anemone (ceriantharia) acquired in 530 m depth in the Gulf of Mexico. The depth (ft) and temperature (Temp, °C) are displayed. Panel (**c**) modified from [90] with permission

2.2.5 Spectral Properties of Natural GFP-Like Proteins

Most of the GFPs isolated from bioluminescent hydrozoa and anthozoa, such as *Renilla* or *Aequorea*, show a narrow emission spectrum peaking at ~508 nm and act as secondary emitters in the chemiluminescence reaction [11, 27]. Exceptions are found in the fluorescent proteins from *Phialidium* sp., where both blue-shifted (498 nm) and red-shifted (537 nm) emission maxima were observed and in *Halistaura* GFP which has also a blue-shifted (497 nm) emission maximum [26, 27, 65]. Regarding the position of excitation maxima, considerable differences were detected among GFPs from bioluminescent cnidaria. They were localized at 465 nm (Halistaura GFP), 485 nm (Phialidium GFP), 498 nm (Renilla GFP), and 525 nm (Phialidium YFP) [26, 27, 65]. The excitation spectrum of avGFP is characterized by a major maximum at 398 nm and a side maximum at 475 nm.

The set of fluorescent proteins from non-bioluminescent anthozoa FPs can be grouped in cyan, green, yellow, and red fluorescent proteins (Fig. 6). Cyan fluorescent proteins (CFPs) show excitation maxima of 400–460 nm and emission maxima of 480–486 nm. GFPs usually have excitation maxima between 480 and 518 nm and show emission spectra peaking between 490 and 522 nm. The excitation of some GFPs such as asFP499 show a second, more or less pronounced, maximum at ~400 nm [5]. This band corresponds to the absorption of the neutral chromophore and contributes to the excitation spectrum via excited state proton transfer [91, 92] (Fig. 6).

A GFP (Dronpa) cloned from a pectinidae coral shows an interesting photoswitching behavior [93]; green fluorescence peaking at 518 nm is emitted when Dronpa is irradiated with light around the excitation maximum at 503 nm. However, the protein undergoes photoconversion to a nonfluorescent state with an absorption maximum at 390 nm when being excited with such wavelengths. Conversely, upon excitation of the nonfluorescent form at 390 nm, the green fluorescent state is almost completely restored. Photoswitchable GFPs were also isolated from a deep sea cerianthid and a ctenophore [74, 90].

So far, only one truly yellow fluorescent protein (YFP) has been cloned from non-bioluminescent anthozoans (*Zoanthus* sp.) showing excitation/emission maxima at 528/538 nm [3]. Finally, the group of naturally red fluorescent proteins have emission maxima ranging from 583 to 611 nm. In this group, excitation spectra are characterized by maxima between 558 and 574 nm. The maturation of both yellow and red fluorescent proteins can produce minor amounts of green fluorescent side products [94]. These green fluorescent states have excitation and emission maxima at 475–500 nm and 500–520 nm, respectively [6, 95]. Their contribution to the total fluorescence depends on the protein. In eqFP611, the residual green fluorescence is less than 1% of the red emission [6], while in dsRed the contribution is considerably higher [94–96]. Fluorescence resonance energy transfer (FRET) can be observed if green and red emitting states of GFP-like proteins are present within a single tetramer [64, 95].

A special case is represented by photoconverting proteins such as Kaede or EosFP [59, 64, 65, 79]. Upon expression, these proteins form a green fluorescent

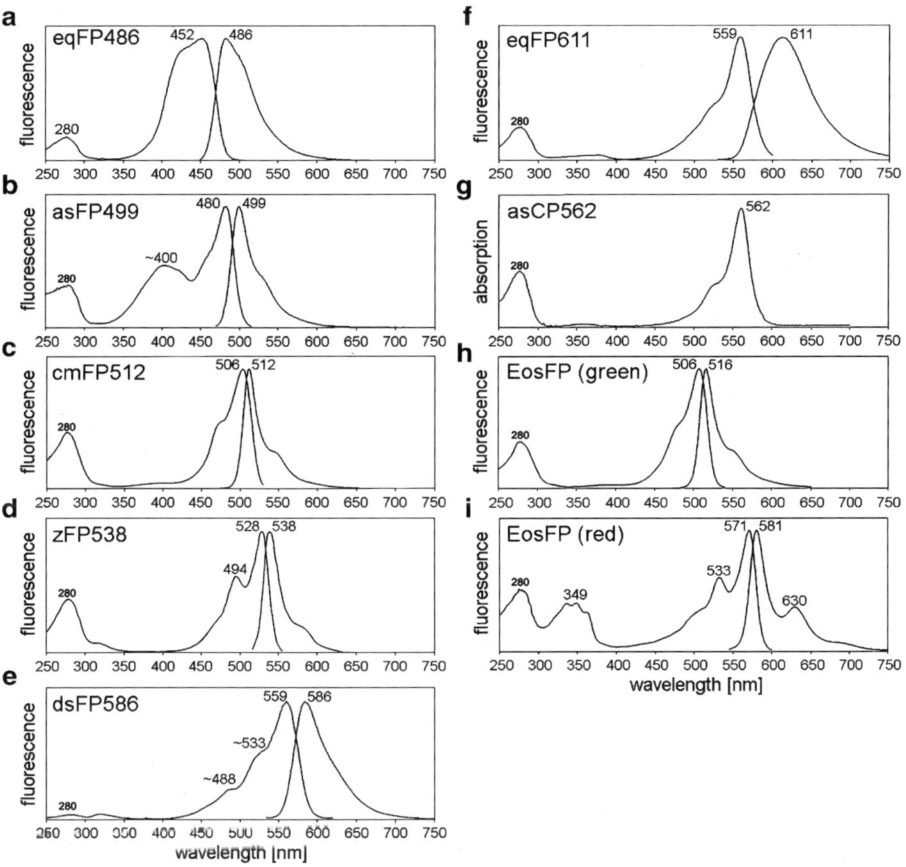

Fig. 6 Spectral properties of major color classes of GFP-like proteins. Panels (**a**)–(**i**) show the excitation and emission spectra of fluorescent proteins or the absorption spectrum in the case of the nonfluorescent chromoprotein asCP562. (**a**) CFP (eqFP486) from *Entacmaea quadricolor* (Wiedenmann, unpublished). (**b**) GFP (asFP499) from *Anemonia sulcata* [5]. (**c**) GFP (cmFP512) from *Cerianthus membranaceus* [67]. (**d**) YFP (zFP538) from *Zoanthus* [3] sp. (**e**) RFP (dsFP586) from *Discosoma* sp. (Wiedenmann, unpublished). (**f**) RFP (eqFP611) from *Entacmaea quadricolor* [6]. (**g**) Pink chromoprotein asCP562 from *Anemonia sulcata* [5]. (**h**) Green-to-red photoconverting protein EosFP from *Lobophyllia hemprichii* before, and (**i**) after photoconversion [64]. zFP538 spectra are a courtesy of M. Matz

state with excitation and emission maxima around 506 and 516 nm, respectively. Irradiation with ~390 nm light irreversibly changes the proteins to red emitters (λ_{max} ~ 581 nm) with an excitation maximum at ~572 nm [64].

Aside from the brightly fluorescent members, the GFP family also comprises a large number of homologous proteins that show intense color but no visible fluorescence [10, 68, 77, 78, 85]. Some of these nonfluorescent chromoproteins can, however, exhibit very weak red fluorescence detectable in a spectrophotometer [67, 68]. For asulCP, a quantum yield of <0.001 was determined. In contrast, with a quantum yield of <0.0001, the blue chromoprotein Rtms5 from *Montipora efflorescens* is de facto nonfluorescent [97].

The chromoproteins asulCP and asCP562 undergo conversion into a red fluorescent state upon excitation with green light ([68]; Wiedenmann and Girod, unpublished). The process can be reversed by irradiation of the red fluorescent protein with blue light [68]. Photoswitching is associated with a *trans–cis* isomerization of the chromophore [98].

2.2.6 Structural Properties

In solution, avGFP exists as monomer at concentrations below 1 mg/ml [27]. At higher concentrations, dimerization occurs with a K_d of 0.1 mM [99]. In contrast, GFPs from *Renilla* sp., *Phialidium gregarium*, and *Halistaura* (*Mitrocoma*) *cellularia* form stable non-dissociatable dimers [27].

Throughout the different color classes of GFP-like proteins, the β-can fold is almost perfectly conserved (Fig. 7). However, most GFP-like proteins from anthozoans exist as homo-tetramers [6, 59, 60, 64, 67]. In some cases, the oligomerization tendency is reduced [6]. Among others, the cyan FP MiCy from a scleractinian coral forms a dimeric association [63].

The rigid β-can fold of coral FPs can help to explain the long half-lives of up to 3 weeks determined in vivo for the corals *L. hemprichii* and *M. cavernosa* [83] (Fig. 7).

The 4-(p-hydroxybenzylidene)-5-imidazolinone structure of the avGFP chromophore appears to be universally involved in the development of chromogenic

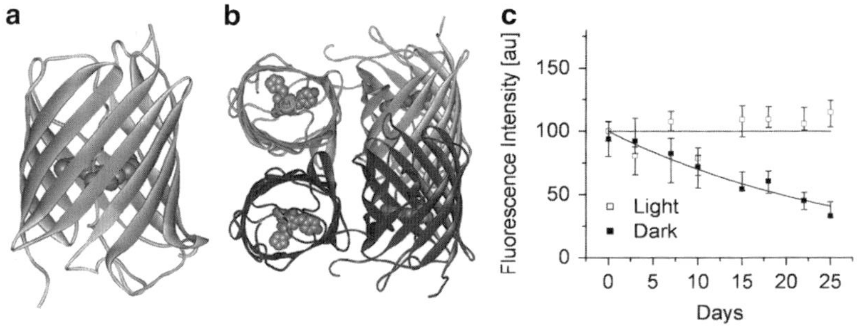

Fig. 7 Structural properties of GFP-like proteins (**a**) The β-can fold of EosFP from the reef coral *Lobophyllia hemprichii*. (**b**) Tetrameric assembly of EosFP. (**c**) Protein turnover measured in vivo and in situ in *L. hemprichii*. The diagram shows the decay of the tissue fluorescence at 581 nm in animals kept in the dark. Under these conditions, green-to-red photoconversion of newly synthesized protein is prevented, and the decrease of red tissue fluorescence can be used as indicator of the decay of the red fluorescent form of EosFP that was present at the beginning of the experiment. No changes in tissue fluorescence can be detected in illuminated corals. The diagrams show the medians; error bars display the first and third quartiles. Data from the dark-treated animals were fitted with an exponential decay. Panel (**c**) reprinted with permission from [83] (Copyright 2007 John Wiley & Sons Ltd)

properties in both fluorescent and nonfluorescent GFP homologs [7]. The conservation of the second and third chromophore-forming amino acids, tyrosine and glycine, and catalytic residues such as arginine in homologous positions to Arg96 in avGFP indicate that the basic mechanisms of chromophore formation are similar. Pronounced differences exist in regard to the modification of the GFP-type chromophore that yields the structures responsible for cyan, yellow, and red fluorescence or nonfluorescent pink to blue colors [7].

Structural features are discussed in greater detail in the chapter "Structure-Function Relationships in Fluorescent Marker Proteins of the GFP Family".

2.2.7 Molecular Evolution of GFP-Like Proteins

At the moment, more than 100 sequences coding for GFP-like proteins are available from databases, many of them awaiting further characterization of the coded protein. The length of the amino acid sequence of these FPs range from 225 (e.g., dsRed) to 238 (avGFP). One exception is the protein cFP484 from *Clavularia* sp. [3] showing a length of 266 amino acids. The length and composition of both the N- and the C-terminus vary considerably among different FPs (Fig. 8). The core protein is characterized by alternating conserved and variable regions. One striking feature is the strict conservation of the tyrosine–glycine motif within the chromophore. Moreover, the amino acids arginine and glutamine in positions homologous to residues Arg96 and Glu222 in avGFP are strictly conserved. These amino acids are thought to be involved in the autocatalytic formation of the GFP chromophore

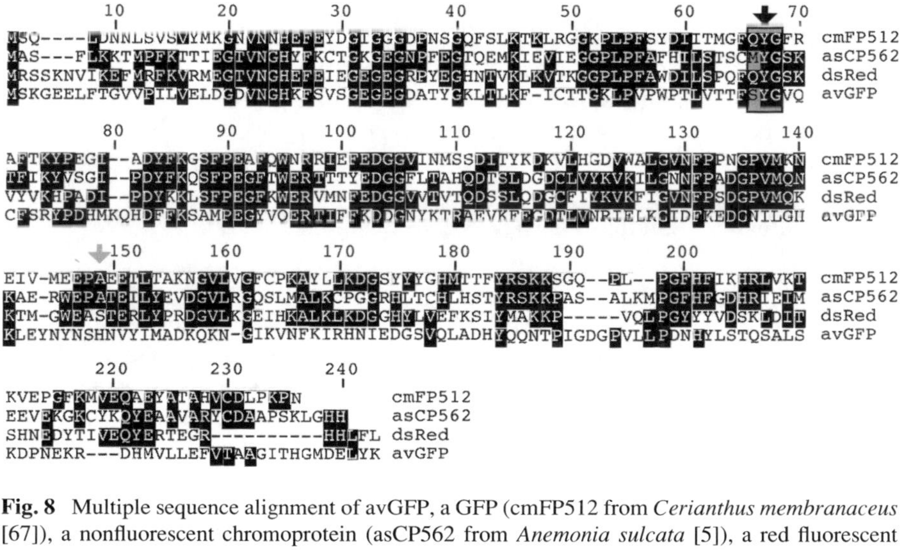

Fig. 8 Multiple sequence alignment of avGFP, a GFP (cmFP512 from *Cerianthus membranaceus* [67]), a nonfluorescent chromoprotein (asCP562 from *Anemonia sulcata* [5]), a red fluorescent protein (dsRed from *Discosoma* sp. [3]). Conserved residues are shaded in *black*. The chromophore-forming amino acids are shaded in *gray* and marked by a *black arrow*. The position corresponding to histidine 148 in avGFP is highlighted by a *gray arrow*

[32, 35, 100–102] (Fig. 6). In contrast, the first residue of the chromophore appears to be rather variable among differently colored GFP-like proteins. Consequently, prediction of the spectral properties based on this position is not reliable. Glutamine can serve as the first chromophore residue in both green and red fluorescent proteins [67] (Fig. 8). An exception is histidine in the first position of the chromophore. In all FPs examined so far, this type of chromophore underwent photoconversion from a green to a red fluorescent state upon irradiation with UV light [59, 64, 65, 103, 104].

In addition, the position corresponding to His148 in avGFP can be indicative of the spectral properties of GFP-like proteins. In highly fluorescent proteins, this residue is a serine, while in nonfluorescent GFP-like proteins this position is occupied by, for instance, alanine, cysteine, or aspartate [3, 73]. However, the proteins eqFP611 [6] and cmFP512 [67] provide exceptions to the rule. These proteins possess amino acids different from serine in the homologous position, aspargine and alanine, respectively, but exhibit a strong red (eqFP611) or green (cmFP512) fluorescence [6, 67, 105].

Molecular phylogeny approaches performed on amino acid sequences of FPs showed that the proteins group often according to the classical taxonomy of the host animals rather than to their spectral properties [5, 10, 58, 65, 106] (Fig. 9). The phylogenetic trees of GFP-like proteins suggest a multiple convergent evolution of red fluorescent proteins and nonfluorescent chromoproteins [10, 65].

The red fluorescent proteins dsRed, Kaede, and EosFP, for instance, show quite similar spectral properties: dsRed has an emission maximum at 583 nm, whereas the emission spectrum of red-converted Kaede and EosFP peaks around 581 nm. Nevertheless, the chromophore structure and underlying formation mechanisms are clearly different and suggest a convergent evolution of these red emitters [106, 107].

Based on the computational reconstruction of ancestral genes, a GFP was proposed to be the common ancestor of differently colored FPs in scleractinian corals [10, 108]. These results show that bioinformatics can help to localize key residues responsible for both biochemical and photophysical properties of FPs.

2.2.8 Biological Function(s) of GFP-Like Proteins

GFPs in bioluminescent cnidarians, as for instance avGFP, are exclusively found in the photogenic tissue of the animals [11, 16, 28]. They can form tight molecular interactions with photoproteins, as for example with luciferase in the case of *Renilla* [109]. The fact that they ubiquitously act as secondary emitters of light generated in the chemiluminescent reaction [11] suggests that their function might be associated with the bioluminescence function. Most cnidarians emit light upon mechanical stimulation [11, 110], which may be interpreted as a deterring response to the action of predators. The sudden flash might dazzle or confuse a nocturnal raider of pelagic jellyfish, allowing the prey to escape into the dark. As most cnidaria are quite well protected by their cnidae charged with potent toxins, they have rather few enemies specialized in feeding on them. A large number of predators of sessile cnidaria,

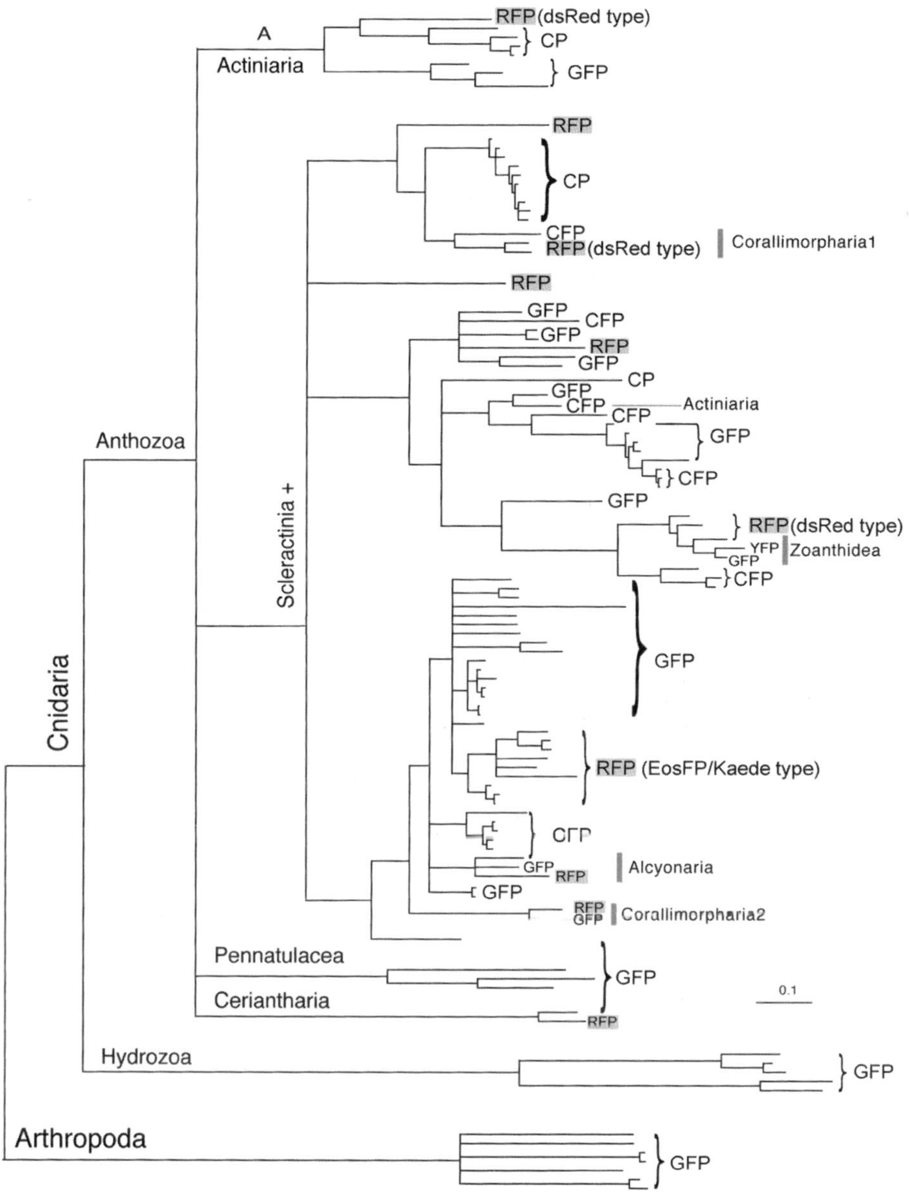

Fig. 9 Bayesian phylogenetic tree of GFP-like proteins from cnidarians showing Arthropoda FPs as outgroup. The emission color of the cyan, green, yellow, and red fluorescent GFP-like proteins are indicated by abbreviations (CFP, GFP, YFP, RFP). Red fluorescent proteins are highlighted by *gray shades*. Nonfluorescent chromoproteins are labeled with CP. The taxonomic origin of the animals from which the proteins were cloned is indicated on the branches. If a protein groups with representatives of a different taxonomic background, the deviating origin is given next to the branch ends. RFPs with different chromorphore types (dsRed type vs EosFP/Kaede type) represent a case of convergent evolution of red fluorescence. Many other cases are likely. Figure modified from [10] with permission

especially of hydrozoans, are found among nudibranchs (sea slugs) [111–113]. It is hard to imagine that these animals can be prevented from feeding by light flashes. Therefore, light emission could probably be used to attract larger predators, subsequently attacking the enemy of the cnidarians [114]. This scenario can be as well assumed for pelagic species. But why use GFPs as secondary emitters? In the case of *Renilla* luminescence, the answer appears to be straightforward: light emission via GFP increases the quantum yield of the chemoluminescence reaction and thereby promotes its effect [109]. Many photoreceptors have a high sensitivity in the green spectral region [115, 116]; therefore, a focused emission of green light might increase the perceptibility of the emitted light as compared to the broad-banded bluish light, which is generated in the chemiluminescent reaction. The optical properties of coastal waters allow most efficiently the penetration of green-yellow light [117], which might also have promoted the evolution of the secondary green emitters to increase the transmission efficiency of the light produced in the bioluminescence reaction.

The great majority of anthozoa species are non-bioluminescent, but many of them contain fluorescent proteins. The wide range of emission colors and the existence of nonfluorescent, yet intensely colored GFP homologs, also poses the question about the biological reasons for this variability [6, 10, 85].

Kawaguti suggested that the green fluorescent pigments convert harmful UV radiation into less energetic light that can be used by symbiotic zooxanthellae [37, 118]. The hypothesis of a photoprotective function seems to be supported by the finding that fluorescent morphs of *A. sulcata* are more abundant in shallow water habitats compared to the nonfluorescent morph *var. rustica* [77]. For the green fluorescent morph *var. smaragdina*, this distribution is probably caused by an enhanced fitness in the competition for space under high light conditions [119]. Similar conclusions were drawn by Salih and coworkers studying fluorescent pigments of Great Barrier Reef corals [56]. They reported the highest number of fluorescent morphs from the shallowest site, thus supporting a photoprotective function.

Moreover, the light regulation of several genes encoding GFP-like proteins suggests function(s) associated with the photobiology of reef corals [85].

Nevertheless, most green, yellow, and red fluorescent proteins as well as GFP-like chromoproteins emit and/or absorb light at wavelengths with sometimes minimal effects on photosynthesis of the symbiotic algae. These optical properties make them rather unsuitable for light shielding purposes [85]. Furthermore, the majority of the proteins show only minor absorption in the UV region (320–400 nm), making an exclusive UV screening function unlikely. Also, the high-level expression in azooxanthellate anthozoans from habitats without light stress questions a general photoprotective function [67, 90]. The great variety of colors and their sometimes convergent evolution point to multiple other uses by the expressing organisms [10]. For GFPs from pelagic copepods, a potential role in camouflage and intraspecific recognition was discussed [65]. Today, convincing explanations of the biological function of this fascinating protein family in non-bioluminescent anthozoa are still missing.

3 Methods: Cloning and Engineering of GFP-Like Proteins

3.1 Cloning of Novel GFP-Like Proteins

The exploration of natural sources of GFP-like proteins is still the most promising approach to discover new lead structures for novel optical tools. Simple tests can ensure that a novel pigment is most likely a GFP-like protein, and spectral measurements can be used to evaluate the potential applicability as a marker protein [4, 67]. However, the cloning of FP-coding cDNAs and the determination of its primary structure stand usually at the beginning of an in-depth characterization. Further experiments are then carried out on the recombinant protein, taking advantage of the relative ease by which the proteins can be expressed in recombinant systems and purified therefrom [3, 6]. This development was promoted by the progress in cloning techniques and the knowledge of the molecular properties of avGFP. The initial isolation of the cDNAs coding for avGFP was performed by probing a cDNA library with radiolabeled oligonucleotides designed on the basis of partial amino acid sequences [30]. Although successful, the handling of radioactivity is inconvenient and the screening of a representative library (~10^6 clones) is painstaking, laborious work. Therefore, currently two major strategies are applied for the cloning of novel FP-coding cDNAs: (a) the PCR amplification of partial cDNA fragments using (degenerate) primers designed against conserved regions of known FP sequences in combination with primers recognizing cDNA adaptors [3, 120], and (b) the direct expression of cDNA libraries in combination with a screening procedure based on the marker gene properties of the target protein [5, 6, 64, 67]. The first method is comparatively fast, and the specific PCR amplification promises success even if only minor amounts of cDNA or samples of lower quality are available. However, this method will fail if the target sequences deviate too much from the template used for primer design. The second method is more laborious and it might fail when the transcript levels of the target gene or the quality of the cDNAs are very low. Further restrictions will arise if the target protein is toxic for the expressing bacteria or the codon usage is incompatible. However, this method offers the great benefit that no sequence information about the novel GFP-like protein needs to be available. Therefore, it is the method of choice when a novel candidate with an unpredictable sequence is to be cloned.

3.2 Engineering of GFP-Like Proteins

Some wild-type proteins such as EosFP proved to be useful as marker proteins in several imaging application without further optimization [64, 121]. However, several features of natural GFP-like proteins hamper their applicability as marker proteins [9]. Common drawbacks are:

- Unspecific aggregation that prevents the correct localization in cells and might have cytotoxic effects

- Oligomerization that can interfere with the functionality of fusion proteins and prevent the correct cellular localization
- Lack of functional expression in mammalian cells requiring culturing temperatures around 37°C
- Fast photobleaching
- Suboptimal fluorescence properties

The fluorescence of the protein chosen for further optimization can often be used as read out for the success of alterations of the amino acid sequence. Consequently, even major engineering projects that might require the alteration of up to 30 residues in more or less unpredictable positions are feasible for skilled experimenters. Currently, three mutagenesis methods are routinely applied for the optimization of GFP-like proteins:

3.2.1 Random Mutagenesis

This method exploits the increased error rates of polymerases under suboptimal reaction conditions. During the amplification of the FP-encoding cDNA in the PCR, errors are introduced randomly in the newly generated DNA fragments. The modified amplicons are subsequently cloned in expression vectors and expressed in bacterial hosts. The fluorescence of the bacterial colonies growing on agar plates can be used as indicator for the effects of certain mutations.

Random mutagenesis proved to be especially helpful to create variants with improved folding and maturation properties [122–124].

3.2.2 Site-Directed Mutagenesis

Today, a considerable number of crystal structures of GFP-like proteins are available, which can help to identify target residues for knowledge-based engineering. The primer overlap extension PCR is a very robust technique to perform directed amino acid exchanges [123, 125]. Oligonucleotide primers can be also designed to randomize a distinct residue so that any possible amino acid will be introduced in this position. The randomization of defined residues can be used, for instance, to verify quickly if a mutation picked up during screening of a random mutant library is indeed the optimal exchange in a certain position. Examples of successful applications of site-directed mutagenesis include the disruption of tetrameric associations or the generation of variants with strongly red-shifted emission [122, 123].

3.2.3 Multisite-Directed Mutagenesis

Up to six oligonucleotide primers can be applied in a single mutagenesis reaction to a mixture of up to several different mutant templates using the QuikChange Multi

Site-Directed Mutagenesis Kit (Stratagene, La Jolla, CA, USA) [124]. Not all of the newly generated DNA fragments will carry all exchanges; many will contain different combinations of the individual mutations. Mutant libraries generated by this method can be subsequently screened in bacteria to spot mutations that act synergistically to improve desired properties, for example, folding of monomeric variants [124].

4 Key Applications

4.1 Key Application of Fluorescent Proteins

Fluorescent proteins, most prominently avGFP, have revolutionized life science research as specific protein labels, markers of gene expression, and reporters of environmental conditions in living cells and tissues [2, 9, 126].

FPs derive their popularity from the fact that in aerobic environments, their fluorophore forms in an autocatalytic reaction without the need for additional cofactors or substrates [2]. Moreover, they can be expressed in principle in virtually any organism. These properties make them excellent genetically encoded markers. The discovery of non-green GFP homologs largely expanded the application potential of the fluorescent protein technology [3–5]. In the following section, major applications of GFP-like proteins are outlined.

4.1.1 Determining Gene Activity

Cell function and differentiation depend on differential gene expression [127]. Important examples of the dramatic effects of the upregulation and downregulation of genes can be found during stem cell differentiation or cancer development. Fluorescent proteins can help to study changes in gene activity. For this purpose, the marker proteins are expressed under control of a promoter system that functions in the same way as the gene of interest (Fig. 10a). Once the cell switches the specific gene on or off, the production of the fluorescent marker is affected. It is important to note that delays in the detection of such changes result from the biochemical properties of the marker protein and the promoter. After the protein is expressed in the cell, it can be visualized only after maturation of the chromophore is completed and sufficient numbers of functional molecules are accumulated to reach the detection threshold [9]. The high stability of the β-can fold and the associated slow turnover of the marker protein can hamper or even prevent the detection of downregulation of a gene. The oligomerization degree of the marker protein is not of primary concern in studies of gene activity. More crucial is the use of fast maturing variants. Destabilized variants were constructed that allow a better monitoring of downregulation processes [128].

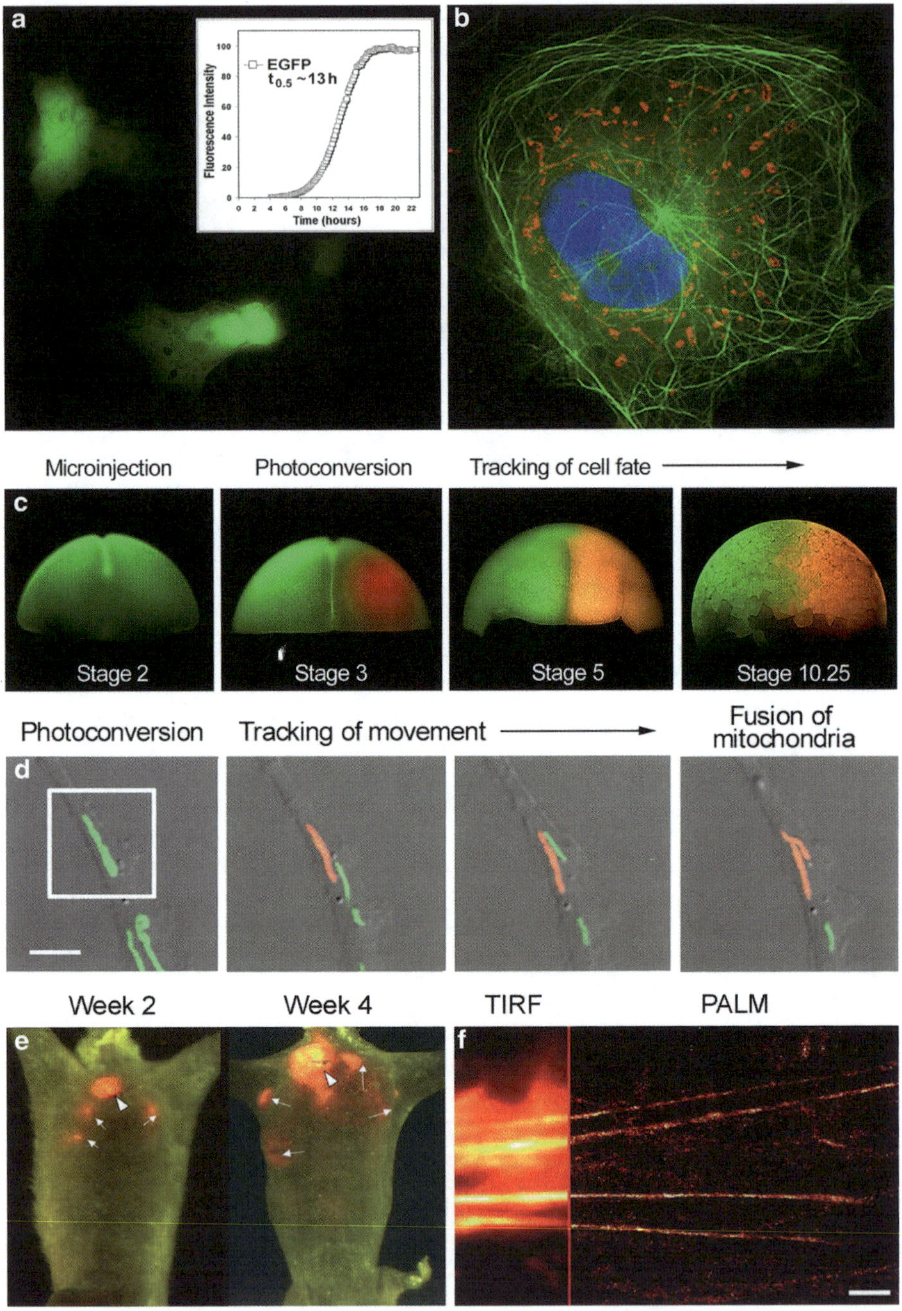

Fig. 10 Applications of GFP-like proteins. (**a**) Cytoplasmatic fluorescence of EGFP expressed under control of a CMV promoter increases with a transition midpoint of 13 h. The cellular expression of FPs can be used as reporter of gene activity. Images courtesy of Franz Oswald, University of Ulm. (**b**) Multicolor labeling of proteins and organelles in HeLa cells. A EGFP-labeled

4.1.2 Tracking of Proteins

Precise knowledge about the localization of proteins in the cell is required to understand their functioning. To study the whereabouts of a protein, the coding DNAs of the protein of interest and the FP are fused. An essential step in the construction of the fusion protein is the removal of the stop codon at the end of the open-reading frame of the first fusion partner. This modification ensures that a tandem of both proteins is produced once the construct is expressed in a cell. The localization of the protein in the cell can subsequently be followed by the fluorescence of the attached marker protein (Fig. 10b). Although some proteins allow the fusion to tetrameric FPs, the application of monomeric variants is recommended to minimize potential mislocalization and functional defects that might be caused by the size of the tetrameric fusion protein complex [9].

4.1.3 Labeling of Subcellular Compartments

Markers are required that highlight distinct subcellular compartments to help the identification of protein localization (Fig. 10a, d). Compartments can be labeled either by FP fusion proteins with a defined localization or using FPs that are tagged with short targeting peptides [64]. Obviously, the FPs used for labeling of the compartment and the protein under study need to have different spectral properties. FPs are also useful when the compartments themselves are to be analyzed, for instance, to study the dynamics of a mitochondrial network [106].

Fig. 10 (continued) tubulin-associated protein highlights parts of the cytoskeleton in *green*; mitochondria show red fluorescence emitted from RFP611, the nucleus is labeled with the blue fluorescent dye DAPI. Reprinted from [123] with permission from Elsevier. (**c**) Monitoring cell fate during embryonic development of *Xenopus laevis* using the green to red photoconvertible protein EosFP. The purified marker was microinjected at stage 2. Local optical marking was achieved by irradiation with ~430 nm light, enabling precise observations of the fate of the highlighted blastomer. Adapted from [121]. Copyright 2009 Wiley-VCH Verlag GmbH & Co. KG, Weinheim, reproduced with permission. (**d**) Tracking of mitochondria labeled with EosFP. The movement of a single mitochondrion was monitored after photoconversion. A fusion event is visible in the panel on the right. Adapted from [106]. Copyright 2006, Expert Reviews Ltd, reproduced with permission. (**e**) Whole body imaging. Tumor cells expressing a red fluorescent protein were implanted in a nude mouse, and tumor growth was documented after 2 weeks (*left panel*) and 4 weeks (*right panel*). The primary tumor is marked by *arrowheads*; *arrows* highlight metastases. Adapted from [129]. Copyright 2005 Wiley-VCH Verlag GmbH & Co. KG, Weinheim, reproduced with permission. (**f**) Super-resolution microscopy using photoactivatable FPs. Aktin fibers were labeled with a EosFP. The *left part* of the image was acquired by resolution achieved by total internal reflection fluorescence (TIRF) microscopy. The *right part* shows the increase in resolution achieved by the PALM method: scale bar 2 μm. Modified from [144]. Copyright 2009 Nature Publishing Group

4.1.4 Tracking of Cells/Whole Body Imaging

The movement of whole cells needs to be followed during cell fate mapping experiments during embryogenesis or to track metastases in tumor models [121, 129] (Fig. 10c, e). During short-time experiments, FPs can be introduced, for instance, in a developing embryo via microinjection. If the progression of a tumor needs to be followed over weeks, the FP-encoding construct has to be stably integrated in the genome of the cell under study.

In many experiments, researchers aim to follow the localization of a few cells in a whole organism. Therefore, the brightness is the most important property that should guide the choice of the marker. Red FPs are particularly desirable in this context because of reduced background fluorescence and less scattering in the red spectral region. Moreover, the green light used for imaging is less cytotoxic compared to UV or blue light.

4.1.5 Multicolor Labeling

Multicolor labeling was greatly facilitated by the advent of red fluorescent proteins such as mCherry, TagRFP, or mRuby [124, 130, 131].

Currently, up to six different fluorescence colors can be imaged in a single cell using a standard fluorescence microscope equipped with suitable filter sets: blue, cyan, green, yellow, red, and far-red. FPs are now available for each of these channels [8]. Consequently, multiplexing experiments became feasible during which several processes can be studied in the same cell (Fig. 10b). Applications such as protein tracking, labeling of whole cells or subcellular compartments, and measurements of gene activity particularly benefit from the recent expansion in the range of FP colors [8].

4.1.6 Monitoring Protein Interactions

The co-localization of two proteins in a distinct subcellular compartment does not prove their molecular interaction. FRET can be exploited to visualize protein interactions [8]. For this purpose, the proteins of interest are fused to two differently colored FPs. If the fusion proteins are localized within a distance of ~100 Å, the energy of the excited chromophore with the blue-shifted emission maximum (the donor) is transferred in a radiationless process to the chromophore with the emission maximum at longer wavelengths (the acceptor). If the two chromophores are further away from each other, no energy transfer takes place and the donor fluoresces normally. The ratio of the fluorescence intensity of the chromophores can be consequently used to monitor protein interactions.

4.1.7 Sensor Applications

Fluorescent proteins have been engineered that respond to changes in cellular environment by altering their fluorescence properties. Examples of a range FP-based sensors include variants that change fluorescence intensity in response to changes in pH or membrane voltage [132–134]. Intracellular calcium concentrations can be determined using so-called camaeleons [135]. These sensors function by changing the FRET efficiency between two differently colored FPs in response to conformational changes induced by binding of Ca^{2+} to a linker containing calmodulin (CaM) and the CaM-binding domain of myosin light chain kinase (M13).

4.1.8 Pulse-Chase Experiments

Tracking of subpopulations proteins, organelles, or cells within a bulk of homogenously labeled structures is enabled using photoactivatable fluorescent proteins. The first variant that became available was the photoactivatable GFP (paGFP) [136]. Its fluorescence increases ~100-fold in response to irradiation with intense blue light. A defined portion of paGFP fusion proteins can be "switched on" within a single cell by irradiation with a focused blue laser beam. Subsequently, the movement of the highlighted molecules can be tracked. Recently, a red version, pamCherry, was introduced [137]. Fluorescent proteins that show a light-induced change in emission colors (photoconversion) are particularly useful for such pulse-chase experiments due to the strong optical contrast that can be generated between the activated and the nonactivated state (Fig. 10c, d). Examples include the cyan-to-green (psCFP) [138], green-to-red Kaede [59], and EosFP [64] photoconverting proteins. The application of relatively high intensities of ~400 nm light induces photoconversion of psCFP. In contrast, Kaede and EosFP can be converted with comparably low amounts of ~400 nm light. Photoconvertible proteins offer also the convenience that the potential target structures are already visible before optical highlighting [9].

4.1.9 Super-Resolution Microscopy

In recent years, concepts were developed that allow imaging of proteins with a resolution beyond the diffraction barrier of conventional light microscopy [139–141]. All super-resolution concepts rely on fluorescent dyes that can be switched between two distinct states [140, 142]. Photoactivated localization microscopy (PALM), for instance, makes extensive use of GFP-like proteins that can be switched on and off or photoconverted in a controlled manner (Fig. 10f). The tandem-dimer variant of the photoconvertible EosFP (tdEosFP) became one of the most widely used tools for PALM [64, 107, 143]. Fusion proteins with tdEosFP are illuminated with low amounts of ~400 nm light to induce a limited number of photoconversion events. The precise loci of the resulting single red fluorescent molecules are obtained by determining the center of the point spread function and logged. Later, the red fluorescent molecules are photobleached to "clear the screen" for the detection of new photoconversion events. After multiple switching and bleaching cycles, the data points are assembled to reconstruct a super-resolution image [141, 144].

Alternatively, green fluorescent labels that can be switched "on" and "off" by alternating irradiation with near-UV and blue light such as Dronpa, rsFastLime, rsCherry, and mIrisFP can be used for the same purpose [93, 145–148]. Together with labels that can be activated by strong irradiation with near-UV or blue light (e.g., paGFP or pamCherry), photoconvertible and photoswitchable FPs have been applied for dual color PALM [137, 143]. Most recently, mIrisFP enabled pulse-chase experiments with super-resolution [148].

5 Conclusions

Nature has created a plethora of fluorescent proteins with useful optical properties. The wide distribution of GFP-like proteins among the animal kingdom suggests that any metazoan organism might contain potential marker proteins; however, anthozoans are surely the most rewarding source for non-green GFP homologs. The overall molecular structure of the fluorescent proteins is essentially identical, with a strongly fluorescent chromophore generated autocatalytically in the interior of the β-can fold.

Molecular phylogeny analyses show that GFP-like proteins tend to evolve along with their "host" organisms; therefore, variations of certain biochemical properties cluster among distinct taxonomic groups. This knowledge will facilitate the search for natural FP variants with altered properties. By now, well-established methods allow the fast isolation of genes encoding novel fluorescent and the subsequent engineering of their products to meet the demands of imaging application in live cells.

The emission colors of natural FPs range from cyan to red, providing a rainbow of marker proteins for multicolor labeling experiments in live cells and tissue. Moreover, photoactivatable proteins were isolated from marine invertebrates that respond to light stimuli by changing their emission color or by reversible switching between a fluorescent and nonfluorescent form. While both types of photoactivatable proteins are key tools in optical microscopy with spatial resolution beyond the diffraction barrier, photoconvertible proteins, in particular, are useful for dynamic tracking of proteins and cells.

In summary, the combination of bioprospecting and protein engineering was hugely successful in generating novel marker proteins and can also be considered the most promising strategy to satisfy the increasing demand for optical tools for biomedical research.

Acknowledgments J.W. acknowledges funding by the Deutsche Forschungsgemeinschaft (DFG) (grant Wi1990/2-1), the Network Fluorescence Applications in Biotechnology and Life Sciences, FABLS, Australia, the Landesstiftung Baden-Württemberg (Elite Postdoc Program), the Natural Environment Research Council, UK (NE/G009643/1), and the University of Southampton. G.U.N. was supported by the DFG and the State of Baden-Württemberg through the Center for Functional Nanostructures, by DFG grant NI 291/9 and by the Baden-Württemberg Stiftung.

References

1. Chalfie M, Tu Y, Euskirchen G, Ward WW, Prasher DC (1994) Green fluorescent protein as a marker for gene expression. Science 263:802–805
2. Tsien RY (1998) The green fluorescent protein. Annu Rev Biochem 67:509–544
3. Matz MV, Fradkov AF, Labas YA, Savitsky AP, Zaraisky AG, Markelov ML, Lukyanov SA (1999) Fluorescent proteins from nonbioluminescent Anthozoa species. Nat Biotechnol 17:969–973
4. Wiedenmann J (1997) The application of an orange fluorescent protein and further colored proteins and the corresponding genes from the species group *Anemonia* sp. (*sulcata*) Pennant, (Cnidaria, Anthozoa, Actinaria) in gene technology and molecular biology. (Die Anwendung eines orange fluoreszierenden Proteins und weiterer farbiger Proteine und der zugehörenden Gene aus der Artengruppe *Anemonia* sp. (*sulcata*) Pennant, (Cnidaria, Anthozoa, Actinaria) in Gentechnologie und Molekularbiologie.). Deutsches Patent und Markenamt. Patent DE 197 18 640
5. Wiedenmann J, Elke C, Spindler KD, Funke W (2000) Cracks in the beta-can: fluorescent proteins from *Anemonia sulcata* (Anthozoa, Actinaria). Proc Natl Acad Sci USA 97: 14091–14096
6. Wiedenmann J, Schenk A, Rocker C, Girod A, Spindler KD, Nienhaus GU (2002) A far-red fluorescent protein with fast maturation and reduced oligomerization tendency from *Entacmaea quadricolor* (Anthozoa, Actinaria). Proc Natl Acad Sci USA 99:11646–11651
7. Nienhaus GU, Wiedenmann J (2009) Structure, dynamics and optical properties of fluorescent proteins: perspectives for marker development. Chemphyschem 10:1369–1379
8. Day RN, Davidson MW (2009) The fluorescent protein palette: tools for cellular imaging. Chem Soc Rev 38:2887–2921
9. Wiedenmann J, Oswald F, Nienhaus GU (2009) Fluorescent proteins for live cell imaging: opportunities, limitations, and challenges. Iubmb Life 61:1029–1042
10. Alieva NO, Konzen KA, Field SF, Meleshkevitch EA, Hunt ME, Beltran-Ramirez V, Miller DJ, Wiedenmann J, Salih A, Matz MV (2008) Diversity and evolution of coral fluorescent proteins. PLoS ONE 3:e2680
11. Morin JG (1974) Coelenterate bioluminescence. In: Muscatine L, Lenhoff HM (eds) Coelenterate biology. Reviews and new perspectives. Academic press, New York, pp 397–438
12. Shimomura O, Johnson FH, Saiga Y (1962) Extraction, purification and properties of aequorin, a bioluminescent protein from the luminous hydromedusan, Aequorea. J Cell Comp Physiol 59:223–239
13. Harvey EN (1925) Studies on Bioluminescence. XVII. Fluorescence and inhibition of luminescence in ctenophores by ultraviolet light. J Gen Physiol 7:331–339
14. Shimomura O, Johnson FH (1969) Properties of the bioluminescent protein aequorin. Biochemistry 8:3991–3997
15. Shimomura O, Johnson FH (1970) Calcium binding, quantum yield, and emitting molecule in aequorin bioluminescence. Nature 227:1356–1357
16. Davenport D, Nicol JAC (1955) Luminescence in hydromedusae. Proceedings of the Royal Society of London Series B, Biological Sciences 144:399–411
17. Hastings JW, Morin JG (1969) Comparative biochemistry of calcium-activated photoproteins from the Ctenophore *Mnemiopsis*, and the Coelenterates *Aequorea, Obelia, Pelagia* and *Renilla*. Biol Bull 137:402
18. Morin JG, Cooke IM (1971) Behavioural physiology of the colonial hydroid Obelia: III. Characteristics of the bioluminescent system. J Exp Biol 54:723–735
19. Morin JG, Hastings JW (1971) Biochemistry of the bioluminescence of colonial hydroids and other coelenterates. J Cell Physiol 77:305–312
20. Wampler JE, Hori K, Lee JW, Cormier MJ (1971) Structured bioluminescence. Two emitters during both the in vitro and the in vivo bioluminescence of the sea pansy, Renilla. Biochemistry 10:2903–2909

21. Wampler JE, Karkhanis YD, Morin JG, Cormier MJ (1973) Similarities in the bioluminescence from the Pennatulacea. Biochim Biophys Acta 314:104–109
22. Cormier MJ, Hori K, Anderson JM (1974) Bioluminescence in coelenterates. Biochim Biophys Acta 346:137–164
23. Cormier MJ, Hori K, Karkhanis YD, Anderson JM, Wampler JE, Morin JG, Hastings JW (1973) Evidence for similar biochemical requirements for bioluminescence among the coelenterates. J Cell Physiol 81:291–297
24. Morin JG, Reynolds GT (1970) Luminescence and related fluorescence in coelenterates. Biol Bull 139:430–431
25. Ward WW (1979) Energy transfer processes in bioluminescence. In: Smith K (ed) Photochemical and photobiological reviews, vol 4. Plenum, New York, pp 1–57
26. Levine LD, Ward WW (1982) Isolation and characterization of a photoprotein, "phialidin", and a spectrally unique green-fluorescent protein from the bioluminescent jellyfish Phialidium gregarium. Comparative Biochemistry and Physiology Part B: Comparative Biochemistry 72:77–85
27. Ward WW (2005) Biochemical and physical properties of green fluorescent protein. In: Chalfie M, Kain SR (eds) Green fluorescent protein: properties, applications and protocols, 2nd edn. Wiley, Hoboken, USA, pp 39–65
28. Titschack Hv (1964) Untersuchungen über das Leuchten der Seefeder Veretillum cynomorium. Vie Milieu 15:547–563
29. Cody CW, Prasher DC, Westler WM, Prendergast FG, Ward WW (1993) Chemical structure of the hexapeptide chromophore of the Aequorea green-fluorescent protein. Biochemistry 32:1212–1218
30. Prasher DC, Eckenrode VK, Ward WW, Prendergast FG, Cormier MJ (1992) Primary structure of the Aequorea victoria green-fluorescent protein. Gene 111:229–233
31. Heim R, Prasher DC, Tsien RY (1994) Wavelength mutations and posttranslational autoxidation of green fluorescent protein. Proc Natl Acad Sci USA 91:12501–12504
32. Cubitt AB, Heim R, Adams SR, Boyd AE, Gross LA, Tsien RY (1995) Understanding, improving and using green fluorescent proteins. Trends Biochem Sci 20:448–455
33. Inouye S, Tsuji FI (1994) Aequorea green fluorescent protein. Expression of the gene and fluorescence characteristics of the recombinant protein. FEBS Lett 341:277–280
34. Yang F, Moss LG, Phillips GN Jr (1996) The molecular structure of green fluorescent protein. Nat Biotechnol 14:1246–1251
35. Ormoe M, Cubitt AB, Kallio K, Gross LA, Tsien RY, Remington SJ (1996) Crystal structure of the Aequorea victoria green fluorescent protein. Science 273:1392–1395
36. Phillips CES (1927) Fluorescence of sea anemones. Nature 119:747
37. Kawaguti S (1944) On the physiology of reef corals VI. Study on the pigments. Palao Trop Biol Stn Stud 2:617–674
38. Marden L (1956) Camera under the sea. National Geographic 109:162–200
39. Wobber D (1967) The miniature world of Wobber – the glowing anemones. Skin Diver Magazine 16:44
40. Read KRH, Davidson JM, Twarog BM (1968) Fluorescence of sponges and coelenterates in blue light. Comp Biochem Physiol A-Mol Integr Physiol 25:873–882
41. Limbaugh C, North WJ (1956) Fluorescent, benthic, Pacific Coast Coelenterates. Nature 178:497–498
42. Catala-Stucki R (1959) Fluorescence effects from corals irradiated with ultra-violet rays. Nature 183:949
43. Catala R (1958) Effets de fluorescence provoque sur des coraux par l'action des rayons ultra-violets. C R Acad des Sciences 247:1678–1679
44. Catala R (1960) Nouveaux organismes marins presentant des effets de fluorescence par l'action des rayons ultra-violets. C R Acad des Sciences 250:1128
45. Catala-Stucki R (1964) Carnival under the sea. R. Sicard, Paris
46. Zahl PA (1963) Fluorescent gems from Davy Jones's locker. National Geographic 124:260–271

47. Schlichter D, Fricke HW (1990) Coral host improves photosynthesis of endosymbiotic algae. Naturwissenschaften 77:447–450
48. Schlichter D, Fricke HW, Weber W (1986) Light harvesting by wavelength transformation in a symbiotic coral of the Red Sea twilightzone. Mar Biol 91:403–407
49. Schlichter D, Fricke HW, Weber W (1988) Evidence for PAR-enhancement by reflection, scattering and fluorescence in the symbiotic deep water coral Leptoseris fragilis. Endocyt C Res 5:83–94
50. Schlichter D, Weber W, Fricke HW (1985) A chromatophore system in the hermatypic, deep-water coral Leptoseris fragilis (Anthozoa: Hexacorallia). Mar Biol 89: 143–147
51. Mazel C (1988) Underwater fluorescence. Sea Frontiers 34:274–279
52. Mazel C (1991) Black night, black light: underwater fluorescence. Ocean Realm, summer: 63–68
53. Mazel CH (1993) Fluorescence in Caribbean coral reef cnidarians. Ph.D. Thesis, Boston University
54. Mazel CH (1995) Spectral measurements of fluorescence emission in Caribbean cnidarians. Mar Ecol Prog Ser 120:185–191
55. Mazel CH (1997) Coral fluorescence characteristics: excitation – emission spectra, fluorescence efficiencies, and contribution to apparent reflectance. Ocean Optics Xiii, Proceedings of the Society of Photo-Optical Instrumentation Engineers (Spie), Ackleson SG, Frouin R (eds), vol 2963, pp 240–245
56. Salih A, Larkum A, Cox G, Kuhl M, Hoegh-Guldberg O (2000) Fluorescent pigments in corals are photoprotective. Nature 408:850–853
57. Dove SG, Hoegh-Guldberg O, Ranganathan S (2001) Major colour patterns of reef-building corals are due to a family of GFP-like proteins. Coral Reefs 19:197–204
58. Labas YA, Gurskaya NG, Yanushevich YG, Fradkov AF, Lukyanov KA, Lukyanov SA, Matz MV (2002) Diversity and evolution of the green fluorescent protein family. Proc Natl Acad Sci USA 99:4256–4261
59. Ando R, Hama H, Yamamoto-Hino M, Mizuno H, Miyawaki A (2002) An optical marker based on the UV-induced green-to-red photoconversion of a fluorescent protein. Proc Natl Acad Sci USA 99:12651–12656
60. Karasawa S, Araki T, Yamamoto-Hino M, Miyawaki A (2003) A green-emitting fluorescent protein from Galaxeidae coral and its monomeric version for use in fluorescent labeling. J Biol Chem 278:34167–34171
61. Kelmanson IV, Matz MV (2003) Molecular basis and evolutionary origins of color diversity in great star coral *Montastraea cavernosa* (Scleractinia: Faviida). Mol Biol Evol 20: 1125–1133
62. Mazel CH, Lesser MP, Gorbunov MY, Barry TM, Farrell JH, Wyman KD, Falkowski PG (2003) Green fluorescent proteins in Caribbean corals. Limnol Oceanogr 48:402–411
63. Karasawa S, Araki T, Nagai T, Mizuno H, Miyawaki A (2004) Cyan-emitting and orange-emitting fluorescent proteins as a donor/acceptor pair for fluorescence resonance energy transfer. Biochem J 381:307–312
64. Wiedenmann J, Ivanchenko S, Oswald F, Schmitt F, Rocker C, Salih A, Spindler KD, Nienhaus GU (2004) EosFP, a fluorescent marker protein with UV-inducible green-to-red fluorescence conversion. Proc Natl Acad Sci USA 101:15905–15910
65. Shagin DA, Barsova EV, Yanushevich YG, Fradkov AF, Lukyanov KA, Labas YA, Semenova TN, Ugalde JA, Meyers A, Nunez JM, Widder EA, Lukyanov SA, Matz MV (2004) GFP-like proteins as ubiquitous metazoan superfamily: Evolution of functional features and structural complexity. Mol Biol Evol 21:841–850
66. Ip DTM, Chan S-H, Allen MD, Bycroft M, Wana DCC, Wong K-B (2004) Crystallization and preliminary crystallographic analysis of a novel orange fluorescent protein from the Cnidaria tube anemone Cerianthus sp. Acta Crystallogr Sect D – Bio Crystallogr D60:340–341

67. Wiedenmann J, Ivanchenko S, Oswald F, Nienhaus GU (2004) Identification of GFP-like proteins in nonbioluminescent, azooxanthellate anthozoa opens new perspectives for bioprospecting. Mar Biotechnol 6:270–277
68. Lukyanov KA, Fradkov AF, Gurskaya NG, Matz MV, Labas YA, Savitsky AP, Markelov ML, Zaraisky AG, Zhao X, Fang Y, Tan W, Lukyanov SA (2000) Natural animal coloration can be determined by a nonfluorescent green fluorescent protein homolog. J Biol Chem 275:25879–25882
69. Martynov VI, Savitsky AP, Martynova NY, Savitsky PA, Lukyanov KA, Lukyanov SA (2001) Alternative cyclization in GFP-like proteins family. The formation and structure of the chromophore of a purple chromoprotein from Anemonia sulcata. J Biol Chem 276: 21012–21016
70. Gurskaya NG, Fradkov AF, Pounkova NI, Staroverov DB, Bulina ME, Yanushevich YG, Labas YA, Lukyanov S, Lukyanov KA (2003) Colourless green fluorescent protein homologue from the non-fluorescent hydromedusa Aequorea coerulescens and its fluorescent mutants. Biochem J 373:403–408
71. Beddoe T, Ling M, Dove S, Hoegh-Guldberg O, Devenish RJ, Prescott M, Rossjohn J (2003) The production, purification and crystallization of a pocilloporin pigment from a reef-forming coral. Acta Crystallogr Sect D – Bio Crystallogr 59:597–599
72. Martynov VI, Maksimov BI, Martynova NY, Pakhomov AA, Gurskaya NG, Lukyanov SA (2003) A purple-blue chromoprotein from Goniopora tenuidens belongs to the DsRed subfamily of GFP-like proteins. J Biol Chem 278:46288–46292
73. Gurskaya NG, Fradkov AF, Terskikh A, Matz MV, Labas YA, Martynov VI, Yanushevich YG, Lukyanov KA, Lukyanov SA (2001) GFP-like chromoproteins as a source of far-red fluorescent proteins. FEBS Lett 507:16–20
74. Haddock SH, Mastroianni N, Christianson LM (2010) A photoactivatable green-fluorescent protein from the phylum Ctenophora. Proceedings of the Royal Society B-Biological Sciences 277:1155–1160
75. Deheyn DD, Kubokawa K, McCarthy JK, Murakami A, Porrachia M, Rouse GW, Holland ND (2007) Endogenous green fluorescent protein (GFP) in amphioxus. Biol Bull 213:95–100
76. Veron JEN (2000) Corals of the world. Australian Institute of Marine Science, Townsville
77. Wiedenmann J, Röcker C, Funke W (1999) The morphs of *Anemonia* aff. *sulcata* (Cnidaria, Anthozoa) in particular consideration of the ectodermal pigments. Verhandlungen der Gesellschaft für Ökologie 29:497–503
78. Leutenegger A, Kredel S, Gundel S, D'Angelo C, Salih A, Wiedenmann J (2007) Analysis of fluorescent and non-fluorescent sea anemones from the Mediterranean Sea during a bleaching event. J Exp Mar Biol Ecol 353:221–234
79. Oswald F, Schmitt F, Leutenegger A, Ivanchenko S, D'Angelo C, Salih A, Maslakova S, Bulina M, Schirmbeck R, Nienhaus GU, Matz MV, Wiedenmann J (2007) Contributions of host and symbiont pigments to the coloration of reef corals. FEBS J 274:1102–1109
80. Gosse PH (1860) Actinologia britannica. A history of the British sea-anemones and corals. Van Voorst, London
81. Andres A (1881) Prodromus neapolitanae actiniarum faunae. Mitt Zool Stn Neapel 2:305–317
82. Andres A (1883) Le Attinie. Accad. Lincei (3. Mem. Class. Scienz. Fis.), ed Atti R), Vol 14, pp 211–674
83. Leutenegger A, D'Angelo C, Matz MV, Denzel A, Oswald F, Salih A, Nienhaus GU, Wiedenmann J (2007) It's cheap to be colorful. Anthozoans show a slow turnover of GFP-like proteins. FEBS J 274:2496–2505
84. Gruber DF, Kao H-T, Janoschka S, Tsai J, Pieribone VA (2008) Patterns of fluorescent protein expression in Scleractinian corals. Biol Bull 215:143–154
85. D'Angelo C, Denzel A, Vogt A, Matz MV, Oswald F, Salih A, Nienhaus GU, Wiedenmann J (2008) Blue light regulation of host pigment in reef-building corals. Mar Ecol Prog Ser 364:97–106

86. Peloux Y (1960) Etude histologique des coraux fluorescents de profondeur. C R Acad des Sciences 250:1129–1130
87. Kawaguti S (1966) Electron microscopy on the fluorescent green of reef corals with a note on mucous cells. Biol J Okayama University 2:11–21
88. Logan A, Halcrow K, Tomascik T (1990) UV excitation-fluorescence in polyp tissue of certain scleractinian corals from Barbados and Bermuda. Bull Mar Sci 46:807–813
89. Wiedenmann J, Lorenz U (2000) Fluoreszenz- und elektronenemikroskopische untersuchungen an der seeanemone Anemonia sulcata Pennant (Cnidaria, Anthozoa). Mikrokosmos 89:1–6
90. Vogt A, D'Angelo C, Oswald F, Denzel A, Mazel CH, Matz M, Ivanchenko S, Nienhaus GU, Wiedenmann J (2008) A green fluorescent protein with photoswitchable emission from the deep sea. PLoS ONE 3:e3766
91. Nienhaus K, Renzi F, Vallone B, Wiedenmann J, Nienhaus GU (2006) Chromophore-protein interactions in the anthozoan green fluorescent protein asFP499. Biophys J 91:4210–4220
92. Nienhaus GU (2010) The "wiggling and jiggling of atoms" leading to excited-state proton transfer in green fluorescent protein. Chemphyschem 11:971–974
93. Ando R, Mizuno H, Miyawaki A (2004) Regulated fast nucleocytoplasmic shuttling observed by reversible protein highlighting. Science 306:1370–1373
94. Verkhusha VV, Chudakov DM, Gurskaya NG, Lukyanov S, Lukyanov KA (2004) Common pathway for the red chromophore formation in fluorescent proteins and chromoproteins. Chem Biol 11:845–854
95. Baird GS, Zacharias DA, Tsien RY (2000) Biochemistry, mutagenesis, and oligomerization of DsRed, a red fluorescent protein from coral. Proc Natl Acad Sci USA 97:11984–11989
96. Cotlet M, Hofkens J, Habuchi S, Dirix G, Van Guyse M, Michiels J, Vanderleyden J, De Schryver FC (2001) Identification of different emitting species in the red fluorescent protein DsRed by means of ensemble and single-molecule spectroscopy. Proc Natl Acad Sci USA 98:14398–14403
97. Prescott M, Ling M, Beddoe T, Oakley AJ, Dove S, Hoegh-Guldberg O, Devenish RJ, Rossjohn J (2003) The 2.2 A crystal structure of a pocilloporin pigment reveals a nonplanar chromophore conformation. Structure 11:275–284
98. Andresen M, Wahl MC, Stiel AC, Grater F, Schafer LV, Trowitzsch S, Weber G, Eggeling C, Grubmuller H, Hell SW, Jakobs S (2005) Structure and mechanism of the reversible photoswitch of a fluorescent protein. Proc Natl Acad Sci USA 102:13070–13074
99. Wang Z, Shah JV, Chen Z, Sun CH, Berns MW (2004) Fluorescence correlation spectroscopy investigation of a GFP mutant-enhanced cyan fluorescent protein and its tubulin fusion in living cells with two-photon excitation. J Biomed Opt 9:395–403
100. Chattoraj M, King BA, Bublitz GU, Boxer SG (1996) Ultra-fast excited state dynamics in green fluorescent protein: multiple states and proton transfer. Proc Natl Acad Sci USA 93:8362–8367
101. Brejc K, Sixma TK, Kitts PA, Kain SR, Tsien RY, Ormo M, Remington SJ (1997) Structural basis for dual excitation and photoisomerization of the Aequorea victoria green fluorescent protein. Proc Natl Acad Sci USA 94:2306–2311
102. Barondeau DP, Putnam CD, Kassmann CJ, Tainer JA, Getzoff ED (2003) Mechanism and energetics of green fluorescent protein chromophore synthesis revealed by trapped intermediate structures. Proc Natl Acad Sci USA 100:12111–12116
103. Mizuno H, Mal TK, Tong KI, Ando R, Furuta T, Ikura M, Miyawaki A (2003) Photo-induced peptide cleavage in the green-to-red conversion of a fluorescent protein. Mol Cell 12: 1051–1058
104. Tsutsui H, Karasawa S, Shimizu H, Nukina N, Miyawaki A (2005) Semi-rational engineering of a coral fluorescent protein into an efficient highlighter. EMBO Rep 6:233–238
105. Nienhaus K, Renzi F, Vallone B, Wiedenmann J, Nienhaus GU (2006) Exploring chromophore-protein interactions in fluorescent protein cmFP512 from *Cerianthus membranaceus*: X-ray structure analysis and optical spectroscopy. Biochemistry 45:12942–12953

106. Wiedenmann J, Nienhaus GU (2006) Live-cell imaging with EosFP and other photoactivatable marker proteins of the GFP family. Expert Rev Proteomics 3:361–374
107. Nienhaus K, Nienhaus GU, Wiedenmann J, Nar H (2005) Structural basis for photo-induced protein cleavage and green-to-red conversion of fluorescent protein EosFP. Proc Natl Acad Sci USA 102:9156–9159
108. Ugalde JA, Chang BS, Matz MV (2004) Evolution of coral pigments recreated. Science 305:1433
109. Ward WW, Cormier MJ (1979) An energy transfer protein in coelenterate bioluminescence. Characterization of the Renilla green-fluorescent protein. J Biol Chem 254:781–788
110. Hastings J, Morin JG (2005) Photons for reporting molecular events: green fluorescent protein and four luciferase systems. Wiley
111. Clark KB (1975) Nudibranch life cycles in the northwest Atlantic and their relationship to the ecology of fouling communities. Helgoländer Meeresundersuchungen 27:28–69
112. Todd CD (1981) he ecology of nudibranch molluscs. Annual Review of Oceanography and Marine Biology 19:141–234
113. Todd CD (1983) Reproductive and trophic ecology of nudibranch molluscs. In: Russell-Hunter WD (ed) The Mollusca. Academic Press, New York, vol 6, Ecology, pp 225–259
114. Haddock SHD, Moline MA, Case JF (2010) Bioluminescence in the sea. Ann Rev Mar Sci 2:443–493
115. Partridge JC, Hilder PE (1990) The colour sensitivity and vision of fishes. In: Herring PJ, Campbell AK, Whitfield M, Maddock L (eds) Light and life in the sea. Cambridge University Press, Cambridge, pp 167–184
116. Partridge JC, Cummings ME (1999) Adaptation of visual pigments to the aquatic environment. In: Archer SN, Djamgoz MBA, Loew ER, Partridge JC, Vallerga S (eds) Adaptive mechanisms in the ecology of vision. Kluwer, Dordrecht
117. Jerlov NG (1976) Marine optics. Elsevier Scientific Publishing Company, Amsterdam
118. Kawaguti S (1969) Effect of the green fluorescent pigment on the productivity of the reef corals. Micronesica 5:313
119. Wiedenmann J, Leutenegger A, Gundel S, Schmitt F, D'Angelo C, Funke W (2007) Long-term monitoring of space competition among fluorescent and nonfluorescent sea anemones in the Mediterranean Sea. J Mar Biol Assoc UK 87:851–852
120. Fradkov AF, Chen Y, Ding L, Barsova EV, Matz MV, Lukyanov SA (2000) Novel fluorescent protein from Discosoma coral and its mutants possesses a unique far-red fluorescence. FEBS Lett 479:127–130
121. Wacker SA, Oswald F, Wiedenmann J, Knochel W (2007) A green to red photoconvertible protein as an analyzing tool for early vertebrate development. Dev Dyn 236:473–480
122. Campbell RE, Tour O, Palmer AE, Steinbach PA, Baird GS, Zacharias DA, Tsien RY (2002) A monomeric red fluorescent protein. Proc Natl Acad Sci USA 99:7877–7882
123. Kredel S, Nienhaus K, Oswald F, Wolff M, Ivanchenko S, Cymer F, Jeromin A, Michel FJ, Spindler KD, Heilker R, Nienhaus GU, Wiedenmann J (2008) Optimized and far-red-emitting variants of fluorescent protein eqFP611. Chem Biol 15:224–233
124. Kredel S, Oswald F, Nienhaus K, Deuschle K, Roecker C, Wolff M, Heilker R, Nienhaus GU, Wiedenmann J (2009) mRuby, a bright monomeric red fluorescent protein for labeling of subcellular structures. PLoS ONE 4:e4391
125. Ho SN, Hunt HD, Horton RM, Pullen JK, Pease LR (1989) Site-directed mutagenesis by overlap extension using the polymerase chain reaction. Gene 77:51–59
126. Shaner NC, Patterson GH, Davidson MW (2007) Advances in fluorescent protein technology. J Cell Sci 120:4247–4260
127. Chalfie M (2009) GFP: lighting up life. Proc Natl Acad Sci USA 106:10073–10080
128. Li X, Zhao X, Fang Y, Jiang X, Duong T, Fan C, Huang CC, Kain SR (1998) Generation of destabilized green fluorescent protein as a transcription reporter. J Biol Chem 273: 34970–34975

129. Yang M, Jiang P, Yamamoto N, Li L, Geller J, Moossa AR, Hoffman RM (2005) Real-time whole-body imaging of an orthotopic metastatic prostate cancer model expressing red fluorescent protein. Prostate 62:374–379
130. Shaner NC, Campbell RE, Steinbach PA, Giepmans BN, Palmer AE, Tsien RY (2004) Improved monomeric red, orange and yellow fluorescent proteins derived from Discosoma sp. red fluorescent protein. Nat Biotechnol 22:1567–1572
131. Merzlyak EM, Goedhart J, Shcherbo D, Bulina ME, Shcheglov AS, Fradkov AF, Gaintzeva A, Lukyanov KA, Lukyanov S, Gadella TW, Chudakov DM (2007) Bright monomeric red fluorescent protein with an extended fluorescence lifetime. Nat Methods 4:555–557
132. Hanson GT, McAnaney TB, Park ES, Rendell ME, Yarbrough DK, Chu S, Xi L, Boxer SG, Montrose MH, Remington SJ (2002) Green fluorescent protein variants as ratiometric dual emission pH sensors. 1. Structural characterization and preliminary application. Biochemistry 41:15477–15488
133. Griesbeck O (2004) Fluorescent proteins as sensors for cellular functions. Curr Opin Neurobiol 14:636–641
134. Tsutsui H, Karasawa S, Okamura Y, Miyawaki A (2008) Improving membrane voltage measurements using FRET with new fluorescent proteins. Nat Methods 5:683–685
135. Miyawaki A, Griesbeck O, Heim R, Tsien RY (1999) Dynamic and quantitative Ca2+ measurements using improved cameleons. Proc Natl Acad Sci USA 96:2135–2140
136. Patterson GH, Lippincott-Schwartz J (2002) A photoactivatable GFP for selective photolabeling of proteins and cells. Science 297:1873–1877
137. Subach FV, Patterson GH, Manley S, Gillette JM, Lippincott-Schwartz J, Verkhusha VV (2009) Photoactivatable mCherry for high-resolution two-color fluorescence microscopy. Nat Methods 6:311
138. Chudakov DM, Verkhusha VV, Staroverov DB, Souslova EA, Lukyanov S, Lukyanov KA (2004) Photoswitchable cyan fluorescent protein for protein tracking. Nat Biotechnol 22:1435–1439
139. Hell SW (2003) Toward fluorescence nanoscopy. Nat Biotechnol 21:1347–1355
140. Hell SW (2007) Far-field optical nanoscopy. Science 316:1153–1158
141. Betzig E, Patterson GH, Sougrat R, Lindwasser OW, Olenych S, Bonifacino JS, Davidson MW, Lippincott-Schwartz J, Hess HF (2006) Imaging intracellular fluorescent proteins at nanometer resolution. Science 313:1642–1645
142. Dedecker P, Flors C, Hotta JI, Uji-I H, Hofkens J (2007) 3D nanoscopy: bringing biological nanostructures into sharp focus. Angew Chem Int Ed Engl 46:8330–8332
143. Shroff H, Galbraith CG, Galbraith JA, White H, Gillette J, Olenych S, Davidson MW, Betzig E (2007) Dual-color superresolution imaging of genetically expressed probes within individual adhesion complexes. Proc Natl Acad Sci USA 104:20308–20313
144. Hedde PN, Fuchs J, Oswald F, Wiedenmann J, Nienhaus GU (2009) Online image analysis software for photoactivation localization microscopy. Nat Methods 6:689–690
145. Egner A, Geisler C, von Middendorff C, Bock H, Wenzel D, Medda R, Andresen M, Stiel AC, Jakobs S, Eggeling C, Schonle A, Hell SW (2007) Fluorescence nanoscopy in whole cells by asynchronous localization of photoswitching emitters. Biophys J 93:3285–3290
146. Stiel AC, Trowitzsch S, Weber G, Andresen M, Eggeling C, Hell SW, Jakobs S, Wahl MC (2007) 1.8 A bright-state structure of the reversibly switchable fluorescent protein Dronpa guides the generation of fast switching variants. Biochem J 402:35–42
147. Adam V, Lelimousin M, Boehme S, Desfonds G, Nienhaus K, Field MJ, Wiedenmann J, McSweeney S, Nienhaus GU, Bourgeois D (2008) Structural characterization of IrisFP, an optical highlighter undergoing multiple photo-induced transformations. Proc Natl Acad Sci USA 105:18343–18348
148. Fuchs J, Bohme S, Oswald F, Hedde PN, Krause M, Wiedenmann J, Nienhaus GU (2010) A photoactivatable marker protein for pulse-chase imaging with superresolution. Nat Methods 7:627–630

Green Fluorescent Protein Photodynamics as a Tool for Fluorescence Correlative Studies and Applications

Giuseppe Chirico, Maddalena Collini, Laura D'Alfonso, Michele Caccia, Stefano Carlo Daglio, and Barbara Campanini

Abstract The use of green fluorescent protein (GFP) mutants has represented a revolution in the application of optical microscopy to cell biology. GFP photodynamics also has been recently addressed in detail due to the crucial role that photoswitchable GFPs have been acquiring in nanoscopy. As a prototype of the class of the reversible photoswitchable GFPs, we discuss here the photophysics of the E222Q mutant of GFPMut2 (Mut2Q).

The fluorescence of the anionic state of this mutant, primed by blue light, is markedly enhanced under the additional excitation at a shorter wavelength and it relaxes within a few milliseconds. By means of modulated double beam fluorescence correlation spectroscopy on protein solutions, two characteristic photoswitching times are found for Mut2Q that lie in the 1–30 ms range and a four states energy diagram can be depicted from these data. By means of experiments on GFPs immobilized in acrylamide gels, we show how the modulated excitation microscopy coupled to the use of photoswitchable proteins can be exploited to obtain the photodynamics parameters on a pixel-by-pixel basis, in view of future applications of this or similar mutants to live cells imaging.

Keywords Cross-correlation · E222Q mutant · Fluorescence correlation spectroscopy · Green fluorescent protein · Photodynamics

Contents

G. Chirico (✉), M. Collini, L. D'Alfonso, M. Caccia, and S.C. Daglio
Dipartimento di Fisica, Università degli study di Milano Bicocca, Piazza della Scienza 3, 21026 Milano, Italy
e-mail: giuseppe.chirico@mib.infn.it

B. Campanini
Dipartimento di Biochimica e Biologia Molecolare, Università degli Studi di Parma, Viale Usberti 23/A, 43100 Parma, Italy

G. Jung (ed.), *Fluorescent Proteins II*, Springer Ser Fluoresc (2012) 12: 35–56,
DOI 10.1007/4243_2011_25
Published online: 28 April 2011

1 Introduction

Green fluorescent protein (GFP) denotes a class of proteins that has revolutionized the use of fluorescence microscopy in biology. The visualization of cells and their constituents is mostly made nowadays through the genetically encoded fusions of fluorescent proteins that permanently transfect a cell line or even a whole eukaryotic organism [1, 2]. However, the possibility to genetically fuse GFP mutants to specific proteins and then to follow their fate within the cell by monitoring the GFP emission is limited by quenching and bleaching, and therefore the scientific community is striving to improve the stability of GFP and to devise new visible-infrared emitting proteins. The GFP itself already displays a remarkable stability against photo-destruction [3] and against denaturing agents [4]. Moreover, its spectroscopic characteristics (absorption and emission wavelengths, quantum yields) can be tuned by specific mutations on the chromophore or in the chromophore pocket creating a multicolor scenario of mutants spanning from blue to near IR emission (from about 350 nm to about 600 nm) [5, 6].

The possibility to tailor the emission wavelength of GFP mutants to the specific needs in imaging or sensing applications is not the only spectroscopic tool: GFP photoactivation is now playing a crucial role in many aspects of the biomedical research, and GFP, GFP-like, and other fluorescent proteins (FPs) represent a large breakthrough in the study of intracellular tracking [7–9].

Already in the very first years of GFP use, Dickinson et al. [10] reported the peculiar feature of a UV-induced photoswitching in GFP proteins trapped in gels after transition to a dark state induced by long-term irradiation (photobleaching). Since then, several photoactivation mechanisms have been reported for GFP and FPs, beside photobleaching and thermally activated blinking, as reviewed by Day and Davidson [11]. These processes consist in changes in the chromophore conformation occurring upon irradiation, which result either in a different absorption and/or emission spectrum [11]. In these years, a variety of names have been used in the literature for these behaviors. The FPs whose fluorescence emission is largely enhanced after irradiation by a high energy (UV–violet) beam can be defined as *photoactivable* FPs, and, among them, the most known and studied is the PA-GFP [12–14]. A drawback of photoactivable FPs could be their low fluorescence before activation which often prevents their localization in cells. The *photoconvertible* FPs, as for example Kaede [15, 16], are proteins which changes their emission wavelength upon irradiation thereby allowing their detection before and after activation in two different channels. For the vast majority of these proteins, the activation is an irreversible process; however, a different class of FPs exists, the *photoswitchable* FPs, for which irradiation at appropriate wavelength(s) can repeatedly convert the chromophore between two states. Examples of such proteins are YQ1, EYQ2, and Dronpa [17–20]. Also wt-GFP can be included in this class with the caveat that its "blinking" on/off behavior is stochastic, whereas the newly developed mutants can be light driven.

The use of these "photoΣ" FPs in fluorescence microscopy allow to attain enhanced spatial resolution, far below the diffraction barrier [14, 21–23], due to the development of techniques such as photoactivated localization microscopy (PALM) [14], fluorescence PALM (FPALM) [24], stochastic reconstruction microscopy (STORM) [25], PALM with independent running acquisition (PALMIRA) [26], stroboscopic PALM (SPALM) [27], and RESOLFT microscopy (reversible saturable optical fluorescence transitions between two marker states) [28].

The photophysical bases of the "FPΣ" behaviors have then been the object of many recent investigations [19, 29–32] on two widely different platforms: proteins trapped in gels through single molecule studies [33] and proteins in solution by means of fluorescence correlation techniques [34].

Although the possibility to single out the properties of a protein at a time is very appealing and can in principle provide detailed knowledge of the fluctuations in the chromophore quantum yield, and the on and off switching times, one cannot exclude some kind of interaction with the host matrix [35]. The observation of proteins while they are flying through an observation volume in solution offers the possibility to follow their photodynamics when highly diluted samples are used and, through correlative methods, allows to single out the kinetic components of the fluorescence fluctuations even in conditions of poor signal/noise ratio [36].

In the last few years, we have been actively involved in this second kind of approach and in trying to transpose it from solution to cell measurements. In the particular case of a switchable FP, the switching itself, regarded so far as a drawback for the use of GFP mutants in microscopy studies, can instead be used advantageously in biological studies, and even more in nanotechnological applications by exploiting the dependence of the switching mechanism on the chemical and physical properties of the environment in which the proteins are embedded. The aim of the present contribution is to give an overview of how GFPs photoswitching can be exploited to obtain technical improvements in fluorescence microscopy, using a time-structured excitation [37]. This can be achieved by focalizing on the sample two laser beams, a pump beam for conventional excitation and a more energetic probe beam which is also modulated in time as a square wave.

The outline of the chapter is as follows. We will first give a short review of the GFP blinking processes. This is not intended as an extensive and comprehensive review of all the large amount of data presented in the literature, but it helps to point out those specific characteristics of the GFP photodynamics that we propose to exploit to develop new methods for microscopy based on time-modulated excitation. We then describe possible protocols to implement fluorescence microscopy based on photoswitching of GFP mutants and time modulation of the excitation intensity. The experimental investigation is paralleled by an effort to develop and theoretically analyze models of energy diagrams [19, 38, 39]. Finally, we discuss the perspectives of this approach to imaging applications.

1.1 GFP Photodynamics

From a biophysical point of view, GFP appeal in biomedical studies is due to its possibility to absorb UV-blue light and emit green light with relatively high efficiency. Mutations of residues within the chromophore or on the protein scaffold have an impressive effect not only on the excitation and emission wavelength but also on the fluorescence quantum yield, ranging from values as large as $\phi = 0.8$–0.2 [3, 11, 21, 40–42]. As briefly discussed in the introduction, the use of GFPs is affected by several photodynamic mechanisms, particularly photobleaching [19, 43], blinking (or flickering) [44], photoactivation [12, 41], and photoswitching [19, 31, 39, 41, 45]. However, most of these properties, when coupled to the sensitivity of the protein chromophore to the external environment (pH, chlorine ions, etc.), can also be exploited to obtain molecular sensors [46, 47] and to enhance the image resolution by single molecule photoactivation [22].

Blinking and bleaching affect the fluorescence signal by reducing the (effective) quantum yield [3] and by inducing fluorescence fluctuations with characteristic times in the range of microseconds to seconds [34, 38, 48–51]. The most limiting process is photobleaching. This mechanism, which can have a different origin depending on the excitation mode, OPE [52] or TPE [43, 48, 53], is recognized in the single molecule experiments as a permanent loss of fluorescence of the observed molecule. In fluorescence correlation spectroscopy (FCS), bleaching induces the shortening of the apparent diffusion times, a change in the autocorrelation function (ACF) amplitude and shape, and an overall decrease in the maximum achievable brightness [48, 54]. The protein spontaneous blinking processes have been ascribed to transitions to dark (neutral or zwitterionic) states reached through tiny conformational changes (*cis–trans* transitions [19, 29]) coupled to changes in the protonation state of the chromophore [34, 40, 49], that also involve the rearrangement of the proton network within the chromophore pocket [49 and references therein, 55]. The blinking efficiency of the particular GFP protein depends on the specific mutations [44] and on the physical (temperature) [56] and chemical (pH and metal ions) [47, 57] parameters of the solution. The dependence of blinking on pH has been investigated extensively by FCS experiments in solution [34, 51, 58], and it has also led to devise intracellular sensors of pH or ions based on GFPs [19, 59, 60].

Among all the processes that at the molecular level are classified as blinking, we focus on the photoswitching processes in GFPs: these can be induced by the mutation of glutamate 222 into glutamine, which is believed to be the key mutation for the induction of stable (i.e., slowly, thermally relaxed) photoswitching in several mutants [3, 19].

The protein on which we will base this discussion is derived from the parent GFPMut2 protein that carries the mutations S65A, V68L, and S72A [61]. The E222Q mutation in GFPMut2 results in a mutant called Mut2Q, which shows photoswitching properties. Differently from other photoswitchable proteins that are characterized by quasi permanent photoactivation, in Mut2Q the activation and deactivation process is characterized by characteristic times of the order of few milliseconds. This rapid activation kinetics might also be exploited in the future to probe cellular processes which fall in this time window.

The spectroscopic properties of the parent protein and its mutant under one-photon (OP) and two-photon (TP) excitation have been reported in a number of previous works [49, 62] and are summarized for discussion purposes in Fig. 1. Briefly, the p*K*a for the chromophore anionic-neutral equilibrium is 6.0 ± 0.1, and the absorption bands fall at ≅390 nm for the chromophore neutral state and ≅490 nm for the anionic state. Excitation at 490 nm leads to fluorescence emission at 508 nm, typical of the anionic form of the chromophore, in a wide range of pH values. In the case of GFPMut2, also excitation at 390 nm produces emission at 508 nm, due to excited state proton transfer (ESPT), whereas this effect is reduced in Mut2Q due to the E222Q mutation [3].

Moreover, the E222Q mutation induces a marked decrease in the chromophore brightness (Fig. 1) particularly under OPE (Fig.1a, b), as also reported by Jung and coworkers for a set of GFP mutants [41, Table 1] encompassing amino acids 203, 65, and 222.

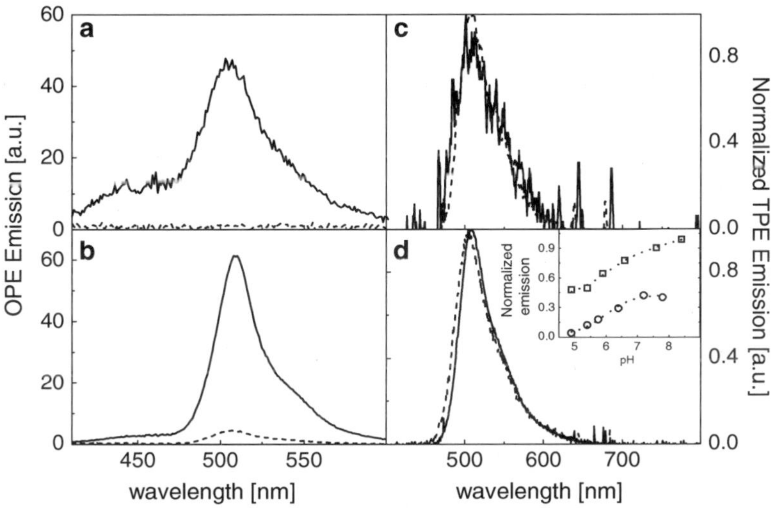

Fig. 1 One-photon primed (λ_{exc} = 390nm) fluorescence emission of the GFPMut2 (*solid line*) and Mut2Q (*dotted line*) mutants in acidic (pH = 5; panel **a**) and basic (pH = 8.5; panel **b**) conditions. The buffer was citrate-phosphate (10–100 mM), protein concentration 0.1 μM. Panels (**c**) and (**d**) report normalized emission spectra at pH = 5 (**c**) and at pH=8.5 (**d**) under TPE excitation at 890 nm. Protein concentration was 10–30 nM, laser intensity was ≅11 MW/cm^2 (the data were Fourier smoothed over 5 points). The *inset* of panel (**d**) reports the pH titration of fluorescence at 510 nm. *Squares* and *circles* refer to GFPMut2 and Mut2Q, respectively. The *solid lines* are the best fit to sigmoidal functions with p*K* = 6.2 ± 0.1 (GFPMut2) and p*K* = 6.0 ± 0.1 (Mut2Q)

1.2 GFP Mutants Single Beam Photodynamics

Photoswitching of GFP involves the use of two laser beams that may conventionally be indicated as pump and probe beams, which respectively induce the ground–excited state transition and photo-induces a change in the chromophore environment. In order to exploit the Mut2Q switching properties, as it is our focus here, it is essential to have first a detailed knowledge of the fluorescence dynamics that can be observed and induced by a single (pump) beam.

The fluorescence dynamics can be studied in solution by means of FCS. The ACFs of the fluorescence intensity are analyzed in terms of a superposition of the diffusive (hyperbolic) and the photodynamics (exponential) components:

$$
\begin{aligned}
G(t) &= G_{\mathrm{D,(OPE,TPE)}}(t) \times \prod_j \left[1 + \frac{A_j}{1-A_j}\exp(-t/\tau_j)\right], \\
G_{\mathrm{D,TPE}}(t) &= \frac{\gamma_{\mathrm{TPE}}}{\langle N \rangle}\left(1+\frac{t}{\tau_{\mathrm{D}}}\right)^{-1}\left(1+\left(\frac{\lambda}{\sqrt{2}\pi\omega_0}\right)^2\frac{t}{\tau_{\mathrm{D}}}\right)^{-0.5}, \\
G_{\mathrm{D,OPE}}(t) &= \frac{\gamma_{\mathrm{OPE}}}{\langle N \rangle}\left(1+\frac{t}{\tau_{\mathrm{XY}}}\right)^{-1}\left(1+\frac{t}{\tau_{\mathrm{Z}}}\right)^{-0.5},
\end{aligned}
\tag{1}
$$

where τ_j is the jth photodynamics relaxation time with fraction A_j, $\langle N \rangle$ is the average number of molecules in the excitation volume, $\gamma_{\mathrm{OPE}} \cong 0.35$ and $\gamma_{\mathrm{TPE}} \cong 0.076$ are geometrical factors, and the diffusion times $\tau_{\mathrm{D}} = \omega_0^2/8D$, for TPE, and $\tau_{\mathrm{XY}} = \omega_0^2/4D$, for OPE, are related to the diffusion coefficient D and to the laser beam waist ω_0. For the axial diffusion time a value $\tau_{\mathrm{Z}} \cong 5\ \tau_{\mathrm{XY}}$ is typically assumed [36].

For our mutants, three photodynamics components are found, typically in the range from one to hundreds of microseconds, attributed to intersystem crossing transition [j = "T" in (1)], to the protonation dynamics [j = "P" in (1)] and to *cis–trans* isomerizations or other photoinduced transitions to high energy states of the chromophore that may also imply structural changes [j = "L" in (1)]. Both the relaxation times and their amplitudes may show dependence on pH, temperature, and excitation intensity, and the characterization of this dependence is relevant for their future applications in cells where the pH may vary substantially, being shifted often to acidic values, and where the temperature is typically 37°C.

The fastest component "T" is the so-called stochastic blinking that has been observed both on immobilized GFP proteins [10, 63] and in diluted solutions [34]. Neglecting this very short (few microseconds) triplet state dynamics, we have found that, for the other components, the relaxation rates range from 5 to 60 kHz (relaxation times ≅ 10–300 μs) depending on the excitation intensity and on pH [49]. Concerning the amplitudes, the "L" component is independent from pH, but increases with the excitation intensity to a value of $A_{\mathrm{L}} \cong 0.45$ at 20 MW/cm^2 (TPE) or 3 kW/cm^2 (OPE). On the contrary, the "P" component strongly depends upon the

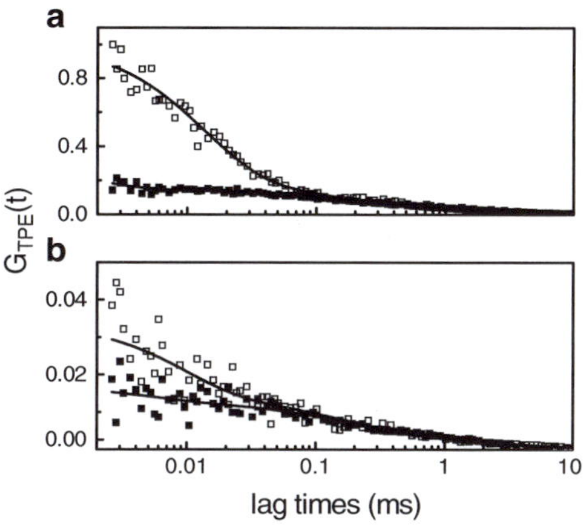

Fig. 2 Normalized ACFs of the fluorescence fluctuations measured on solutions of Mut2 (panel **a**) and Mut2Q (panel **b**) for pH ≅ 5 (*open symbols*) and pH ≅ 8.6 (*filled symbols*) by TPE at $\lambda = 890$ nm. The *solid lines* are the best fit to (1) (TPE)

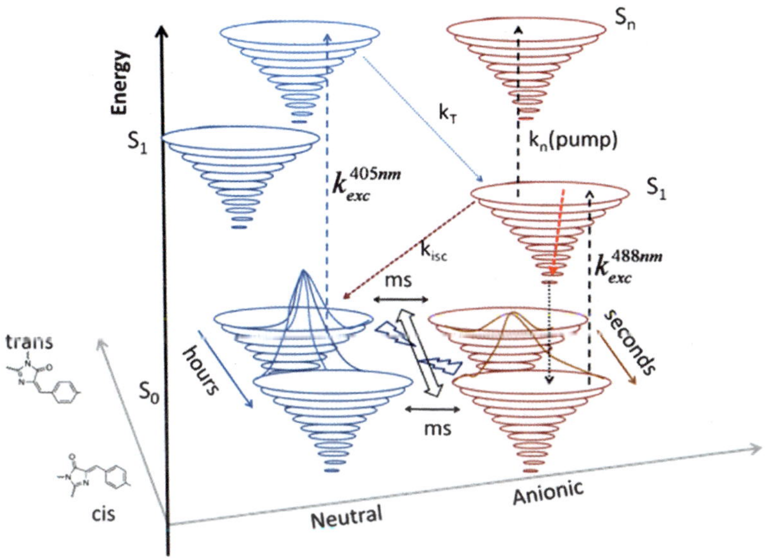

Scheme 1 The *circles* represent the vibrational sublevels in the electronic states. Only the most relevant energetic barriers are indicated in the scheme. The *trans* configuration of the neutral chromophore represents the dark state of the GFP [19]. The *solid lines* correspond to thermally activated transitions between the ground states. The *dashed lines* represent the photoinduced transitions and correspond to the 490 nm absorption band [S_0 (anionic)→S_1 (anionic)], to the photoswitched recovery of the fluorescence by the depletion of the trap neutral *trans* state by the (405 nm) probe beam (S_1→S_2). The ESPT mechanism (not shown) should involve S_1 (neutral) and S_1 (anionic) levels

pH ranging from $A_P \cong 0$ at alkaline values up to $A_P \cong 0.8$ at pH ≅ 5, independent from the excitation intensity [49]. Such a marked contribution of the photodynamics to the fluorescence correlation curves is evident at acidic pH values, as reported in Fig. 2.

A recently proposed scheme for the chromophore photodynamics [19] takes into account both the protonation of the *cis* and *trans* forms and a photoinduced isomerization process which is much faster than the thermally driven *trans–cis* isomerization. A possible outline of this complex scenario is given in Scheme 1.

The photoinduced isomerization and the neutral–anionic transitions are relatively fast processes lying in the range of the millisecond [38, 39]. The thermal recovery of the *cis* form of the chromophore at the ground state occurs instead over longer times, from tens of seconds to hours [19, 39]. These processes deal with the relaxation between ground electronic states [19, 34, 64–66]. However, there may be photoinduced transitions from excited states of the chromophore. One of these transitions is the so-called excited state electron transfer (ESPT) that involves directly the E222 residue, is the origin of the large Stokes shift observed in many GFP mutants [64, 66, 67], and it should be strongly reduced in Mut2Q mutant [3].

1.3 GFP Mutants Dual Beam Excitation Photodynamics

Under two laser beams excitation, a reversible photoswitching effect is observed in E222Q mutants [19, 38, 68, 69]. In these experiments, a protein solution is shined continuously with a 488 nm laser beam that primes the emission of the anionic form of the chromophore, while a second laser beam tuned in the range 405–420 nm is superimposed. The 508 nm emission is found to be substantially larger than the sum of the emissions primed by one beam at a time, suggesting that a photoactivated process has occurred, as shown, for example, in Fig. 3a for Mut2Q. A characterization of this process can be obtained by estimating the increase in the fluorescence emission of the protein under dual beam excitation and its characteristic time.

Concerning the fluorescence enhancement, we found that the activation of the fluorescence by the blue probe laser is a property of the E222Q mutation, since the parent protein GFPMut2 does not show any enhancement or delayed emission under 405 nm irradiation (Fig. 3a, inset), but at most a limited bleaching. The maximum relative enhancement $\Delta F/F_0$ can be obtained by fitting the fluorescence emission as a function of I_{probe}, at selected values of I_{pump}, to the trial function $\langle F \rangle = F_0 + \Delta F \times I_{\text{probe}} \left(I_{\text{s1}} + I_{\text{probe}} + (I_{\text{probe}}/I_{\text{s2}})^2 \right)^{-1}$, where I_{S1} and I_{S2} are saturation intensities that account for the triplet state population and for the transitions from the excited singlet to higher energy states [38]. From the best fit of the data, we can estimate a maximum value of $\Delta F/F_0 \cong 3.5$–4 as shown in Fig. 3b. The fluorescence enhancement amplitude is of the same order of magnitude of that found for wt-GFP [2, 12, 70]. Nevertheless, the limited bleaching induced by the probe beam in this mutant makes it potentially amenable to cell applications [38].

A characterization of the characteristic times related to this process can be achieved by modulating the probe (or the pump) beam at a frequency ν of the same order of magnitude of the characteristic rate of the process itself. From the

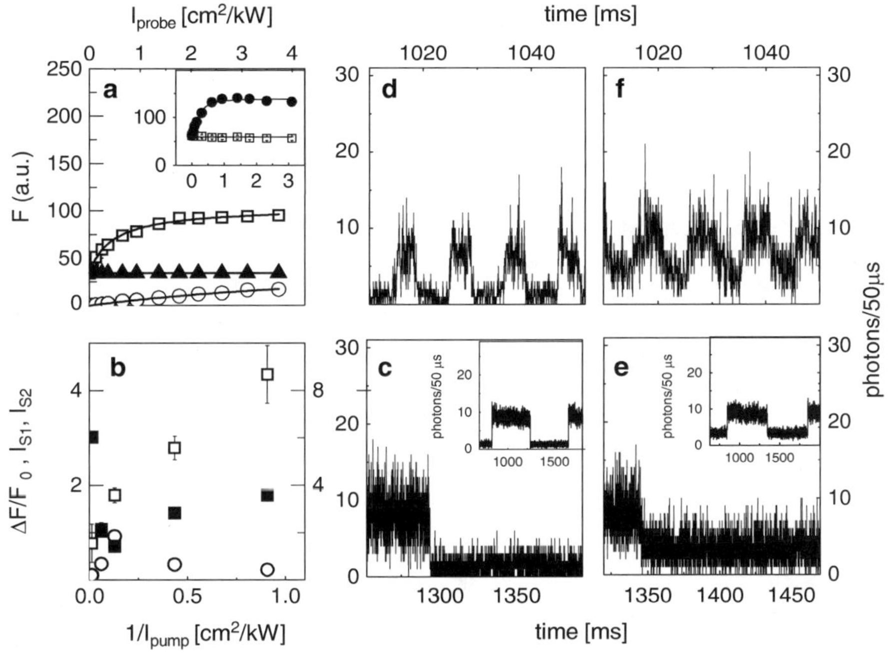

Fig. 3 Panel (**a**) reports the fluorescence emission ($\lambda_{em} = 515/30$ nm) of Mut2Q at pH 8.4 versus the probe beam intensity: probe excitation only (405 nm) *open circles*; pump excitation only (488 nm, 1.1 kW/cm2) *full triangles*; dual beam excitation *open squares*. *Inset* of Panel (**a**): fluorescence emission under dual beam excitation versus the probe intensity ($I_{pump} = 1.7$ kW/cm2) for GFPMut2 (*open squares*) and Mut2Q (*full circles*). Data refer to the same protein concentration. The *solid lines* are best fit curves as described in the text, and the best fit parameters are reported as a function of $1/I_{pump}$ in panel (**b**): $\Delta F/F_0$ (left axis, *open squares*), I_{s1} (right axis, *open circles*) and I_{s2} (right axis, *filled circles*). Panels (**c**–**f**): Fluorescence traces collected from dilute solutions (20 nM) of GFPMut2 (**c**, **d** panels) and Mut2Q (**e**, **f** panels) proteins under the action of the pump (488 nm, continuously irradiating) and probe (405 nm, modulated) beams. The solution pH was 8.6. The probe beam was alternated on the sample by means of an electromechanical shutter at $\nu = 1$ Hz (panels **c** and **e**) and $\nu = 100$ Hz (panels **d**, **f**). The *insets* in panels (**c**) and (**e**) report a longer stretch of the traces (smoothed by adjacent averaging over 5 points)

fluorescence traces reported in Fig. 3e, f, it is apparent that the Mut2Q mutant displays a particularly fast (millisecond) kinetic response to changes in the blue excitation light intensity, resulting in a rounded transition of the signal for the Mut2Q mutant only when the blue beam is switched on and off. This effect is more pronounced under high modulation frequencies, $\nu \cong 100$–300 Hz, but also at modulation frequencies as low as $\nu = 1$ Hz, it is possible to note that the on–off (or off–on) transitions for Mut2Q are not as sharp, at the 50 μs time resolution, as for the GFPMut2 parent protein (Fig. 3c, d).

The process outlined above and represented by the data reported in Fig. 3 can be analyzed by means of fluctuation spectroscopy and ACFs decays. In general, for steady single (pump) beam excitation, the fluorescence signal can be written as:

$$F(t,T) \cong k_R T \eta \sigma \langle I_{pump} \rangle \int W(\mathbf{r})\ \mathrm{S}_1(\mathbf{r},t)\,d\mathbf{r}, \tag{2}$$

where k_{R} is the radiative constant, T is the sampling time (supposed to be much smaller than the characteristic fluorescence fluctuation time), η is the detection efficiency, σ is the absorption cross section, $W(\mathbf{r})$ is the normalized intensity of the excitation profile, $\langle I_{\mathrm{pump}} \rangle$ is the average intensity of the steady pump beam, $\mathrm{S}_1(\mathbf{r}, t)$ is the number concentration of the molecules in the S_1 excited state. The fluctuations of the fluorescence signal may come from the number fluctuations, the changes in the probability P_1 of the molecules to be in the S_1 state, or the changes in the pump intensity, $I_{\mathrm{pump}}(t)$. Let us first consider the simplest case: no photodynamics exists in the time range explored by the FCS ACFs, and the modulation frequency of the pump excitation fall in ranges where no coupling with the photodynamics may exist. Equation (2) becomes then:

$$F_{NP}(t,T) \cong \eta k_R \sigma T I_{pump}(t) \int W(\mathbf{r}) N(\mathbf{r},t)\,d\mathbf{r}, \tag{3}$$

where $N(\mathbf{r}, t)$ is the total number of fluorescent molecules. The fluorescence ACF can be described by:

$$g(\tau) = \frac{\langle \delta F_{NP}(t+\tau,T)\delta F_{NP}(t,T)\rangle_t}{\langle F(t,T)\rangle_t^2} \cong \frac{\langle N(t+\tau)N(t)\rangle_t}{\langle I_{pump}N\rangle_t^2}\langle \delta I_{pump}(t+\tau)\delta I_{pump}(t)\rangle_t$$
$$+ \frac{\langle I_{pump}(t+\tau) I_{pump}(t)\rangle_t}{\langle I_{pump}N\rangle_t^2}\langle \delta N(t+\tau)\delta N(t)\rangle_t, \tag{4}$$

where $N(t) = \int W(\mathbf{r})N(\mathbf{r},t)\,d\mathbf{r}$ and $\delta N(t) = \int W(\mathbf{r})\ \delta N(\mathbf{r},t)\,d\mathbf{r}$.

If we now define the number and pump intensity correlation functions as:

$$\begin{aligned} g_N(\tau) &= \left\langle \int \langle W(\mathbf{r})\ \delta N(\mathbf{r},t)\ W(\mathbf{r}')\ \delta N(\mathbf{r}',t+\tau)\rangle d\mathbf{r}\,d\mathbf{r}' \right\rangle / \langle N\rangle^2, \\ g_I(\tau) &= \langle \delta I_{pump}(t+\tau)\delta I_{pump}(t)\rangle / \langle I_{pump}\rangle_t^2 \end{aligned} \tag{5}$$

and rearrange terms in (4), we can write that the normalized fluorescence correlation function as:

$$g(\tau) = g_I(\tau) + (1 + 4g_I(\tau))g_N(\tau). \tag{6}$$

From this expression, it is easy to identify two limiting behaviors. When highly diluted solutions are investigated under modulated excitation (of either the pump or the probe beam), the fluorescence fluctuations correlation function is dominated by the number fluctuations [i.e., $g_N(\tau)$], and the ACF is well approximated by the form:

$$g(N \simeq 0)(\tau) \cong (1 + 4g_I(\tau))g_N(\tau). \quad (7)$$

On the other hand, for relatively concentrated solutions the limiting behavior is:

$$g(N \gg 1)(\tau) \cong g_I(\tau). \quad (8)$$

Similar results are obtained if we take into account the protein photodynamics whose characteristic times are assumed to be much smaller than the typical modulation time of the pump beam intensity. In this case, the $g_N(\tau)$ function should be simply replaced with the whole ACF decay as reported in (1). If we are specifically interested in studying the effect of the beam intensity modulation on the protein photodynamics, we should then measure at $\langle N \rangle \cong$ 20–40 molecules per excitation volume.

However, if the internal photodynamics contains relaxation times of the order of the pump beam intensity modulation time, we should consider a direct coupling between the photodynamics and the pump beam modulation that brings us to reconsider the parameters entering (3). In this case, the variable over which to compute the $g_I(\tau)$ correlation function becomes $\int_{-\infty}^{t} I_{pump}(x)P_1(r, t - x; I_{probe})dx$, where $P_1(\mathbf{r}, t; I_{probe})$ represents the probability, per unit of excitation intensity, to find a molecule that sits at $\mathbf{r}$ at the time t in the S_1 excited state. The shape of the $P_1(\mathbf{r}, t; I_{probe})$ function can be inferred by the analysis, in terms of eigenvalues and eigenvectors, of the linear system of equations that describe the evolution of the energy levels populations [38].

It must be further noticed that by modulating the excitation beam, we are sampling the dynamic response of the quantum system. In other words, we are performing a Laplace analysis of the response function, $P_1(\mathbf{r}, t; I_{probe})$, of the quantum system to a pulsed excitation. In fact, if we suppose that the quantum system response is a single exponential with relaxation rate Γ, i.e., $P_1(\mathbf{r}, t; I_{probe}) \cong \exp(-\Gamma t)$, independently of the position within the excitation volume (which is an hypothesis requiring additional considerations), and if we suppose that the excitation (pump only at this stage) changes with time as $I_{pump}(x) = A\cos(\Omega t)$, the intensity over which to compute the fluctuations becomes:

$$\int_{-\infty}^{t} I_{pump}(x)P_1(\mathbf{r}, t - x; I_{probe})\mathrm{d}x = \frac{A}{\sqrt{\Omega^2 + \Gamma^2}}\cos\left(\Omega t - \frac{\Omega}{\Gamma}\right). \quad (9)$$

The intensity ACF is then given by:

$$g_I(\tau) = \frac{1}{\Omega^2 + \Gamma^2}\cos^2(\Omega\tau) = \frac{m^2(\Omega)}{2}\cos(\Omega\tau). \quad (10)$$

Therefore, we can measure the relaxation rate Γ from the zero lag time extrapolation of the correlation function, $g_I(0)^{-1} = \Omega^2 + \Gamma^2$.

It is interesting to note that by performing a cross-correlation of the protein (endowed with photodynamics) and the excitation signal, or, even better, a signal coming from a solution with no photodynamics in the analyzed range of frequencies, we would be able to measure the relaxation rate Γ of the P_1 function not only from the amplitude modulation but also from the dephasing term $\cong \Omega/\Gamma$.

The above formalism can be applied to the characterization of photoswitching processes, both in solutions and on images. In our case, the probe beam is modulated, and we can apply the same formalism as in (9) and (10) if we exploit the hypothesis that the $P_1(\mathbf{r},t; I_{\text{probe}})$ function can be expanded in series of eigenvalues that depend linearly on I_{probe} [38]. Any relaxation component will then provide a demodulation of the zero lag time correlation function as that reported in (10). If we use an average of 30–40 molecules per excitation volume, the correlation function is given by (8) and (10), and the dependence of the zero lag time extrapolation, $g_I(0)$, on the modulation frequency can be analyzed accordingly, as shown for example in Fig. 4. In real experiments, the situation is more complex due to the presence of a background (with average value $\langle B \rangle$), and the demodulation factor changes to the form:

$$g_I(\tau) = \frac{m^2(\Omega)}{2} \frac{\langle F^2 \rangle}{\langle F^2 \rangle + \langle B^2 \rangle} \cos^2(\Omega\tau). \tag{11}$$

Moreover, if more than one relaxation component is found each with number fractions f_i in the range accessible to FCS (relaxation rates <1 MHz), the intensity correlation function becomes [38]:

$$g_I(\tau) = \frac{1}{2} \frac{\langle F^2 \rangle}{\langle F^2 \rangle + \langle B^2 \rangle} \times \left[\left(\sum_i f_i m_i^2(\Omega) \right)^2 + \left(\sum_i f_i m_i(\Omega) \sqrt{1 - m_i^2(\Omega)} \right)^2 \right] \cos^2(\Omega\tau). \tag{12}$$

The zero lag time extrapolation of the ACF has been measured under modulated pump (488 nm) or probe (405 nm) irradiation at a modulation frequency in the range $0.5 < \Omega/2\pi < 100$ Hz. The results of the experiments on Mut2Q are reported in Fig. 4a. The analysis of the dependence of $g_I(0)$ as a function of the modulation frequency according to (12) indicates the presence of two relaxation components with relaxation times of the order of few tens of milliseconds, as shown in Table 1. These characteristic times depend on the solution pH while are not affected appreciably by the solution viscosity and temperature [38].

An energy diagram model that includes at least a trap state and a higher energy singlet state over which the excitation of the chromophore can cycle has been suggested to rationalize the data reported in Fig. 4, and thoroughly discussed in terms of equilibrium solutions and eigenvalues, or relaxation times [38]. The main feature of this scheme is the presence of a trap state from which the chromophore

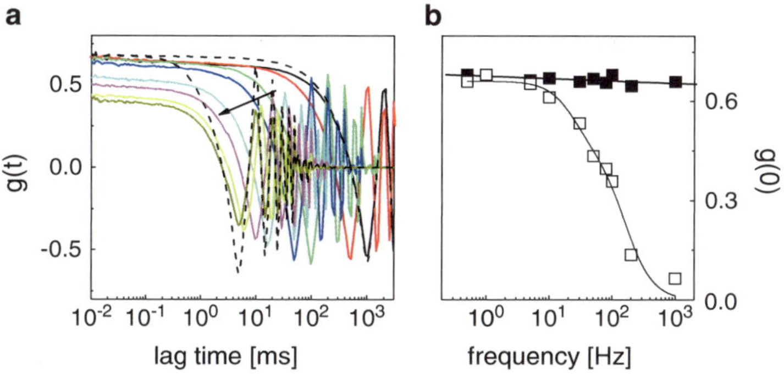

Fig. 4 Panel (**a**): ACFs of Mut2Q solutions (40 molecules per excitation volume), under modulated probe beam excitation ($\lambda_{probe} = 420$ nm, $I_{probe} = 1.6$ kW/cm^2 $I_{pump} = 16.3$ kW/cm^2) at modulation frequencies varying from 0.5 to 100 Hz (0.5, *black*; 1, *red*; 5, *green*; 10, *blue*; 30, *cyan*; 50, *magenta*; 80, *yellow*, and 100 Hz, *dark olive*). The *dashed black lines* report the ACFs of the same solution under pump modulation. Panel (**b**): zero lag time correlation function, $g(0)$, as a function of the modulation frequency for the cases of probe modulation (*open squares*) and pump modulation (*filled squares*). The *lines* are the best fit to the data of (12), with $\tau_1 = 1.3$ ms, $f_1 = 68\%$, and $\tau_2 = 32.8$ ms

Table 1 Analysis of the dependence of the modulation factor, $g_I(0)$, on the modulation frequency at two different pH values for Mut2Q, $I_{pump} = 1.7$ kW/cm^2

pH	f_1	$\tau_{1,R}$ (ms)	$\tau_{2,R}$ (ms)	$\langle\tau\rangle$ (ms)
6.2	0.70 ± 0.04	8.2 ± 2.0	113.9 ± 44	40.1 ± 14.0
9.0	0.69 ± 0.02	1.3 ± 0.2	27.7 ± 3.5	9.55 ± 1.23

can escape with the help of a more energetic radiation (420 nm typically). In [38], the possibility was envisaged that the action of the probe energetic beam was to promote a transition to a higher energy state, S_2 from which a rapid decay to S_1 and then to S_0 could follow. This assumption was also supported by similar behaviors found in many organic compounds and fluorescent proteins, in which the phenomenon is known as (photoinduced) reverse intersystem crossing [71 76]. However, it must be considered the possibility that configuration switchings such as the *cis–-trans* transitions proposed by several groups [19, 77–79] for the ground state equilibrium (e.g., see Scheme 1) might also occur at the excited state, and therefore the S2 state (Scheme 1) might be a different chromophore configuration rather than a distinct electronic state.

1.4 Dual Beam Modulation Imaging

How can we exploit the millisecond activation time of the fluorescence displayed by the E222Q mutants? A dependence of the fluorescence enhancement and of the activation times on the pH has been proved in vitro [38], and it might be also translated to the cell environment. However, a more basic application consists in the exploitation of the fluorescence enhancement and of the relaxation times to perform fluorescence imaging at increased signal/noise ratio. In fact, both the activation time and the enhancement factor could be measured on immobilized proteins and used as a contrast parameter.

This kind of applications must deal with the (minor but finite) photobleaching of the GFP mutants. A simple way to limit such an effect is to fast scan the sample at a scanning frequency $v = f_S$ that is slightly detuned with respect to a multiple of the (pump or probe) beam modulation frequency, f_L. In this way, the phase of the beam modulation wave on each pixel is scanned linearly, and a limited number of repetitions of each line can provide the response to a full period of light beam modulation. This principle is outlined and exemplified in Fig. 5a, b: a single line is scanned repeatedly 180 times, while it is continuously irradiated by the 488 nm beam and by the modulated 420 nm beam. In this case, the x-axis is the actual scanning axis and the y-axis can be taken as a time axis. From the analysis of this time axis (as detailed below), an image of the line can be reconstructed where the pixel content is either the photoswitching amplitude or the time. A full reconstructed image, called beating mode (BM) image, can then be built by repeating this procedure for all the single lines of the conventional image.

More in detail, the line scan shown in Fig. 5 has been obtained by modulating the probe beam by means of an electronic shutter or a mechanical chopper at a frequency in the range 1–500 Hz (9.78 Hz in Fig. 5b). Depending on the shift between the shutter (modulation) frequency f_s and the line scanning rate f_L, in each line scan the probe switching occurs on different portions of the line and of its pixels (see scheme in Fig. 5a).

By analyzing the rows and columns of the BM image, we can derive the information on both the fluorescence enhancement and the activation time on each pixel of the line. The analysis of the line in the x-direction (Fig. 5c) provides us with an indication of the actual size of the point spread function on the sample. In fact, the on–off transition appears to be smoothed due to the extension of the beam well over the size of the pixel, which is actually the step size of the x-motion of the scanner (100 nm in the case presented in Fig. 5). From the values of the on–off transition decays along the x-direction, it is possible to estimate for the PSF a width of 215 $\pm$ 25 nm, in reasonable agreement with the value expected for the numerical aperture, NA $=$ 1.2, and the excitation wavelength, $\lambda_{exc} = 420$ nm, used here. The analysis of the BM image along the y-axis provides instead a time-resolved scanning of the probe on and off switching. The time resolution is computed from the difference between f_S and f_L, respectively, according to the equations:

$$
\begin{aligned}
f_S &= f_L M + \delta, \\
\tau_y &= \frac{\delta}{f_S},
\end{aligned}
\tag{13}
$$

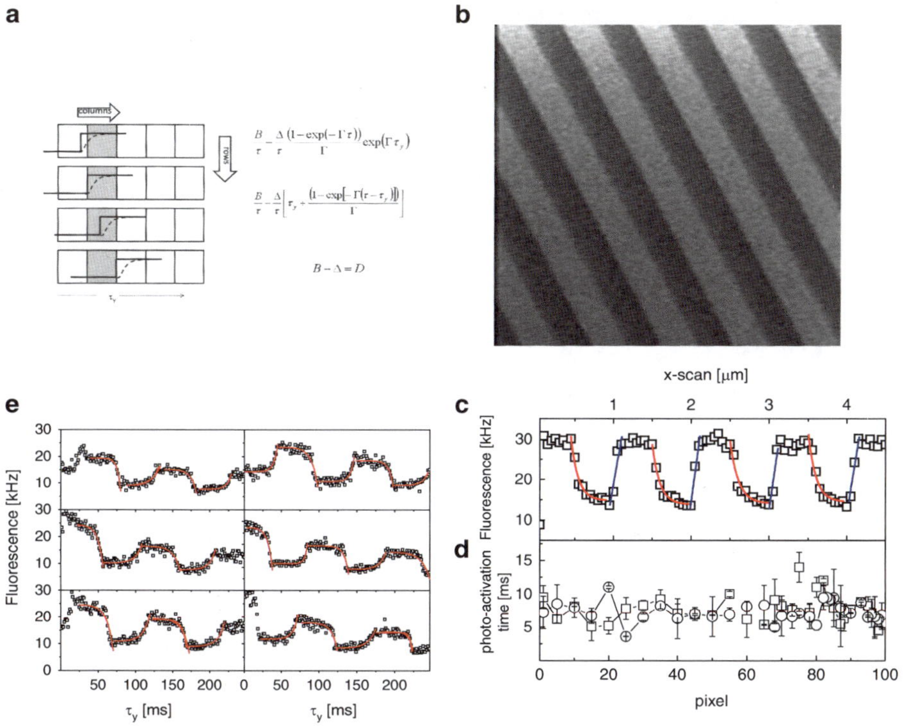

Fig. 5 Beating mode imaging scheme. Panel (**a**) sketches the principle of the beating mode scanning in which the scanning along the line (*x*-direction) performs the conventional line scanning, while the *y*-axis scanning implements a high resolution time scanning along the line. The time resolution is determined by the shift between the shutter and the line scan frequencies (see text). The equations reported indicate the average rate per pixel at the different scanning steps. Panel (**b**): a sample image obtained by continuously scanning a single line of a sample of Mut2Q loaded acrylamide gel, under modulate probe beam excitation, $f_S = 9.78$ Hz and $f_L = 1.9608$ Hz, pixel size 100 nm, time along the *y*-axis $\tau_y = 1.25$ ms. Panel (**c**): scanning along the *x*-axis of a line of panel (**b**). The *red curves* are exponential decay fits to the data, and the *blue* ones are linear fit to the stretches of data. Panel (**d**): activation times measured by fitting the growth (*squares*) and the decay (*circles*) of the *y*-scan along each pixel of the line as reported in panel (**e**). Panel (**e**): samples of *y*-scan along selected pixels (spaced by 16 columns each). The *red lines* are the best fit of the growth and decay observed to the function $A + B{*}\exp(x/\tau)$

where *M*, an integer, is the number of lines of the BM image and $\delta < f_L$ is the scanner dead time. For example, for $f_S = 9.78$ Hz and $f_L = 1.9608$ Hz, the resolution is $\tau_y = 1.25$ ms.

Sample traces along the *y*-axis of the BM image are reported in Fig. 5e. It is noteworthy that the on and off transitions are smooth and last for a large number of pixels that correspond to times of the order of tens of milliseconds. It is worth noting that, with the same kind of excitation, the analysis of the BM image acquired from immobilized samples of fluorescein provides sharp transitions that last for few pixels (data not shown) that correspond closely to the ratio between the pixel dwell time and the y-axis time resolution [38]. The smoothness of the transitions is then to be ascribed to the photoswitching kinetics.

From the traces, it is then possible to estimate the switching on and off times on single pixels of the BM image. The analysis of the fluorescence trace should be performed according to the equations developed in [38]:

$$\langle F\rangle(y) = \begin{cases} B - \left[\frac{\Delta}{\tau}\frac{(1-\exp(-\Gamma\tau))}{\Gamma}\right]\exp(\Gamma y\tau_y)\ y<0 \\ B - \frac{\Delta}{\tau}\left[\tau_y y + \frac{(1-\exp(-\Gamma(\tau - y\tau_y)))}{\Gamma}\right] 0 \le y < \frac{\tau}{\tau_y} \\ B - \Delta = D\ \frac{\tau}{\tau_y} \le y \end{cases} \tag{14a}$$

for the on to off transition and

$$\langle F\rangle(y) = \begin{cases} D + \left[\frac{\Delta}{\tau}\frac{(1-\exp(-\Gamma\tau))}{\Gamma}\right]\exp(\Gamma y\tau_y) y<0 \\ D + \frac{\Delta}{\tau}\left[\tau_y y + \frac{(1-\exp(-\Gamma(\tau - y\tau_y)))}{\Gamma}\right] 0 \leqslant y < \frac{\tau}{\tau_y} \\ D + \Delta = B\frac{\tau}{\tau_y} \leqslant y \end{cases} \tag{14b}$$

for an off to on transition. In the above equations, D and B indicate the levels of emission under the pump only and the pump&probe beams, respectively, and $\Delta = B - D$. The times τ and τ_y indicate the pixel dwell time and the time resolution along the y-scan as results from (13). Γ indicates the effective photoswitching rate. In (14), the condition $y = 0$ corresponds to the situation when the probe laser switching occurs synchronously with the line scan (i.e., the situation depicted in the second row from top on Fig. 5a).

Equation (14) is too complex to allow a routine analysis of the y-scans of all the pixels along a line of the image. However, since $\tau \cong 5\ \tau_y$, while the decay observed for $y < 0$ may last for several tens of pixels (depending obviously on the value of the photoswitching time), we can adopt as a trial fitting function a simple exponential:

$$f(y) = a + b\exp[(y - y_0)\Gamma]. \tag{15}$$

A simulation, reported in Fig. 6, motivates this choice at least for the present case in which the available y-time resolution is $\cong$1 ms, the dwell time is 5 ms, and the average fluorescence activation rate is $\Gamma \cong 50$ Hz.

The analysis according to (15) performed on a full line acquired on a uniform sample of immobilized Mut2Q proteins provides the plot reported in Fig. 5d. This kind of plot can be used to build, line by line, an image in which the pixel content is the fluorescence on switching (or off) times, similarly to what previously reported for the fluorescence average enhancement factor [38]. The use of both the fluorescence enhancement and the switching time and the dependence of these

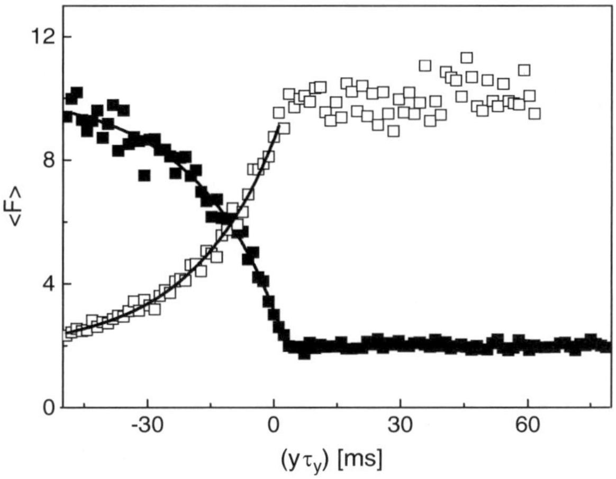

Fig. 6 Simulation of the decay that can be observed along the *y*-axis of each pixel of the line acquired in the BM image (see Fig. 5). A 5% Gaussian noise was added to the data. The simulation switching time was $\Gamma = 50$ Hz. The best fit values of the two lines to (15) reported in the plot are $\Gamma = 50 \pm 2.5$ Hz and 49 ± 5 Hz for the on–off and the off–on transition, respectively

parameters on the solution pH would, in principle, offer also the possibility to follow locally (i.e., on the pixels) the pH.

2 Conclusions and Perspectives

Reversible photoswitching is a key tool to perform functional and enhanced resolution optical imaging [80]. Here, we have discussed how time-structured excitation of one E222Q mutant of GFP (GFPMut2) can be exploited to provide information on the protein photodynamics itself, and how it could be transposed to imaging applications where the image contrast is provided by the fluorescence enhancement or the photoswitching time. Far from being conclusive, the discussion of the examples reported in this chapter opens the possibility of new technical developments in which photoswitching may play a role in enhancing image signal/noise also offering the way to measure locally chemical and physical parameters.

Acknowledgments This work has been also funded by the PRIN MIUR fund 2008 to M.C. (2008JZ4MLB) and by the FP7 program ENCITE (contract no. 201842) to G.C.

References

1. Chalfie M, Tu Y, Euskirchen G, Ward WW, Prasher DC (1994) Green fluorescent protein as a marker for gene expression. Science 263:802–805
2. Tsien RY (1998) The green fluorescent protein. Annu Rev Biochem 67:509–544
3. Jung G, Zumbusch A (2006) Improving autofluorescent proteins: comparative studies of the effective brightness of green fluorescent protein (GFP) mutants. Micr Res Tech 69:175–185

4. Hsu ST, Blaser G, Jackson SE (2009) The folding, stability and conformational dynamics of beta-barrel fluorescent proteins. Chem Soc Rev 38:2951–2965
5. Miyawaki A, Sawano A, Kogure T (2003) Lighting up cells: labelling proteins with fluorophores. Nat Cell Biol Suppl S1–7
6. Lippincott-Schwartz J, Patterson GH (2003) Development and use of fluorescent protein markers in living cells. Science 300:87–91
7. Chudakov DM, Belousov VV, Zaraisky AG, Novoselov VV, Staroverov DB, Zorov DB, Lukyanov S, Lukyanov KA (2003) Kindling fluorescent proteins for precise in vivo photolabeling. Nat Biotechnol 21:191–194
8. Post JN, Lidke KA, Rieger B, Anrdt-Jovin DJ (2005) One- and two-photon photoactivation of a paGFP-fusion protein in live Drosophila embryos. FEBS Lett 579:325–330
9. Kim PK, Mullen RT, Schumann U, Lippincott-Schwartz J (2006) The origin and maintenance of mammalian peroxisomes involves a de novo PEX16-dependent pathway from the ER. J Cell Biol 173:521–532
10. Dickson RM, Cubitt AB, Tsien RY, Moerner WE (1997) On/off blinking and switching behavior of single molecules of green fluorescent protein. Nature 388:355–358
11. Day RN, Davidson MW (2009) The fluorescent protein palette: tools for cellular imaging. Chem Soc Rev 38:2887–2921
12. Patterson GH, Lippincott-Schwartz J (2002) A photoactivatable GFP for selective photolabeling of proteins and cells. Science 297:1873–1877
13. Lippincott-Schwartz J, Patterson GH (2008) Fluorescent proteins for photoactivation experiments. Methods Cell Biol 85:45–61
14. Betzig E, Patterson GH, Sougrat R, Lindwasser OW, Olenych S, Bonifacino JS, Davidson MW, Lippincott-Schwartz J, Hess HF (2006) Imaging intracellular fluorescent proteins at nanometer resolution. Science 313:1642–1645
15. Ando R, Hama H, Yamamoto-Hino M, Mizuno H, Miyawaki A (2002) An optical marker based on the UV-induced greento- red photoconversion of a fluorescent protein. Proc Natl Acad Sci USA 99:12651–12656
16. Ando R, Mizuno H, Miyawaki A (2004) Regulated fast nucleocytoplasmic shuttling observed by reversible protein highlighting. Science 306:1370–1373
17. Lippincott-Schwartz J, Altan-Bonnet N, Patterson GH (2003) Photobleaching and photoactivation: following protein dynamics in living cells. Nat Cell Biol Suppl S7–14
18. Cinelli RAG, Pellegrini V, Ferrari A, Faraci P, Nifosi R, Tyagi M, Giacca M, Beltram F (2001) Green fluorescent proteins as optically controllable elements in bioelectronics. Appl Phys Lett 79:3353–3355
19. Bizzarri R, Serresi M, Cardarelli F, Abbruzzetti S, Campanini B, Viappiani C, Beltram F (2010) Single amino acid replacement makes Aequorea victoria fluorescent proteins reversibly photoswitchable. J Am Chem Soc 132:85–95
20. Habuchi S, Ando R, Dedecker P, Verheijen W, Mizuno H, Miyawaki A, Hofkens J (2005) Reversible single-molecule photoswitching in the GFP-like fluorescent protein Dronpa. Proc Natl Acad Sci USA 102:9511–9516
21. Hell SW, Jakobs S, Kastrup L (2003) Imaging and writing at the nanoscale with focused visible light through saturable optical transitions. Appl Phys A Mater Sci Process 77:859–860
22. Hofmann M, Eggeling C, Jakobs S, Hell SW (2005) Breaking the diffraction barrier in fluorescence microscopy at low light intensities using reversibly photoswitchable proteins. Proc Natl Acad Sci USA 102:17565–17569
23. Subach FV, Patterson GH, Manley S, Gillette JM, Lippincott-Schwartz J, Verkhusha VV (2009) Photoactivatable mCherry for high-resolution two-color fluorescence microscopy. Nat Methods 6:153–159
24. Hess ST, Girirajan TP, Mason MD (2006) Ultra-high resolution imaging by fluorescence photoactivation localization microscopy. Biophys J 91:4258–4272
25. Rust MJ, Bates M, Zhuang X (2006) Sub-diffraction-limit imaging by stochastic optical reconstruction microscopy (STORM). Nat Methods 3:793–795

26. Egner A, Geisler C, von Middendorff C, Bock H, Wenzel D, Medda R, Andresen M, Stiel AC, Jakobs S, Eggeling C, Schönle A, Hell SW (2007) Fluorescence nanoscopy in whole cells by asynchronous localization of photoswitching emitters. Biophys J 93:3285–3290
27. Flors C, Hotta J, Uji-i H, Dedecker P, Ando R, Mizuno H, Miyawaki A, Hofkens J (2007) A stroboscopic approach for fast photoactivation-localization microscopy with Dronpa mutants. J Am Chem Soc 129:13970–13977
28. Hofmann M, Eggeling C, Jakobs S, Hell SW (2005) Breaking the diffraction barrier in fluorescence microscopy at low light intensities by using reversibly photoswitchable proteins. PNAS 102:17565–17569
29. Luin S, Voliani V, Lanza G, Bizzarri R, Amat P, Tozzini V, Serresi M, Beltram F (2009) Raman study of cromophore states in photochromic fluorescent proteins. J Am Chem Soc 131:96–103
30. Andresen M, Stiel AC, Trowitzsch S, Weber G, Eggeling C, Wahl MC, Hell SW, Jakobs S (2007) Structural basis for reversible photoswitching in Dronpa. Proc Natl Acad Sci USA 104:13005–13009
31. Andresen M, Wahl MC, Stiel AC, Gräter F, Schäfer LV, Trowitzsch S, Weber G, Eggeling C, Grubmüller H, Hell SW, Jakobs S (2005) Structure and mechanism of the reversible photoswitch of a fluorescent protein. Proc Natl Acad Sci USA 102:13070–13074
32. Stiel AC, Trowitzsch S, Weber G, Andresen M, Eggeling C, Hell SW, Jakobs S, Wahl MC (2007) 1.8 Å bright-state structure of the reversibly switchable fluorescent protein Dronpa guides the generation of fast switching variants. Biochem J 402:35–42
33. Moerner WE (2002) Single-molecule optical spectroscopy of autofluorescent proteins. J Chem Phys 117:10925–10937
34. Haupts U, Maiti S, Schwille P, Webb WW (1998) Dynamics of fluorescence fluctuations in green fluorescent protein observed by fluorescence correlation spectroscopy. Proc Natl Acad Sci USA 95:13573–13578
35. Moerner WE, Fromm DP (2003) Methods of single-molecule fluorescence spectroscopy and microscopy. Rev Sci Inst 74:3597–3619
36. Haustein E, Schwille P (2007) Fluorescence correlation spectroscopy: novel variations of an established technique. Annu Rev Biophys Biomol Struct 36:151–169
37. Marriott G, Mao S, Sakata T, Ran J, Jackson DK, Petchprayoon C, Gomez TJ, Warp E, Tulyathan O, Aaron HL, Isacoff EY, Yan Y (2008) Optical lock-in detection imaging microscopy for contrast-enhanced imaging in living cells. Proc Natl Acad Sci USA 105:17789–17794
38. Quercioli V, Bosisio C, Daglio SC, Rocca F, D'Alfonso L, Collini M, Baldini G, Chirico G, Bettati S, Raboni S, Campanini B (2010) Photoinduced millisecond switching kinetics in the GFPMut2 E222Q mutant. J Phys Chem B 114:4664–4677
39. Jung G, Brauchle C, Zumbusch A (2001) Two-color fluorescence correlation spectroscopy of one chromophore: application to the E222Q mutant of the green fluorescent protein. J Chem Phys 114:3149–3156
40. Zimmer M (2002) Green fluorescent protein (GFP): applications, structure, and related photophysical behavior. Chem Rev 102:759–781
41. Jung G, Wiehler J, Zumbusch A (2005) The photophysics of green fluorescent protein: influence of the key amino acids at positions 65, 203, and 222. Biophys J 88:1932–1947
42. Heilemann M, Margeat E, Kasper R, Sauer M, Tinnefeld P (2005) Carbocyanine dyes as efficient reversible single-molecule optical switch. J Am Chem Soc 127:3801–3806
43. Patterson GH, Piston DW (2000) Photobleaching in two-photon excitation microscopy. Biophys J 78:2159–2162
44. Seefeldt B, Kasper R, Seidel T, Tinnefeld P, Dietz K-J, Heilemann M, Sauer M (2008) Fluorescent proteins for single-molecule fluorescence applications. J Biophoton 1:74–82
45. Eggeling C, Hilbert M, Bock H, Ringemann C, Hofmann M, Stiel AC, Andresen M, Jakobs S, Egner A, Schoenle A, Hell SW (2007) Reversible photoswitching enables single-molecule

fluorescence fluctuation spectroscopy at high molecular concentration. Micr Res Tech 70:1003–1009
46. Bizzarri R, Arcangeli C, Arosio D, Ricci F, Faraci P, Cardarelli F, Beltram F (2006) Development of a novel GFP-based ratiometric excitation and emission pH indicator for intracellular studies. Biophys J 90:3300–3314
47. Bizzarri R, Serresi M, Luin S, Beltram F (2009) Green fluorescent protein based pH indicators for in vivo use: a review. Anal Bioanal Chem 393:1107–1122
48. Schwille P, Haupts U, Maiti S, Webb WW (1999) Molecular dynamics in living cells observed by fluorescence correlation spectroscopy with one- and two-photon excitation. Biophys J 77:2251–2265
49. Bosisio C, Quercioli V, Collini M, D'Alfonso L, Baldini G, Bettati S, Campanini B, Raboni S, Chirico G (2008) Protonation and conformational dynamics of GFP mutants by two-photon excitation fluorescence correlation spectroscopy. J Phys Chem B 112:8806–8814
50. Malvezzi-Campeggi F, Jahnz M, Heinze KG, Dittrich P, Schwille P (2001) Light-induced flickering of DsRed provides evidence for distinct and interconvertible fluorescent states. Biophys J 81:1776–1785
51. Wang ZF, Shah JV, Chen ZP, Sun CH, Berns MW (2004) Fluorescence correlation spectroscopy investigation of a GFP mutant-enhanced cyan fluorescent protein and its tubulin fusion in living cells with two-photon excitation. J Biomed Opt 9:395–403
52. Jung G, Werner M, Schneider M (2008) Efficient photoconversion distorts the fluorescence lifetime of GFP in confocal microscopy: a model kinetic study on mutant Thr203Val. ChemPhysChem 9:1867–1874
53. Chirico G, Cannone F, Baldini G, Diaspro A (2003) Two-photon thermal bleaching of single fluorescent molecules. Biophys J 84:588–598
54. Kim SA, Heinze KG, Bacia K, Waxham MN, Schwille P (2005) Two-photon cross-correlation analysis of intracellular reactions with variable stoichiometry. Biophys J 88:4319–4336
55. Nifosì R, Tozzini V (2006) Cis–trans photoisomerization of the chromophore in the green fluorescent protein variant E2GFP: a molecular dynamics study. Chem Phys 323:358–368
56. Wong FHC, Banks DS, Abu-Arish A, Fradin C (2007) A molecular thermometer based on fluorescent protein blinking. J Am Chem Soc 129:10302–10303
57. Cannone F, Milani R, Chirico G, Diaspro A, Krol S, Campanini B (2007) Voltage regulation of single green fluorescent protein mutants. Biophys Chem 125:368–374
58. Paradise A, Levin MK, Korza G, Carson JH (2007) Significant proportions of with reduced intracellular nuclear transport proteins mobilities resolved by fluorescence correlation spectroscopy. J Mol Biol 365:50–65
59. Wang Z, Shah JV, Berns MW, Cleveland DW (2006) In vivo quantitative studies of dynamic intracellular processes using fluorescence correlation spectroscopy. Biophys J 91:343–351
60. Reiff DF, Ihring A, Guerrero G, Isacoff EY, Joesch M, Nakai J, Borst A (2005) In vivo performance of genetically encoded indicators of neural activity in flies. J Neurosci 25:4766–4778
61. Cormack BP, Valdivia RH, Falkow S (1996) FACS-optimized mutants of the green fluorescent protein (GFP). Gene 173:33–38
62. Abbruzzetti S, Grandi E, Viappiani C, Bologna S, Campanini B, Raboni S, Bettati S, Mozzarelli A (2005) Kinetics of acid-induced spectral changes in the GFPmut2 chromophore. J Am Chem Soc 127:626–635
63. Chirico G, Diaspro A, Cannone F, Collini M, Bologna S, Pellegrini V, Beltram F (2005) Selective fluorescence recovery after bleaching of single E(2)GFP proteins induced by two-photon excitation. ChemPhysChem 6:328–335
64. Chattoraj M, King BA, Bublitz GU, Boxer SG (1996) Ultra-fast excited state dynamics in green fluorescent protein: multiple states and proton transfer. Proc Natl Acad Sci USA 93:8362–8367

65. Brejc K, Sixma TK, Kitts PA, Kain SR, Tsien RY, Ormo M, Remington SJ (1997) Structural basis for dual excitation and photoisomerization of the Aequorea victoria green fluorescent protein. Proc Natl Acad Sci USA 94:2306–2311
66. Lossau H, Kummer A, Heinecke R, Pollinger-Dammer F, Kompa C, Bieser G, Jonsson T, Silva CM, Yang MM, Youvan DC, Michel-Beyerle ME (1996) Time-resolved spectroscopy of wild-type and mutant green fluorescent proteins reveals excited state deprotonation consistent with fluorophore-protein interactions. Chem Phys 213:1–16
67. Stoner-Ma D, Jaye AA, Ronayne KL, Nappa J, Meech SR, Tonge PJ (2008) An alternate proton acceptor for excited-state proton transfer in green fluorescent protein: rewiring GFP. J Am Chem Soc 130:1227–1235
68. Jung G, Wiehler J, Steipe B, Bräuchle C, Zumbusch A (2001) Single-molecule microscopy of the green fluorescent protein using simultaneous two-color excitation. ChemPhysChem 6:392–396
69. Cinelli RAG, Pellegrini V, Ferrari A, Faraci P, Nifosì R, Tyagi M, Giacca M, Beltram F (2001) Green fluorescent proteins as optically-controllable elements in bioelectronics. Appl Phys Lett 79:3353–3355
70. Yokoe H, Meyer T (1996) Spatial dynamics of GFP-tagged proteins investigated by local fluorescence enhancement. Nat Biotech 14:1252–1256
71. Redmond RW, Kochevar IE, Krieg M, Smith G, McGimpsey WG (1997) Excited state relaxation in cyanine dyes: a remarkably efficient reverse intersystem crossing from upper triplet levels. J Phys Chem A 101:2773–2777
72. Widengren J, Seidel CAM (2000) Manipulation and characterization of photo-induced transient states of merocyanine 540 by fluorescence correlation spectroscopy. Phys Chem Chem Phys 2:3435–3441
73. Huang Z, Ji D, Xia A (2005) Fluorescence intensity and lifetime fluctuations of single Cy5 molecules immobilized on the glass surface. Colloids Surf A Physicochem Eng Asp 257–258:203–209
74. Larkin JM, Donaldson WR, Foster TH, Knox RS (1999) Reverse intersystem crossing from a triplet state of rose Bengal populated by sequential 532- + 1064-nm laser excitation. Chem Phys 244:319–330
75. English DS, Harbron EJ, Barbara PF (2000) Probing photoinduced intersystem crossing by two-color. Double resonance single molecule spectroscopy. J Phys Chem A 104:9057–9061
76. Ringemann C, Schönle A, Giske A, von Middendorff C, Hell SW, Eggeling C (2008) Enhancing fluorescence brightness: effect of reverse intersystem crossing studied by fluorescence fluctuation spectroscopy. ChemPhysChem 9:612–624
77. Palm GJ, Zadnov A, Gaitanaris GA, Stauber R, Pavlakis GN, Wlodawer A (1997) The structural basis for spectral variations in green fluorescent protein. Nat Struct Biol 4:361–365
78. Weber W, Helms V, McCammon JA, Langhoff PW (1999) Shedding light on the dark and weakly fluorescent states of green fluorescent proteins. Proc Natl Acad Sci USA 96:6177–6182
79. Agmon N (2007) Kinetics of switchable proton escape from a proton-wire within green fluorescence protein. J Phys Chem B 111:7870–7878
80. Heilemann M, Dedecker P, Hofkens J, Sauer M (2009) Photoswitches: key molecules for subdiffraction-resolution fluorescence imaging and molecular quantification. Laser Photon Rev 3:180–202

Part II
Quantification of Basic Physiological Parameters

The Proton Sensitivity of Fluorescent Proteins: Towards Intracellular pH Indicators

Ranieri Bizzarri

Abstract On account of the presence of an ionizable phenol group on their chromophore, most fluorescent proteins display pH-sensitive absorption and fluorescence emission bands. This property has great implications for most applications that involve FPs as intracellular markers of target proteins. The photophysical modulation of pH sensitivity relies upon a complex network of exchange reaction pathways involving the chromophore and funneling protons in and out the protein structure. In parallel with the fast-growing use of FPs as intracellular probes, many groups reported on the use of FP variants as genetically-encodable pH indicators of intracellular compartments that cannot be probed using conventional pH-sensitive dyes. These pH indicators shed light on many cell functions strongly coupled with local pH. Here, we shall review both the properties that make FPs so efficient as intracellular pH indicators and the probes that are more utilized by the scientific community.

Keywords Excited-state proton transfer · GFP chromophore ionization · Intracellular pH · pH indicator · Proton pathways

Contents

R. Bizzarri
IIT@NEST, Center for Nanotechnology Innovation, NEST, Scuola Normale Superiore and Istituto Nanoscienze-CNR, Piazza S. Silvestro 12, 56127 Pisa, Italy
e-mail: r.bizzarri@sns.it

G. Jung (ed.), *Fluorescent Proteins II*, Springer Ser Fluoresc (2012) 12: 59–98,
DOI 10.1007/4243_2011_30 © Springer-Verlag Berlin Heidelberg 2011
Published online: 2 August 2011

1 Introduction

Intracellular pH (pH_i) plays many critical roles in the biochemistry of the cell thereby representing a fundamental modulator of cell function [1]. This relevance stems from the acid–base characteristics of many residues in biomolecules, particularly enzymes, which rely upon pH control to perform their activity at optimal level.

pH_i affects cell metabolism and growth [2, 3], ion transport and homeostasis [4–7], endocytosis [8], and muscle cellular contractility [9]. pH_i alterations may directly lead to cell apoptosis and cancer [10–12], to the impairment of posttranslational modifications and processing of secreted proteins [13], to a mislocalization of the biosynthetic cargo [14], and to severe defects in functionality of the organelles [15]. Modification of the pH_i of the secretory compartment seems related to several tumors [16]. The nervous system is influenced by pH_i, because it affects neuronal excitability, cell–cell coupling, and signal cascades [17]. The lumen of presynaptic vesicles displays an acidic interior. During exocytosis, i.e., the fusion of vesicles with the plasma membrane, the synaptic vesicles undergo a pH jump as their inner part comes in contact with the neutral extracellular medium. Note that exocytosis is at basis of the release of neurotransmitters stored in presynaptic vesicles [18].

The pH_i of eukaryotic cells is in most cases within the 7.1–7.4 range, being regulated by: (1) the flux of acid/base moieties (H^+, OH^- or HCO_3^-) across the surface membrane by means of specific ion-coupled transporters such as the NHE proteins [5], and (2) by acid-generating processes, such as aerobic and anaerobic metabolism [19]. Nonetheless, pH_i is not spatially uniform and its value depends strongly on the nature and function of subcellular domains. Some organelles, like endosomes and plant vacuoles, have intracompartmental pH between 4 and 6; the progressive luminal acidification of endosomes is essential for the distribution and degradation of internalized ligands into lysosomes [20, 21]. Consistently, cellular dysfunction may be associated with abnormal pH values in organelles.

The fundamental role of pH_i in modulating cell functions implies that accurate pH_i measurement can provide critical information to study physiological and pathological processes in living organisms. The need of high spatiotemporal resolution while at the same time avoiding invasive measurements and preserving cell viability restricts the selection of the analytical probes to fluorescence indicators [22]. High-resolution quantitative fluorescence imaging microscopy has evolved to a very sensitive analytical tool in cell and molecular biology [23]. The development of fluorescence microscopy was paralleled by the engineering of fluorescent indicators for *in cellulo* and in vivo use. Owing to its relevance for cell biology, pH_i has been one of the first parameters addressed by fluorescent indicators in the microscopy of living cells [24, 25]. Although synthetic probes can efficiently report on pH_i, indicators constituted by Fluorescent Protein mutants have quickly become the best selection to monitor proton concentration variations within cells [26].

The Green Fluorescent Protein (wild type GFP, abbreviated as wtGFP) is an autofluorescent protein characterized by violet–blue light absorption and bright green light emission. wtGFP was discovered in 1962 as an accessory protein of the bioluminescence system of the hydroid jellyfish *Aequorea victoria*, encompassing also the protein *Aequorin* [27, 28]. Thirty years later the cloning [29], and the successful heterologous expression of the wtGFP gene [30], proved that the fluorescent emission is *genetically encoded* into the primary sequence of the protein. On account of this outstanding property, wtGFP can be used as intrinsic intracellular reporters of target proteins by simple genetic fusion followed by expression into cells. In recent years, many other organisms were found to possess emissive proteins structurally similar to wtGFP, although endowed with slightly different optical properties (e.g., different emission wavelength) [31–33]. Furthermore, the primary sequence of wtGFP and other ancestor fluorescent proteins can be modified thereby manipulating their optical properties; as a consequence, hundreds (and their number is currently expanding) of *Fluorescent Proteins* (FPs), characterized by a wide range of photophysical properties [33], are now available to researchers. The introduction of FPs revolutionized cell biology studies and pioneer scientists Osamu Shimomura, Martin Chalfie, and Roger Tsien received the Nobel Prize in Chemistry in 2008.

The main advantages of fluorescent protein-based indicators over simple organic dyes are that they can be fused to protein of interests and possibly thereby targeted to specific subcellular compartments as they are genetically encoded. FP indicators can be designed to respond to a variety of biological events and signals, they can be introduced in many tissues and organisms, and they seldom cause photodynamic toxicity [34–36].

In this chapter, we shall describe the pH-dependent optical properties that make some variants excellent fluorescent indicators of pH_i, describing also their biological applications. Note that we shall limit our review to mutants obtained from wtGFP and thus belonging to the *Aequorea* family (GFPs). Indeed, popular FPs recovered from other organisms display generally poor pH-responsiveness [32]; insofar only two efficient pH indicators not based on *Aequorea* proteins have been described [37, 38]. The reader is referred to these papers for additional information.

The pH-dependent optical characteristics of GFPs will be reviewed in Sect. 2. The engineered GFP-based indicators for pH_i measurement will be described in Sect. 3. In Sect. 4, we shall discuss the most common microscopy setups that are used to image pH_i through engineered GFP indicators. In the last chapter, we shall debate the future perspectives about the development and applications of these remarkable indicators.

2 The Basis of pH Sensitivity in Fluorescent Proteins

2.1 The Y66 GFP Chromophore: Intrinsic Protonation and Optical Characteristics

Upon the formation of the native β-barrel tertiary structure of GFP, the Y66 chromophore (hereafter denoted as *Chro*) is spontaneously generated through three chemical steps [39]. At first, the tripeptide sequence S65–Y66–G67 undergoes intramolecular cyclization to yield a five-member ring β-amino lactame (Scheme 1, **i**). Then, in the presence of water-soluble oxygen, an oxidation step takes place (Scheme 1, **ii**). Eventually, a water molecule is released and the hyperconjugated imidazolinone structure of *Chro* is formed (Scheme 1, **iii**). Note that the rate-limiting step of the overall process is the oxidation reaction, requiring at least 30 min to reach completion [40].

The *intrinsic* chromophore protonation properties were thoroughly investigated by using several synthetic *Chro* analogs. *Chro* bears two main protonation sites: the phenol function of Tyr^{66} and the N^{66} nitrogen in the imidazolinone ring (Scheme 2). Two protonation sites imply four protonation states: double protonated (net charge: +1, hereafter denoted as **P**), neutral (net charge: 0, hereafter denoted as **N**), zwitterionic (net charge: 0, hereafter denoted as **Z**), and anionic (net charge: −1, hereafter denoted as **P**) (Scheme 2). These four protonation states have fairly different optical properties (Fig. 1a). The **P**, **N**, and **D** forms absorb, respectively,

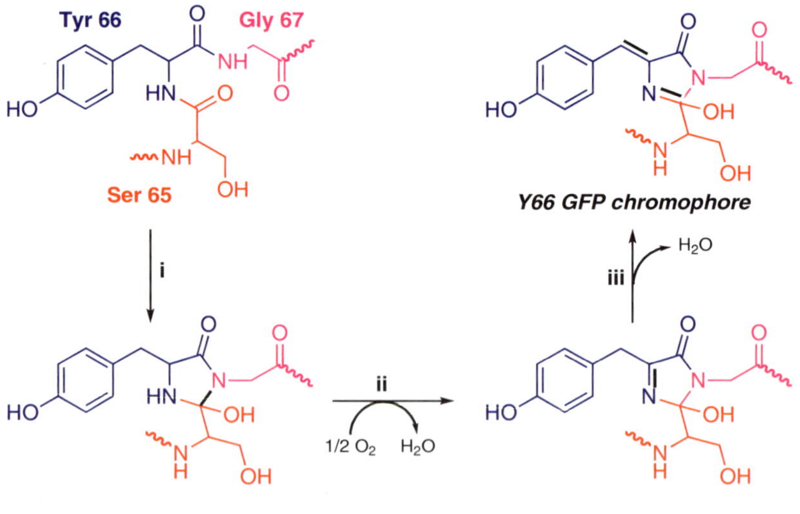

Scheme 1 Formation of Y66 GFP chromophore from the autocatalytic post-translational processing of the Ser^{65}-Tyr^{66}-Gly^{67} aminoacid sequence [39]

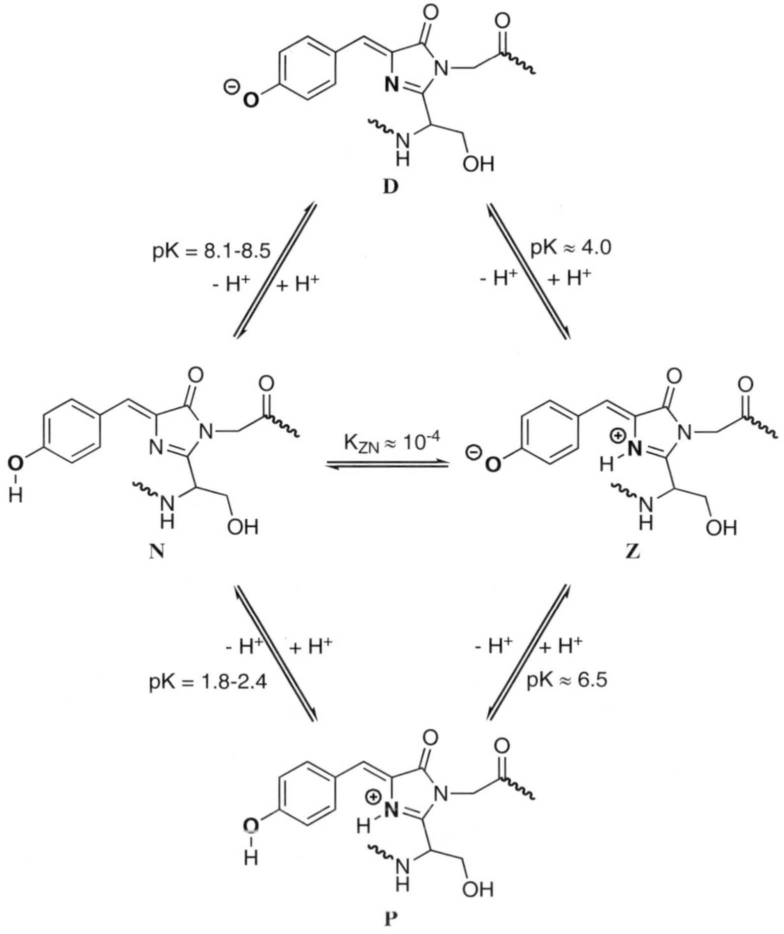

Scheme 2 The four accessible protonation states **P** (protonated), **N** (neutral), **Z** (zwitterionic), and **D** (anionic) of synthetic GFP chromophore and their associated protonation reactions. For each reaction, the equilibrium constant (pK_a) is reported

at 387–393,[1] 368–372[a], and 425–428[a] nm in water [41–43], showing also considerable solvatochromism when other solvents are considered [43]. The red shifted absorption of **P** and **D** states compared to **N** is attributed to extended electronic conjugation which lowers the $S_0 \rightarrow S_1$ transition energy [44]. Owing to its negligible population at any pH (see below), the absorption spectrum of **Z** is not directly measurable; nonetheless, a zwitterionic-mimicking derivative was found to absorb

[1]Minor variability depends on the structure of the synthetic *Chro* analog.

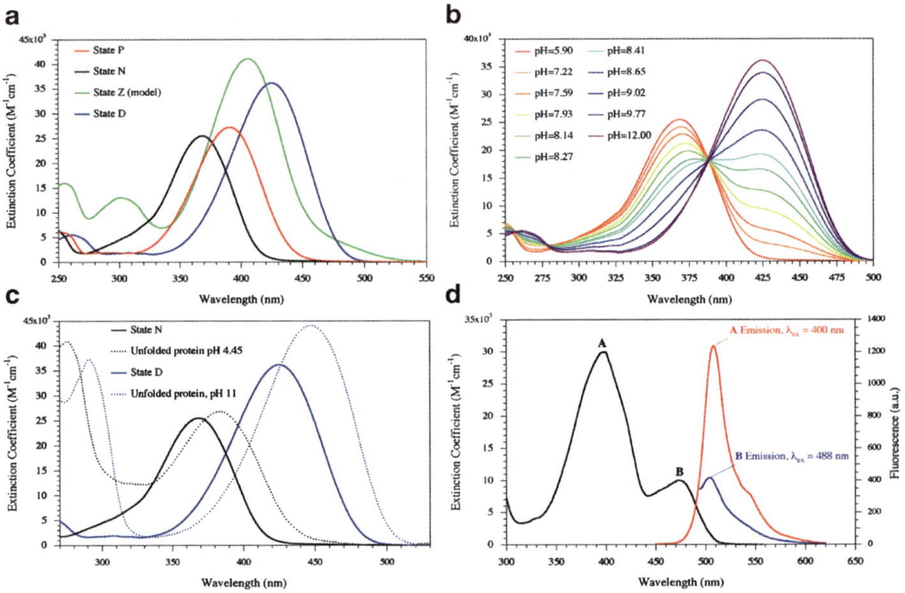

Fig. 1 pH-dependent optical properties of GFP Y66 chromophore. (**a**) Absorption spectra of the four protonation states of synthetic *Chro* (for nomenclature and structural reference see Scheme 1). (**b**) pH-dependent interconversion between **N** and **D** state of synthetic chromophore as monitored by absorption spectroscopy. (**c**) Comparison between absorption spectra of synthetic chromophore and unfolded protein at pH 4.45 and pH 11: state **N** and **D** are, respectively, observed. (**d**) Absorption (*left* scale) and emission (*right* scale) spectra of F64L GFP, a mutant that is optically identical to wtGFP; the neutral *Chro* state (**A**) and the anionic *Chro* state (**B**) are indicated; note that the absorption spectra of **N** and **D** states in the denatured protein are added for comparison

at 406 nm (Fig. 1a) [43], in good agreement with the value theoretically calculated for **Z**.

The absorption differences between the protonation states of *Chro* allowed for the calculation of the ionization pK_a values by means of spectrophotometric titrations (Fig. 1b) [42, 43]. In synthetic *Chro* analogs, N^{66} has pK_a = 1.8–2.4[2] (Scheme 2, **P**↔**N**), whereas the phenol group has pK_a = 8.1–8.5 (Scheme 2, reaction **N**↔**D**). The pK_a of **P**↔**Z** was determined to be around 6.5 by using the already mentioned zwitterionic-mimicking *Chro* analog [43]. These three pK_a values allow for the full quantitative description of the thermodynamic system of *Chro* protonation (Scheme 2). Importantly, this calculation shows that **Z** always accounts for less than 0.01% of its isoelectric counterpart **N**. Hence, in the physiological range of pH only the **N**↔**D** ionization is active (Fig. 1b).

Strikingly, none of the protonation forms of the synthetic GFP chromophore displays appreciable fluorescence emission (quantum yields $<10^{-3}$). Several

[2]Minor variability depends on the addition of a small amount of methanol/ethanol to improve *Chro* solubility.

studies linked the absence of emission to the presence of a fast and efficient deactivation channel involving intramolecular *hula-twist* torsion of the chromophore in the excited state [45]. Indeed, *Chro* fluorescence is restored at very low temperatures and in high-viscosity media that strongly reduce molecular mobility thereby hampering the torsional deactivation channel of the excited state [41, 45–47].

The absorption properties of the synthetic chromophore are conserved in the *unfolded* state of the protein or in a protein fragment containing the chromophore, although detectable red shift of maxima to 384 and 448 nm occurs for the **N** and **D** states, respectively (Fig. 1c). Note that the absorption of **D** state in the unfolded protein at pH 11 is commonly used as standard to calculate the degree of chromophore maturation in the folded protein state [48, 49]. Similarly, *Chro* in the unfolded protein is nearly nonfluorescent at room temperature, but it becomes emissive at 77 K [41].

2.2 Wild-Type GFP: Optical States and Mechanism of Fluorescence Emission

At first, we shall consider the "archetype" of the fluorescent protein family characterized by Y66 chromophore, i.e., *wild-type GFP* (wtGFP). wtGFP is folded between pH 4 and 11; in this range, only the stable neutral and ionic *Chro* states contribute to the optical properties. To avoid notation confusion between pure *Chro* and protein *Chro*, and following a long-standing tradition in the field of FP photophysics, the neutral and anionic states of the wtGFP chromophore will be hereafter identified as state **A** and **B**, respectively.

The tertiary β-barrel structure of fluorescent proteins has a striking effect upon the optical properties of *Chro* [50, 51]. Concerning absorption, state **A** of wtGFP is red-shifted by 30 nm compared to **N** state, peaking at 398 nm (Fig. 1d). The absorption red shift of state **B** compared to **D** is even larger: 50 nm, because **B** peaks at 475 nm (Fig. 1d). These major optical changes are attributable to the complex network of interactions, mostly H-bonds, which are established between the chromophore and the surrounding residues. Indeed, X-ray data of wtGFP displayed that *Chro* is surrounded by four entrapped water molecules and several charged and polar residues such as Gln^{69}, Gln^{94}, Arg^{96}, His^{148}, Thr^{203}, Ser^{205}, and Glu^{222} [52–55]. Additionally, **A** and **B** states display different conformations of the residues around the chromophore [53]. A proton relay system connecting the phenol group to residue Glu^{222} is present in **A**, whereas it is blocked in **B** owing to the rotation of Thr^{203} by 120° to establish a strong H bond with the phenolate anion [53]. The **A/B** population ratio is around 6/1, as calculated from the extinction coefficients of these states; notably, this ratio is nearly unchanged by external pH in the 5–11 pH range (an argument that will be further discussed in Sect. 2.3) [56].

The rigid tertiary structure of wtGFP is responsible for the significant protein fluorescence emission ($\Phi_A = 0.78$ [57], $\Phi_B = 0.79$ [48]), which is also very resistant to classical quenching agents [58, 59]. In spite of their large difference in absorption maxima, however, **A** and **B** are characterized by minor differences of emission maxima and shapes (Fig. 1d). This surprising similarity of fluorescent emissions stems from a subtle photophysical effect that was fully elucidated only after pump-probe experiments targeting the excited state depletion at short (ps) time-scale [60, 61]. Photon absorption by **B** results in the formation of the excited state **B*** whose depletion is mainly due to fluorescence emission, peaking at 503 nm. When **A** is excited, instead, two competing photoprocesses are activated (beside some thermal deactivation back to ground state): (a) the direct fluorescence emission from **A*** (in the 440–480 nm range) and (b) proton transfer from A* to Glu^{222}, presumably through a proton relay system involving one water molecules and the hydroxyl function of Ser^{205} [60–62]. The latter process is called *Excited State Proton Transfer* (ESPT) and is triggered by the increased acidity of the phenol group in the excited state [44] a well-known phenomenon in classical photochemistry. ESPT takes place in a few picoseconds and it represents a much more efficient depletion channel of **A*** than direct fluorescence emission [63]. By ESPT, **A*** states evolves to an intermediate excited form **I***, where *Chro* is anionic like in **B*** but its surrounding residues are in the *relaxed form* typical of **A**, because the timescale of the **A*** → **I*** photoprocess (ps) is too fast to allow for the protein structural rearrangements that are *thermodynamically driven* by chromophore deprotonation (e.g., flipping of the lateral chain of Thr^{205}) [60, 64]. Hence, **I*** emits at wavelengths, similar to $\mathbf{B}^*$, although their emission spectra are not fully superimposable (Fig. 1d). After emission, **I*** is converted to **I**, which is a metastable ground-state moiety where anionic *Chro* is embedded in an environment typical of the neutral **A** form. **I** quickly evolves to A, which is about 7.6 kJ/mol lower in energy [65], by receiving back the proton of Glu^{222} possibly through the same internal protonation relay involving Ser^{205} and a bound water molecule [53]. A more detailed kinetic analysis revealed that the **I** moiety actually comprises two metastable intermediates $\mathbf{I_1}$ and $\mathbf{I_2}$, although $\mathbf{I_1}$ is very short lived (3 ps) compared to $\mathbf{I_2}$ (0.4 ns) [66].

The main optical characteristics of wtGFP are shared by most FPs from *Aequorea Victoria* containing the Y66 chromophore, although the mutation pattern has always some effect on the absorption and emission wavelength of **A** and **B** as well as the fluorescence quantum yields and lifetime. Unfortunately, a detailed structure–function photophysical model to guide protein engineering for obtaining desired optical properties is not yet available. Nonetheless, the absorption and emission of **B** are more sensitive to environmental effects within the chromophore pocket than its protonated counterpart **A**. For example, the presence of aromatic residues near *Chro* leads to significant red shifts of both **B** absorption and emission. A particularly relevant case is represented by the mutants where Thr^{203} is replaced by Tyr: here **B** absorption is shifted to 510–525 nm and it becomes almost superimposable to its emission (515–527 nm); this large red-shift is attributable to π–π^* interaction between the aromatic ring of *Chro* and Tyr^{203}, which adopt a parallel configuration to each other.[55, 67] Because of such optical properties, Tyr^{203} mutants are collectively denoted as the Yellow Fluorescent Protein class (YFPs). YFPs represent the most optically red-shifted mutant class belonging to the

Aequorea family. Larger red-shifts are observed only when an additional double bond conjugated to the original imidazolinone *Chro* is generated in mutants derived from different sea organisms. The very high extinction coefficients (>75,000 M^{-1} cm^{-1}), emission quantum yields ($\Phi > 0.8$), and fluorescence lifetimes ($\tau > 3$ ns) make YFPs among the best candidates for high-resolution and single-molecule imaging studies [68, 69].

The optical properties of the **A** states can be in principle modulated by careful engineering of the ESPT efficiency, i.e., of the molecular pathway that conveys the proton from **A*** to a proton acceptor and vice versa. This is usually done by adding polar/ionizable residues in the chromophore surroundings [70]. Hampering the ESPT channel leads to mutants whose **A** state has a significant blue-shifted emission compared to wtGFP, although the quantum yield of **A*** fluorescence is usually low [62].

2.3 pH Dependence of Optical Response in GFP Mutants

In a billion years, natural selection has engineered wtGFP to be almost unresponsive to external pH changes to avoid slight fluctuations of the jellyfish bioluminescence that could hamper its (yet unknown) biological role. Actually, wtGFP shows detectable deviations from the 3:1 optical ratio between the **A** and **B** bands only at pH below pH 4.5 and above pH 10 [56] where the protein starts unfolding. This means that, in the physiological pH range, the ionization of *Chro* phenol does not involve a *net proton transfer* to or from the external solution, but only an *internal* proton displacement to or from other ionizable residues surrounding *Chro*.

GFP mutants engineered by scientists have no natural role to play and therefore the ionization of *Chro* phenol is often very sensitive to the external proton concentration in the 5–10 pH range, in spite of the *Chro*-insulating β-barrel tertiary structure [71–73]. In such cases, the phenol group of *Chro* is connected to the external solution by means of "proton wires" constituted by several residues networked by H-bonds. Agmon et al. recently demonstrated also the existence of "proton-collecting antenna" systems, i.e., several carboxylates within H-bond networks on the surface of proteins, which convey protons to the orifice of an internal proton wire leading to the protein's active site [74]. Note that the pH dependence of *Chro* ionization, and the existence of pH-interconverting different optical states, is a prerequisite for GFP-based pH intracellular indicators [71, 73, 75].

In many GFP mutants, the pH-dependent optical response is analogous to that of the isolated *Chro*, and follows a strict single-site protonation behavior (Fig. 2a, b, c). Here absorption spectra show the progressive interconversion of neutral *Chro* (absorbing at 390–420 nm) into anionic *Chro* (absorbing at 475–520 nm) (Fig. 2a). The fluorescence spectrum at each pH is determined by the relative populations of neutral and anionic *Chro*; in absence of ESPT neutral *Chro* is usually very dim, whereas anionic *Chro* is always bright (Fig. 2c).

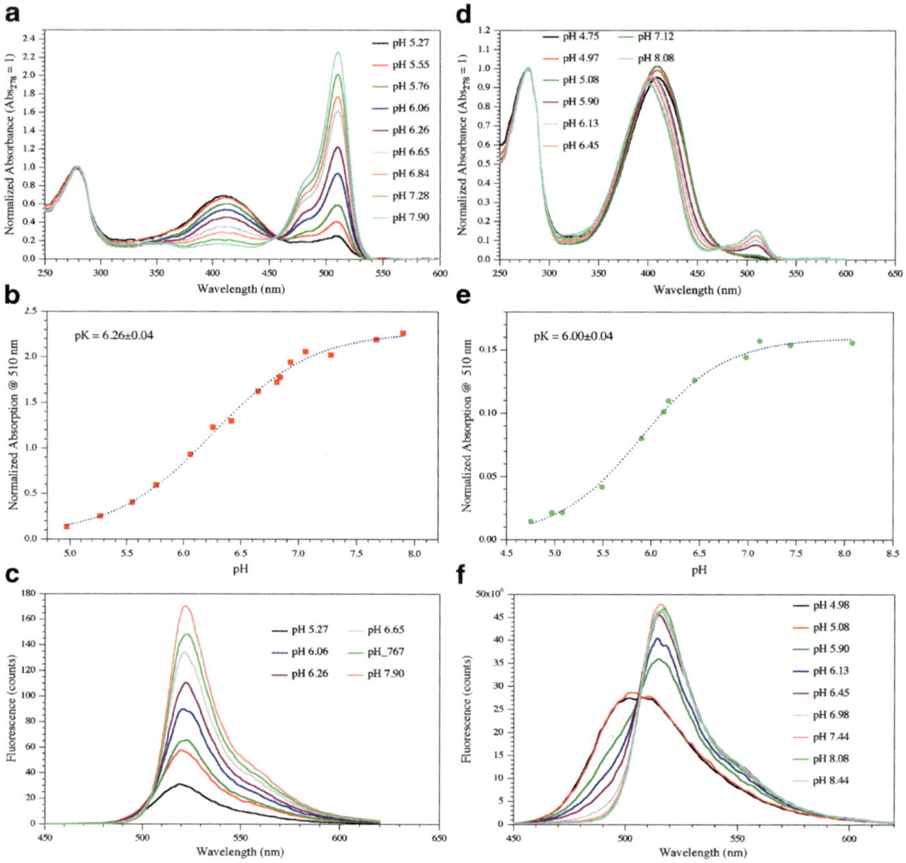

Fig. 2 pH titrations of representative FP mutants. (**a**) pH-dependent interconversion between neutral and anionic *Chro* in S65A/V68L/S72A/T203Y GFP (Mut2Y [76]) as monitored by absorption spectroscopy. (**b**) Absorption maxima at 510 nm of Mut2Y is plotted vs. pH to show the typical sigmoid trend of a single-site pH titration ($pK_a = 6.26$). (**c**) pH-dependent interconversion between neutral and anionic *Chro* in Mut2Y as monitored by fluorescence emission by exciting at 470 nm, where the anionic *Chro* is predominantly excited. (**d**, **e**, **f**) Same as in *panels* (**a**, **b**, **c**) but Mut2Y is replaced by F64L/T203Y (E[1]GFP, [76], $pK_a = 6.0$) and its fluorescence is excited at 395 nm (**f**) where the only neutral *Chro* absorbs; note the large shift in emission wavelength maxima as pH is increased

In other mutants, the absorption trend with pH is more complicated (Fig. 2d, e, f). Plotting the absorbance of anionic *Chro* vs. pH, a nice single-ionization curve is obtained, as expected (Fig. 2e) [62, 76, 77]. This plot allows for the calculation of pK_a, which we should interpret as the thermodynamic value associated with *Chro*'s deprotonation. At pH » pK_a, however, the absorption band corresponding to neutral *Chro* is still intense, although it is slightly wavelength-shifted compared to that found at acidic pH (Fig. 2d); this phenomenon is not consistent with the simple ionization scheme of *Chro*, and consequently the experimental pK_a cannot be attributed uniquely to *Chro*. Additionally, these mutants display detectable emission also from neutral state(s) [62]. The latter effect appears to be pH-dependent, leading to emission spectra that are red-shifted by raising the solution pH (Fig. 2f).

We shall now describe a comprehensive model (named 2S-model) that rationalizes most experimental findings related to the pH dependence of the optical response in pH-sensitive GFPs as well as in unresponsive wtGFP by inclusion of a *second, optically inactive, protonation site* thermodynamically coupled or uncoupled with *Chro* ionization [76].

The first assumption in our model is that, likewise wtGFP, the principal optical characteristics of any Y66 GFP variant stem from the ionization of the phenol group on the chromophore and therefore from the ground state populations of neutral and anionic forms of *Chro*. In certain conditions, these populations can be modulated by external pH. We shall not include in the treatment structurally/environmentally "non-relaxed" anionic species **I** (if present), as they negligibly contribute to the protein absorption spectra on account of their metastable character and short lifetime. Note, also, that many GFP mutants (e.g., those devoid of Thr^{203}) usually display no structural relaxation upon chromophore protonation/deprotonation, thus yielding **I*** = **B***.

The second assumption is that the optical behavior of the protein can, in some cases, be affected by a second *ionizable site* XH (henceforth denoted as X when deprotonated), which resides on the lateral chain of a nearby residue to *Chro*. *Chro* and XH (or X) are supposed to exchange protons at ground state (and perhaps also at excited state), possibly through a few proton relays belonging to an internal H-bonding network including the chromophore. Actually, many protonatable groups are likely exchanging protons with the chromophore (e.g., His^{148}, Glu^{222}, Arg^{96}). We shall see that only one can modulate, through its own ionization, the proton equilibrium of *Chro*. This assumption is verified experimentally [76]. Both *Chro* and XH are supposed to be in direct proton communication with the external solution through some "proton antenna" entry or exit pathways such as those described by Agmon et al. [74].

Two protonation sites imply four distinct ground states that we denote as **A′**, **A**, **B**, and **B′**. **A′** and **B′** (Scheme 3). **A** and **B** are mixed states: **A** pertains to protonated *Chro* and X, whereas **B** corresponds to anionic *Chro* and XH (Scheme 3). This notation is coherent with the reported protonation scheme of wtGFP [53, 60, 61], where the **A** state is characterized by neutral chromophore and anionic Glu^{222} (that plays the role of "coupled" X residue, as we shall show in the following), whereas **B** has the opposite configuration. We further assume that: (1) each state maintains its optical characteristics when the pH is varied, and (2) a linear relationship holds between absorbance/fluorescence and concentration (linearity is usually ensured in intracellular microscopy as measurements are performed on diluted solutions and/or very thin observation volumes). Thus, the absorption or fluorescence of the protein at a given pH is given by the sum of the populations of each state multiplied by its molar absorbance or fluorescence yield, respectively. Following the mathematical treatment of Ullmann [78], we can write:

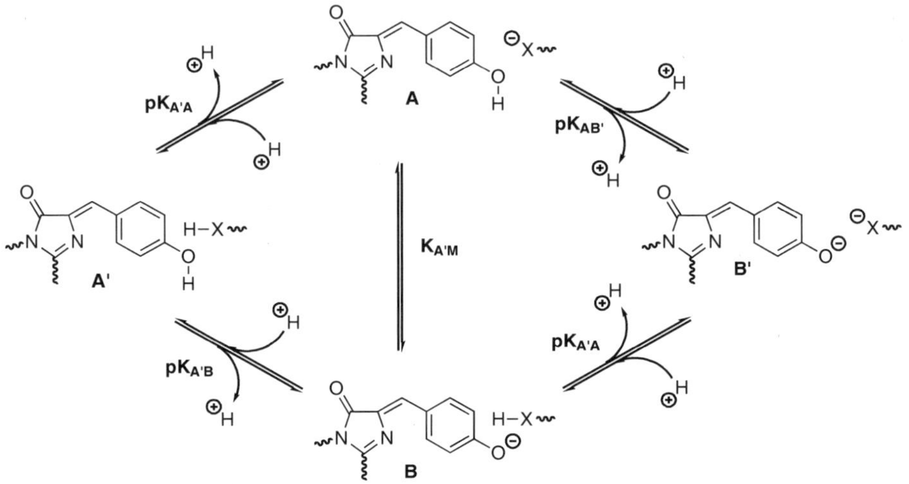

Scheme 3 2S model of pH-dependent optically active ground states in GFPs [76]. XH is an ionizable residue nearby the chromophore. The four possible states **A′**, **A**, **B**, and **B′** are reported together with their associated protonation reactions. If X and the chromophore are uncoupled, only the [**A′**↔**B** + H$^+$] equilibrium determines the optical properties of the protein. When strong anticooperative coupling takes place (owing to unfavorable thermodynamics given by the close proximity of two negative charge – **B′** state), only competing [**A′**↔**B** + H$^+$] and [**A′**↔**A** + H$^+$] processes are observed in the pH range of protein folding. The mixture of **B** and **A** obtained at high pH constitutes the "apparent" **M** state in our notation [76]

$$S = C_0\left[S_{A'}+S_A\cdot 10^{(pH-pK_{A'A})}+S_B\cdot 10^{(pH-pK_{A'B})}+S_{B'}\cdot 10^{(2pH-pK_{A'A}-pK_{A'B}+W)}\right] \times\left[1+10^{(pH-pK_{A'A})}+10^{(pH-pK_{A'B})}+10^{(2pH-pK_{A'A}-pK_{A'B}+W)}\right]^{-1} \quad (1)$$

where S is the absorption or the fluorescence of the protein at a given observation (excitation) wavelength; C_0 is the protein concentration; $S_{A'}$, S_A, S_B, and $S_{B'}$ are the molar absorbance/fluorescence of each state at the observation (excitation) wavelength; $pK_{A'A}$ and $pK_{A'B}$ are the pK_a values pertaining to the **A′**↔**A** and **A′**↔**B** proton dissociation equilibria, respectively (Scheme 3); W is the *interaction free energy* between the ionization reactions of *Chro* and XH; W is positive when the coupling is *cooperative* (i.e., deprotonation of one site favors the deprotonation of the other one), whereas it is negative when anticooperative coupling takes place (i.e., deprotonation of one site hampers the deprotonation of the other one). From the mathematical treatment by Ullmann [78] and (1) we calculated how the molar fraction of each state (Fig. 3a, c, e) and the signals of neutral and anionic *Chro* (Fig. 3b, d, f) change with pH for some prototypical cases, i.e., $W = 0$ (Fig. 2a, b), $W \ll 0$ and $pK_{A'B} > pK_{A'A}$ (Fig. 3c, d), and $W \ll 0$ and $pK_{A'B} < pK_{A'A}$ (Fig. 3e, f).

A first remarkable case occurs for $W = 0$. This means that XH/X and *Chro* are decoupled, i.e., the ionization of the first *does not influence* the ionization thermodynamics of the second. Under the reasonable hypothesis that XH influences

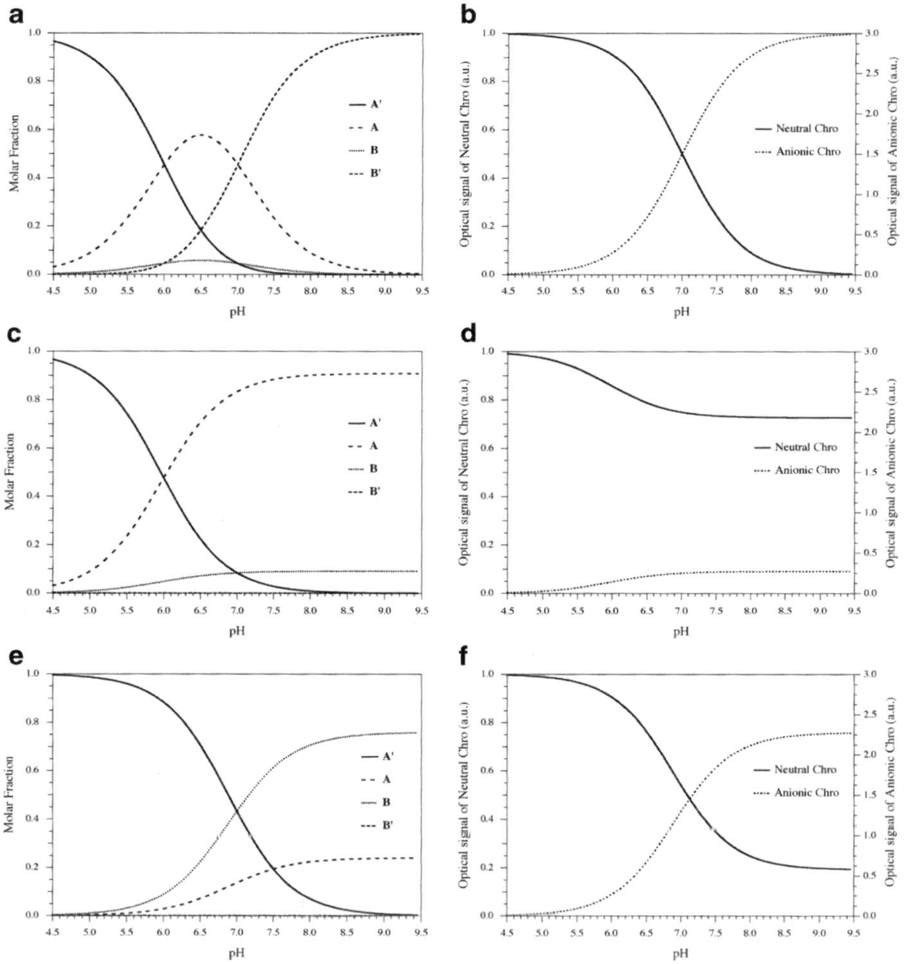

Fig. 3 pH-dependence of molar fractions of **A′**, **A**, **B**, and **B′** states (*panels* **a**, **c**, **e**) and optical signals of neutral and anionic *Chro* states (*panels* **b**, **d**, **f**) for prototypical cases relevant to the 2 S model [76]. (*panels* **a**, **b**): W = 0, $pK_{A'A} = 6$, $pK_{A'B} = 7$, and $S_{A'} = S_A = 1$, $S_B = S_{B'} = 1$; (*panels* **c**, **d**): W = –1,000, $pK_{A'A} = 6$, $pK_{A'B} = 7$, and $S_{A'} = 1$, $S_A = 0.8$, $S_B = S_{B'} = 3$; (*panels* **e**, **f**): W = –1,000, $pK_{A'A} = 7.5$, $pK_{A'B} = 7$, and $S_{A'} = 1$, $S_A = 0.8$, $S_B = S_{B'} = 3$

neither the optical properties of *Chro* states (i.e., $S_{A'} = S_A$ and $S_B = S_{B'}$), (1) becomes:

$$S = C_0 \cdot \frac{S_{A'} + S_B \cdot 10^{(pH - pK_{A'B})}}{1 + 10^{(pH - pK_{A'B})}} \tag{2}$$

Equation (2) represents the typical isotherm for a single-proton dissociation equilibria and generates a sigmoidal dependence of protein absorption and fluorescence on pH (Fig. 2b). The absorption of synthetic *Chro* (or *Chro* in unfolded protein) in water follows (2) at pH above 4 (below the N^{66}-protonation equilibrium must be included, Sect. 2.1). In protein, (2) predicts that a rise in the solution pH leads to the full conversion of the *Chro* neutral states (here **A**′ and **A**) into the anionic state (here **B** and **B**′) (Fig. 2a); the opposite holds when pH is reversed. At pH = $pK_{A'B}$, the optical signal is exactly at half of its global variation. Differentiation of (2) shows also that the maximum signal sensitivity to pH changes is found in the pH range [$pK_{A'B}$ − 1;$pK_{A'B}$ + 1], where the fluorescence response to pH is nearly linear (maximum deviation: 16%). This "chromophore-like" optical behavior *is shared by many GFP mutants* (e.g., some YFPs [55, 79], citrine [80], Mut2 class [72]); $pK_{A'B}$, however, is 1–3 pH units lower than the pK_a of the isolated chromophore. Preliminary results on the thermodynamics of GFP protonation (R. Bizzarri, unpublished data) suggest that this difference is related to the stronger negative ionization entropy of the chromophore outside the protein environment (−97.7 J/mol K for synthetic *Chro* in water) compared to inside the β-barrel (−14.4 J/mol K for EYFP). As for most organic acids, deprotonation of the chromophore in water is associated with the structural reorganization of the H_2O molecules around the newly formed negative charge, a process that leads to a significant entropy decrease of the system. Ionization of *Chro* within the folded protein does induce some rearrangements of nearby residues (e.g., Thr^{203} flipping in T203 mutants) but apparently at lower entropy expense.

The second remarkable case occurs when $W \ll 0$, i.e., the ionization of XH strongly hampers the deprotonation of *Chro*. This behavior is attributable to all GFP mutants whose Glu^{222} is not forced in a single protonation state by strong H-bonds with surrounding residues [76]; in such cases, X = Glu^{222} and the strong anticooperative coupling presumably stems from the strong negative energy implied in the creation of two close negative charges on *Chro* and Glu^{222} upon concomitant deprotonation. Consistently, in these mutants the fully deprotonated **B**′ state cannot be reached within the stability range of the protein (note that **B**′ becomes significantly populated at pH ≈ [$pK_{A'A}$ + $pK_{A'B}$−W], Fig. 3c, e). Assuming $W \ll 0$ (so as to discard the **B**′ population at any meaningful pH) and introducing the new ionization constant $K_{A'M}$ and molar absorption/fluorescence S_M:

$$K_{A'M} = K_{A'A} + K_{A'B} \tag{3}$$

$$S_M = \frac{S_A + S_B(K_{A'B}/K_{A'A})}{1 + (K_{A'B}/K_{A'A})} \tag{4}$$

Equation (1) rearranges into:

$$S = C_0 \cdot \frac{S_{A'} + S_M \cdot 10^{(pH - pK_{A'M})}}{1 + 10^{(pH - pK_{A'M})}} \tag{5}$$

Equation (5) is formally equivalent to (2), and it describes the sigmoidal evolution of the optical response between the **A**′ state (prevalent at pH < $pK_{A'M}$) and the new

M state (prevalent at pH > $pK_{A'M}$). The sigmoidal behavior is depicted in Fig. 3d, f. Note that the **M** state *does not coincide* with **B**, i.e., the **M** state does not pertain to the pure anionic form of protein *Chro*. Actually, **M** is an apparent state that corresponds to a *mixed form* of **A** (neutral *Chro*) and **B** (anionic *Chro*); the optical characteristics of **M** are given by (4), whereas the thermodynamics of **A**′↔**M** apparent ionization is expressed by (3) [76]. This explains the non-negligible presence of neutral chromophore in the absorption and/or fluorescence spectra of the "coupled" mutants at pH well above the protein chromophore's pK_a, although they display pH-titration behavior typical of a single ionization state (Fig. 2e). Interestingly, the relative molar ratio between neutral (state **A**) and anionic (state **B**) *Chro* at high pH is determined by the comparison between $pK_{A'A}$ and $pK_{A'B}$: for $pK_{A'A} < pK_{A'B}$, **A** is the prevalent form at high pH (Fig. 3c), whereas **B** becomes prevalent in the opposite case (Fig. 3e). Thus, we can play with the acidity of the second ionizable site to induce or hamper *Chro* deprotonation by rising pH.

We should stress that also wtGFP belongs to this protein family, as previously suggested by Scharnagl [81]. wtGFP differs from other "coupled" mutants *only* for a much lower $pK_{A'M}$, which is around 4.5–4.9 (other mutants display $pK_{A'M}$ from 5.5 up to 7.2). This explains why two states with different *Chro* protonation, **A** and **B**, can be constantly seen in the optical spectra of wtGFP at any pH without invoking an unlikely "proton-shield" from the external solution. Accordingly, when Glu^{222} is replaced by Gln in wtGFP the protonation behavior changes drastically: the optical response becomes sensitive to external pH change in the physiological range [65, 82].

Assuming $pK_{A'M} = 4.9$ and a population ratio **B**/**A** = 6, (3) and the trivial identity **B**/**A** = $K_{A'B}/K_{A'A}$ make us calculate $pK_{A'B} = 5.75$ and $pK_{A'A} = 4.97$. This means that the origin of the lower $pK_{A'M}$ in wtGFP is fully attributable to the "easier" ionization of Glu^{222} carboxylic group (a case resembling that of Fig. 3c, d). Thus, natural selection has engineered Glu^{222} to play multiple pivotal roles in wtGFP: (1) it helps the maturation of chromophore [39];(2) it deprotonates much more easily than the phenol group in *Chro* (Glu side chain: $pK_a = 4.07$, *Chro*: $pK_a = 8.3$) and, by means of the anticooperative coupling effect, it strongly lowers the optical pK_a to a value outside the physiological range thereby stabilizing the protein emission; (3) it is a good recipient for the *Chro's* proton upon ESPT, conferring significant emission to the protein's neutral *Chro*. Note that the second and third properties positively cooperate with the biological function of wtGFP: in fact, the pH-independent neutral/anionic **M** state implies that the protein fluorescence can be excited in a very large wavelength range (390–500 nm, Fig. 1d), thus making wtGFP a very efficient FRET acceptor for the bioluminescence of its partner *Aequorin* (which bioluminesces between 400 and 550 nm, [57]) in the jellyfish cells.

2.4 *Ratiometric Optical Response*

Equation (2), or its equivalent (5), shows that the optical signal depends on the total protein concentration C_0. This poses a limit on the use of GFP mutants as intracellular indicators of pH. Indeed, in living specimens it is rather complicated to control protein expression: this makes very difficult to determine whether fluorescence changes observed in different cells (or intracellular environments) are originated by a change in pH of the concentration of the fluorescent indicator protein. The fluorescence response of the indicator may also be affected by the cell thickness in thin regions as the cell periphery, as the probe excitation is proportional to the optical path length. Calibration methods making use of purified protein solutions of known concentration are often cumbersome or inaccurate [83]; fluctuation methods such as FCS yield molar brightness and protein concentration only when the indicator is very diluted in the intracellular context (usually below 30–50 nM) and require a complex fluorescence acquisition apparatus.

A straightforward way to circumvent all these problems is represented by GFP mutants whose optical response can be monitored in the so-called *ratiometric configuration*. These proteins display opposite optical trends of fluorescence excitation (*excitation ratiometric indicators*) and/or emission (*emission ratiometric indicators*) in two distinguishable optical intervals: the signal ratio in these two intervals drops the dependence from C_0 while retaining the reliance on the concentration of the analyte under observation (here: pH). The theoretical analysis of a ratiometric pH indicator, in a form equivalent to that described by Grynkiewicz for Ca^{2+}-sensors [84], is presented in the following.

Let us consider a GFP whose fluorescence obeys (2) or (5) and two sets of excitation–emission wavelengths or wavelength intervals (λ_{x1},λ_{e1}) and (λ_{x2},λ_{e2}). Taking the fluorescence ratio in these two intervals we have:

$$\frac{F(\lambda_{x1},\lambda_{e1})}{F(\lambda_{x2},\lambda_{e2})}=\left(\frac{F_{\infty}(\lambda_{x1},\lambda_{e1})+F_0(\lambda_{x1},\lambda_{e1})\cdot 10^{(pK-pH)}}{F_{\infty}(\lambda_{x2},\lambda_{e2})+F_0(\lambda_{x2},\lambda_{e2})\cdot 10^{(pK-pH)}}\right) \tag{6}$$

where F_0 and F_∞ represent the molar brightness at very low pH (and it replaces the notation $S_{A'}$) and at very high pH (and it replaces the notation S_B or S_M), respectively. pK_a must be identified with $pK_{A'M}$ for "coupled" mutants or with $pK_{A'B}$ otherwise. Equation (6) can be recast as:

$$R[1,2]=\frac{F(\lambda_{x1},\lambda_{e1})}{F(\lambda_{x2},\lambda_{e2})}=R_0[1,2]\cdot\left(\frac{R_f[1,2]+10^{(pK'-pH)}}{1+10^{(pK'-pH)}}\right) \tag{7}$$

where

$$R_0[1,2]=\frac{F_0(\lambda_{x1},\lambda_{e1})}{F_0(\lambda_{x2},\lambda_{e2})} \tag{8}$$

$$R_f[1,2]=\frac{F_{\infty}(\lambda_{x1},\lambda_{e1})}{F_0(\lambda_{x1},\lambda_{e1})}\cdot\frac{F_0(\lambda_{x2},\lambda_{e2})}{F_{\infty}(\lambda_{x2},\lambda_{e2})} \tag{9}$$

$$pK_a{}'=pK_a-\log\left[\frac{F_{\infty}(\lambda_{x2},\lambda_{e2})}{F_0(\lambda_{x2},\lambda_{e2})}\right] \tag{10}$$

R[1,2] is the ratiometric fluorescence signal relevant to the optical setup [1,2]. Ratiometric indicators *by excitation* display $\lambda_{x1} \neq \lambda_{x2}$ and $\lambda_{e1} = \lambda_{e2}$; those by emission have $\lambda_{x1} = \lambda_{x2}$, $\lambda_{e1} \neq \lambda_{e2}$.

The value of R_0 is linked to instrumental characteristics such as the excitation intensity and the detector efficiency because it represents a fluorescence ratio taken adopting two specific excitation/emission optical sets (8). Instead, R_f and pK_a' depend only on the photophysical/thermodynamic properties of the fluorescent protein and on the selection of the excitation/emission sets (λ_{x1},λ_{e1}) and (λ_{x2},λ_{e2}) (9)–(10). Importantly, (7) shows the same functional dependence of the measurable quantity from pH as (2) and (5). Thus, a plot of R[1,2] vs pH is sigmoidal with R_0 as lower asymptote (the *ratiometric offset*), R_f as amplitude (the *dynamic range*), and pK_a' as inflection point (the *ratiometric pK$_a$*) (Fig. 4a). Different from (2) and (5), however, (7) does not retain the dependence from the concentration C_0. In addition, the ratio R[1,2] is independent from any geometrical feature, such as cell or specimen thickness, as well as

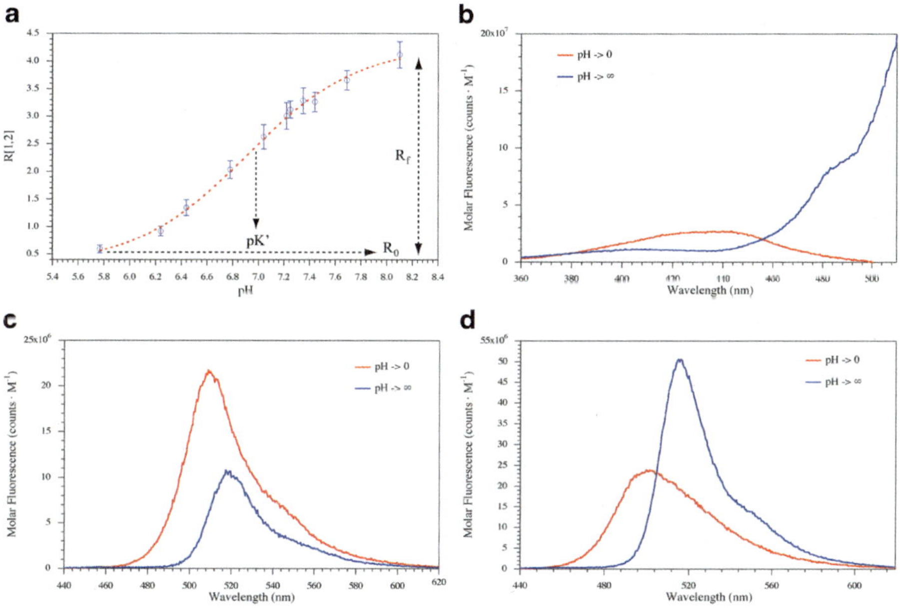

Fig. 4 Ratiometric properties of E[1]GFP and E[2]GFP. (**a**) Universal calibration curve for Tat-E[2]GFP obtained in U2-OS cells by adopting the excitation ratiometric scheme $\lambda_{x1} = 488$ nm, $\lambda_{x2} = 458$ nm, and $\lambda_e = 500$–600 nm; the three parameters relevant for ratiometric measurements, i.e., R_0, R_f, and pK_a' are shown in the graphic. (**b**) Extrapolated excitation spectra collected at 523 nm [76] of E[2]GFP at pH → 0 (*red*) and at pH → ∞ (*blue*) to show the intrinsic optical properties of **A**′ and **M** states (for notation, see text and Scheme 3); note the isosbestic point near 460 nm that is very close to the 458-nm emission line of Ar-laser. (**c**) Same as in *panel* (**b**), but the extrapolated fluorescence emissions upon 395-nm excitation are now considered; note the emission shift that stems from different ESPT pathways for **A**′ and **A** (see text). (**d**) Same as in *panel* (**c**), but E[1]GFP is now considered

from general fluorescence variations due to photobleaching effects. In principle, (7) *describes a universal calibration curve under the selected excitation/observation conditions.*

The necessary condition $R_f \neq 1$ to meaningful pH-dependent R[1,2] explains why only fluorescent probes characterized by multiple excitation or emission maxima that show opposing changes in fluorescence excitation or emission in response to pH [see (9)] are utilizable as ratiometric pH indicators. Nonetheless, it is worth noting that the optical setup (λ_{x2},λ_{e2}) affects strongly the pK_a' value, i.e., the mid-point of fluorescence ratiometric response upon pH (10). Careful selection of this setup is required to tailor the pH indicator to the desired biological application(s), as maximum sensitivity of ratiometric signal to pH occurs in the range [pK_a'−1;pK_a' + 1], where R[1,2] vs pH is nearly linear.

Ideally, all GFP mutants should be excellent excitation ratiometric pH indicators, on account of the large absorption and excitation difference between the neutral and anionic chromophore (Fig. 1d and 2a, d). Some variants show also significant alteration of the emission spectra upon pH, and could in principle be used as emission ratiometric indicators (Fig. 2f). Unfortunately most GFPs display poor emissivity at low pH from the neutral chromophore and cannot be employed to monitor ratiometrically pH with good accuracy. High-yield ESPT from the neutral excited state seems actually a pre-requisite for a GFP-based ratiometric pH-indicator, although some authors described mutants whose intrinsic emission of neutral *Chro* is enough to provide adequate signal to noise ratio in pH measurements [62].

We noticed earlier that "uncoupled" mutants are characterized by Glu^{222} fixed in the protonated state by the stabilizing action of some surrounding residues: these GFPs lack the primary acceptor of ESPT and are usually dimly fluorescent at low pH. We have experimentally observed that many "coupled" variants, instead, display significant fluorescence from **A**′, and sometimes even from **A** (e.g., wtGFP, characterized by an impressive $\Phi_{\mathbf{A}} = 0.78$ [57]), on account of an efficient ESPT mechanism (Fig. 2f). Notably, Glu^{222} is protonated in **A**′: this means that one or more additional residues, for instance His^{148}, must act as ESPT recipient. This subtle difference between **A**′ and **A** is at basis of a minor, but well-detectable, wavelength shift in the emission spectra of the two states (Fig. 1h). We shall show at Sect. 3.2 that such shift can be proficiently used to engineer GFP-based emission ratiometric indicators of pH.

3 GFP-Based pH Indicators for Intracellular Use

In 1998, four different papers came out describing for the first time the pH_i determination in different intracellular compartment by means of pH-sensitive GFP mutants [71, 75, 85, 86]. Since then, several GFPs have been proposed to the attention of the scientific community as genetically encodable intracellular pH indicators. A recent estimate based on simple bibliographical keywords sets to about 100 the number of published papers reporting on pH measurements by GFP indicators. Many of these papers provided new quantitative insight on subtle biochemical processes (for example see [87, 88]) and there are no doubts that in the near future, new scientific evidences will be collected with the help of this kind of pH indicators.

A simple classification divides GFP-based pH indicators into two families: nonratiometric and ratiometric indicators. In the following, we shall review the components of each family. The reader is also referred to [26] for further details.

3.1 Nonratiometric Indicators

The members of this family are GFP variants that show strong pH-dependence of their optical properties but whose poor emission from the neutral chromophore makes them unsuitable for ratiometric measurements. As we already stated at Sect. 2.3, GFP mutants where Glu^{222} is forced in the protonated state regardless of the pH generally display these optical characteristics. Nonratiometric pH indicators are conventionally used to report *changes* of pH_i rather than absolute pH_i values, owing to the variable, and not easily measurable, intracellular GFP expression (Sect. 2.4).

In 1998, the group of Brooks Robey [75], Verkman [71], and Tsien [85] introduced the F65L/S65T GFP mutant, better known as Enhanced GFP (EGFP), as nonratiometric efficient pH_i indicator. EGFP displays a significantly green-emissive anionic state ($\varepsilon_{488} = 60{,}000\ M^{-1}\ cm^{-1}$, $\Phi = 0.7$, $\lambda_e = 509$ nm, [76]), although its pK_a around 6 prevents the pH monitoring of alkaline subcellular component such as mitochondria [71]. Actually EGFP is characterized by a weak coupling between Glu^{222} and *Chro* that leads to about 12% of **A** state at high pH [76, 89]; thus, for most applications EGFP can be safely considered an uncoupled variant. EGFP was applied to probe pH changes in the cell cytoplasm [71, 75, 85], Golgi apparatus [71, 85], and synaptic vesicle cycling at nerve terminals [90].

By means of several rounds of random mutagenesis on wtGFP, Rothman and coworkers generated in 1998 and 2000 two pH-sensitive GFPs, named "Ecliptic" and "Superecliptic" pHlourins (EcGFP and sEcGFP), which since then appeared more suitable than EGFP to monitor intracellular pH changes on account of their higher pK_a values ($pK_a = 7.1$–7.2, [86, 90]). EcGFP shows the S147D/N149Q/T161I/S202F/Q204T/A206T GFP mutation pattern. Inspection of reported optical properties clearly suggests that EcGFP is characterized by coupled protonation between *Chro* and presumably Glu^{222}. The absorption spectra of EcGFP maintains a predominant neutral state at $pH > pK_a$ [91]. Additionally, FCS of EcGFP evidenced a pH-independent internal proton transfer at high pH [91], which may be attributed to the proton exchange reaction between *Chro* and XH in the **M** state (Sect. 2.3). EcGFP is conventionally used as a nonratiometric sensor by selective excitation of the **B** state at 476 nm: the fluorescence disappears at low pH as **B** is no more populated. Consistently with other "coupled" variants, however, the neutral states of EcGFP maintain non-negligible fluorescence and this protein can be also used as ratiometric indicator.

sEcGFP adds the F64L/S65T mutations to the EcGFP sequence [90], and the anionic chromophore of sEcGFP is about nine times brighter than in EcGFP. Although no absorption spectra of sEcGFP are available, this fluorescence enhancement is likely related to the S65T replacement, which is known to inhibit strongly Glu^{222} ionization thereby favoring **B** population [73, 76].

In the first report by Rothman [86], EcGFP (and its ratiometric counterpart RaGFP, Sect. 3.2) was fused to the luminal C-terminus of vesicle-associated membrane protein (VAMP)-synaptobrevin (the full construct was named synapto-pHlourin) to monitor synaptic vesicle exocytosis and recycling. In the lumen of vesicles the pH is low, and EcGFP excitation at 476 nm yields a very dim emission; however, upon exocytosis EcGFP comes into contact with cytoplasmic pH thereby boosting its fluorescence. On account of this property, EcGFP and its brighter alternative sEcGFP have been proposed as general markers of cell exocytosis [92]. EcGFP and sEGFP are broadly used to monitor presynaptic secretory vesicles, vesicle fusion events, exocytosis and endocytosis (see, for example: [87, 93, 94]).

As already discussed at Sect. 2.2, the T203Y substitution in the GFP sequence leads to significant red-shifted absorption and emission of the **B** state because of a π–π stack interaction between *Chro* and Tyr^{203} [95]; T203Y GFPs are collectively denoted as Yellow Fluorescent Proteins (YFPs). YFP mutant S65G/S72A/T203Y, usually indicated as Enhanced YFP (EYFP), is pH responsive and is characterized by an uncoupled chromophore, owing to the stabilization of protonated Glu induced by the S65G replacement [55, 95]. Accordingly, the neutral chromophore of EYFP is almost non-emissive [55]. Instead, the anionic chromophore of EYFP is very bright ($\varepsilon_{514} = 84{,}000\ M^{-1}\ cm^{-1}$, $\Phi = 0.61$, $\lambda_e = 527$ nm, [96]). On account of its rather high $pK_a = 7.1$, EYFP was shown to be very effective to probe pH changes in the cytosolic, Golgi, and mitochondrial environment by the Tsien lab [85]. EYFP is also a good selection for multicolor experiments in tandem with pH-unresponsive cyan or green mutants. Unfortunately, the emission of EYFP is severely quenched by the chloride ion [55, 79], which can be abundant in some cell types and subcellular organelles. Attempts at reducing the chloride sensitivity of EYFP by introducing mutations in its well characterized ion-binding pocket yielded chloride-resistant mutants characterized with rather low pK_a [80], which seem unsuitable for pH monitoring in vivo.

3.2 Ratiometric Indicators

In the same study that introduced EcGFP, Rothman and coworkers described the first GFP-based ratiometric pH indicator by excitation, named Ratiometric pHlourin (RaGFP) [86]. RaGFP shows the following mutational pattern: E132D/S147E/N149L/N164I/K166Q/I167V/R168H/L220F. Surprisingly, RaGFP is characterized by a reversed pH dependence of its chromophore states; in fact, rising pH leads to the parallel decrease of the anionic band at 476 nm and the increase of the neutral band at 395 nm. The photophysical reasons underlying this odd behavior have not been elucidated yet. In a typical experiment, RaGFP is imaged by taking the ratio of 500–550 nm emission excited at 410 and 470 nm [88, 97]. The group of Schulte carried out a thorough ratiometric analysis on this indicator and reported it to be associated with $pK_a' = 6.9$ and $R_f = 8.8$ for $\lambda_{x1} = 415$ nm, $\lambda_{x2} = 475$ nm, and $\lambda_e = 508$ nm [38]. Hence, RaGFP is capable of probing pH in the physiological range with high sensitivity. Accordingly, RaGFP has been applied to monitor the pH in a variety of intracellular locations, such as the cytoplasm [97, 98], peroxisomes [88], mitochondria [98], axons of motoneurons [99], bacterial flagellar motor [100], and endosomes and the *trans*-Golgi network [101]. Likewise EcGFP, RaGFP was conjugated with VAMP-synaptobrevin (ratiometric synapto-pHlourin): this construct allowed the monitoring of the actual pH in presynaptic secretory vesicles [86].

As anticipated at Sect. 3.1, EcGFP pHlourin has been proposed as an emission ratiometric pH indicator [38]. Upon excitation at 400 nm **A**′ state does not display ESPT and its emission reflects the intrinsic decay of the neutral excited state at 460–480 nm. Conversely, **A** state is characterized by efficient ESPT, emitting at 510 nm. Thus, a pH-dependent emission shift is obtained by excitation at 400 nm [38]. Adopting the ratiometric sets ($\lambda_x = 400$ nm, $\lambda_{e1} = 511$ nm, $\lambda_{e2} = 464$), Schulte and coworkers determined $R_f = 28$ and $pK_a' = 7.6$ [38].

In 2002, Remington and coworkers published remarkable results about four new ratiometric pH indicators, named deGFPs [77]. deGFPs are characterized by the S65T and H148C(G) and/or T203C replacements and display pK_a values from 6.8 to 8.0 [77]. The chromophore of deGFPs is likely anticooperatively coupled with Glu^{222}, as their absorption spectra retain large absorption of the neutral pH at $pH \gg pK_a$ [77]. Likewise EcGFP, excitation of the **A**′ state of deGFPs at 400 nm does not activate ESPT but the intrinsic emission of the neutral excited *Chro* at 460 nm ($\Phi = 0.05$–0.1); instead, 400 nm-excitation of **A** triggers efficient ESPT yielding green fluorescence ($\Phi = 0.05$–0.1) [62]. X-ray analysis of the crystal structure of deGFP1 highlighted a pH-induced structural rearrangement in the chromophore pocket, presumably at basis of the observed dual emission characteristics [77]. The crystal structure at low pH evidenced the interruption of the proton relay towards Glu^{222} typical of wtGFP, thus explaining the inactivation of ESPT. Raising pH was shown to trigger a backbone motion placing Tyr^{145} and Ser^{147} in a favorable position to form hydrogen bonds with the phenol group of *Chro*; this configuration generates a novel proton relay encompassing Ser^{147} and two water molecules and allows ESPT towards the bulk solvent. On account of these and other findings, Remington and co-workers were the first to propose the modulation of ESPT rate by rationale modification of the primary sequence as a method to change the emission properties of GFP mutants [62, 102, 103].

Although all deGFPs can be applied as excitation ratiometric pH-indicators, deGFP4 variant was tested for emission ratiometry adopting the optical set $\lambda_x = 365$ nm, $\lambda_{e1} = 475$–525 nm, $\lambda_{e2} = 385$–470 nm. deGFP4 was shown to possess a much greater dynamic range [77] compared to popular emission ratiometric pH-indicator SNARF-1 dye [104, 105]. The SNR of deGFP4, however, appeared worse likely because of the dim intrinsic fluorescence at low pH. In the same study, deGFP4 was used to monitor pH changes in the cytoplasm of PS120 cells. Additionally, deGFP4 was tested as ratiometric indicator by emission under two-photon excitation by replacing illumination at 365 nm with 810 nm strong pulsed laser excitation. The use of two-photon-excitable indicators will be discussed in some detail at Sect. 4.2.

In 2006 and 2009, our group introduced the YFP mutants F64L/T203Y (named E^1GFP) and F64L/S65T/T203Y (named E^2GFP) as novel efficient and versatile ratiometric pH indicators for intracellular study [106, 107]. Like deGFPs, E^1GFP and E^2GFP shows proton-dependent optical behavior strongly affected by *Chro* and Glu^{222} anticooperative coupling [76]. Indeed, the neutral form of the chromophore is evident in the absorption spectra of both proteins at high pH (Fig. 2d), indicating active **A**↔**B** internal equilibrium typical of the apparent **M** state. We calculated that **A** state encompasses about 95 and 60% of the overall protein population at high pH in E^1GFP and E^2GFP, respectively [76].

The spectroscopic data of E^1GFP and E^2GFP and their $pK_{A'M}$ values are reported in Table 1. Notably, these two proteins are characterized by significant quantum yield of emission from the **A′** state, probably on account of efficient ESPT towards His^{148} (R. Bizzarri, unpublished results); both the absorption and the excitation of the **A′** state are red-shifted compared to typical neutral *Chro* absorption. The **B** state has strong absorption and emission (Table 1). These date support the use of these two proteins as excitation ratiometric indicators by illumination of **A′** (low pH predominant) and **B** state. Actually, E^2GFP represents a very attractive selection when imaging setups equipped with popular Ar-laser excitation source, owing to its isosbestic point located at 460 nm that can be addressed by the 458-nm Ar emission line (Fig. 4b). Indeed, we demonstrated that when ratiometric λ_{x2} corresponds to an isosbestic point, (10) predicts a minimal deviation of pK_a' from the actual thermodynamic value, which is close to the optimal value 7 in this protein [106]. When we adopted $\lambda_{x1} = 488$ nm, $\lambda_{x2} = 458$ nm (two Ar-laser lines) and $\lambda_e = 500$–600 nm, we obtained $pK_a' = 6.9$ associated with a large dynamic range of measurement in vivo ($R_f > 3$, Fig. 4a). E^2GFP was successfully employed to monitor cellular alkalinization upon mitosis and subsequent acidification upon entry in the G1 phase [106]. A fusion construct of E^2GFP and Tat, the transactivator

Table 1 Protein absorption and emission wavelengths, quantum yields, and pK_a as determined from steady state absorption/emission measurements at different pH. $\lambda_{A'}$, λ_A, λ_B wavelengths at maxima of **A′**, **A**, and **B** absorption, respectively; $\lambda_{emA'}$, $\lambda_{emA/B}$: wavelengths at maxima of the **A′** and **M** emission, respectively; $\Phi_{A'}$, Φ_A, Φ_B quantum yields of emission of **A′**, **A**, and **B**, respectively

Protein	$\lambda_{A'}$ (nm)	$\Phi_{A'}$	λ_A (nm)	Φ_A	λ_B (nm)	Φ_B	$\lambda_{emA'}$ (nm)	λ_{emM} (nm)	pK_a
E^1GFP[a]	410	0.11	400	0.15	509	0.88	500	516	6.00
E^2GFP[b]	419	0.10	401	0.05	515	0.88	510	523	6.78

[a]Base mutation pattern: F64L/T203Y
[b]Base mutation pattern: F64L/S65T/T203Y

protein of HIV-1, allowed us to determine the pH in nucleolar and promyelocytic leukemia protein (PML) bodies at high spatial resolution [106]. Interestingly, we observed a significant pH change upon Tat relocation that might be biologically related to the control of HIV-1 transcription [108]. It is worth mentioning that chloride (and other halides) ions are static quenchers of E^2GFP fluorescence [109], but the mathematical dependence of fluorescence emission from quencher concentration leads to unaffected ratiometric measurements [106]. Actually, the chloride dependence of E^2GFP and one slightly different variant was exploited to engineer an efficient pH/chloride tandem intracellular indicator [110].

Inspection of Table 1 shows that also **A** state of both proteins display non-negligible emission yield at wavelengths distinguishable to that of the **A′** state. The emission displacement can be rationalized by considering that ESPT of **A** is presumably directed to anionic Glu^{222} and it takes part in a different H-bond network around the chromophore. This ESPT rewiring allows the use of E^1GFP and E^2GFP as emission ratiometric pH indicators by exciting selectively the neutral states (Fig. 4c, d). In E^2GFP, $\lambda_x = 403$ nm and collection in the two ranges $\lambda_{e1} = 515$–600 nm $\lambda_{e2} = 475$–515 nm (where **A′** and **A** are predominant, respectively) shifted pK_a' to 7.48, making this setup valuable for pH determination in alkaline subcellular regions. The higher quantum yield of **A** and the larger emission wavelength shift in E^1GFP (Table 1), however, make this protein more useful for emission ratiometry. The optical setup $\lambda_x = 403$ nm, $\lambda_{e1} = 505$–600 nm $\lambda_{e2} = 460$–505 nm was found to yield $pK_a' = 6.68$ and $R_f = 8.57$ [107]. Accordingly, Tat-E^1GFP was applied to monitor vesicle pH in real-time and at high spatial resolution during multi-step internalization through the intracellular endocytic network (Fig. 5) [107].

Ratiometric indicators need not to be constituted by a single GFP molecule. Indeed, two or more GFPs can be linked together affording a fluorescent system composed by unit with distinguishable optical properties. Awaij and coworkers developed in 2001 two ratiometric pH sensors, named GFpH and YFpH, by fusing the pH-insensitive variant GFPuv, excitable at 380 nm, and a classical pH-sensitive GFP excitable at longer wavelengths [111]. GFpH is constituted by GFPuv and EGFP. On account of this, GFpH works well as excitation ratiometric pH indicators for mildly acidic subcellular components adopting the $\lambda_{x1} = 480$ nm, $\lambda_{x2} = 380$ nm, and $\lambda_e = 520$ nm optical scheme ($pK_a' = 6.2$ and $R_f \approx 5$) [111].

YFpH is composed by GFPuv and EYFP. Notably, the excitation at 380 nm leads to an emission shift of YFpH from 509 to 527 nm as the pH was raised. This effect was interpreted as FRET between GFPuv and EYFP, as EYFP is nonfluorescent when excited at 380 nm. Conversely the emission wavelength of YFpH upon 380-nm excitation remained stable when with pH was nearly unchanged, as such emission comes mostly from EYFP. Hence, YFpH was proposed as a ratiometric pH indicator by both excitation ($\lambda_{x1} = 480$ nm, $\lambda_{x2} = 380$ nm and $\lambda_e = 520$) and emission ($\lambda_x = 380$ nm, $\lambda_{e1} = 525$ nm and $\lambda_e = 510$) [111]. The pK_a' of YFpH was shown to be around 6.5–6.8 in both cases [111]. GFpH and YFpH were used to visualize pH changes in the cytosol/nucleus of living cells and during internalization caused by endocytosis upon agonist stimulation [111]. A pH indicator similar

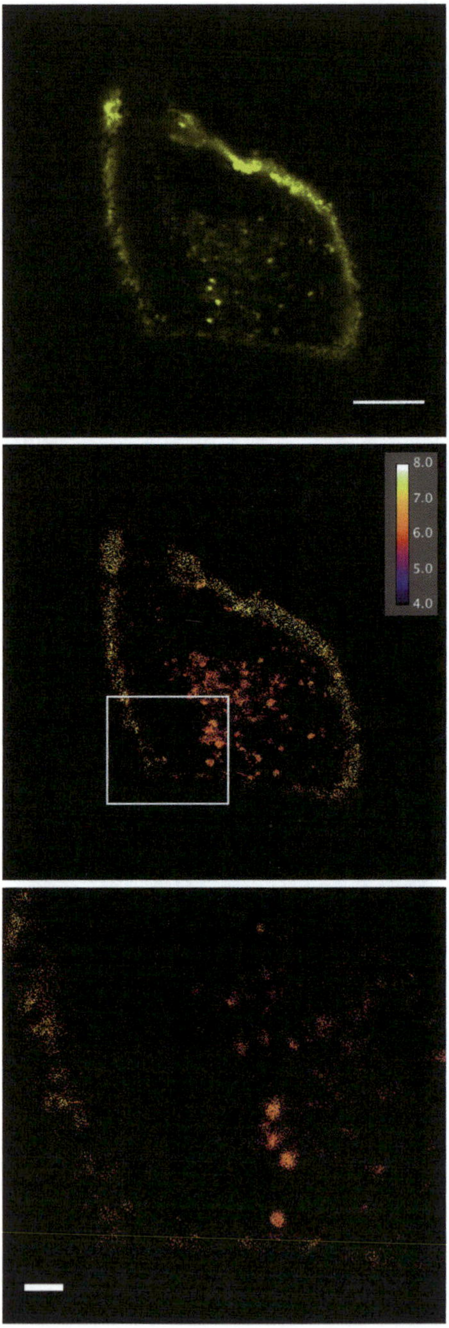

Fig. 5 Ratiometric pH_i map by emission of Tat-E^2GFP internalized in endocytic vesicles collected by confocal microscopy [107]. *Upper panel*: fluorescence emission (scale of *yellows*). *Central panel*: ratiometric map in pseudocolor scale at medium magnification (*Scale bar*: 10 μm). *Lower panel*: high-resolution ratiometric map in pseudocolor scale of the region enclosed by a white rectangle in *central panel* (*Scale bar*: 1 μm). The ratiometric maps clearly display the progressive acidification from the membrane to the inner part of the cell, which is a hallmark of the dynamic evolution of organelle contents from endosomes and pinosomes to degradative lysosomal compartments

to YFpH was realized by conjugation of Cyan Fluorescent Protein (CFP), the cyan-emitting, poorly pH-sensitive [85] Y66W GFP mutant, together with EYFP. Note that in the vast majority of FRET experiments involving *Aequorea* proteins, CFP and EYFP are the two optical partners, owing to their large optical complementarity [112, 113]. CFP-EYFP was proficiently used to detect changes in pH secondary to H^+ efflux into the basolateral space of MDCK cells [114].

In 2004, Pozzan and coworkers introduced mtAlpHi [10], an excitation ratiometric indicator tailored for measurements in high-pH subcellular regions. mtAlpHi was obtained by replacing the calmodulin linker of the Ca^{2+}-sensor camgaroo-II [80] with a portion of Aequorin. mtAlpHi was used to monitor pH changes occurring in a variety of physiological and non-physiological situations in mitochondria [10]. Notably, mtAlpHi (and its parent construct camgaroo-II) comprises a YFP variant that is poorly sensitive to the quenching effect of chloride ions [80].

4 Imaging Intracellular pH

4.1 Confocal Microscopy Setups

Recent advances in microscopy methods have paved the way to the use of many fluorescent indicators to provide high-resolution intracellular measurements in physiological and nonphysiological conditions. A typical microscope apparatus for intracellular fluorescence measurements can be divide into three major components: (1) the excitation sources; (2) the actual microscope; (3) the detector(s). A further component is represented by the optical system(s) that discriminates the excitation light from the emission of the specimen.

The major breakthrough in fluorescence imaging of biological samples was represented by the diffusion of confocal microscopes [104, 115, 116]. In a confocal microscope, fluorescence from out-of-focus planes is almost completely removed by a *pinhole*, which is optically-conjugated to the objective focus and placed just before the detector [117]. This configuration greatly improves the image quality and the axial resolution compared to conventional (e.g., wide-field) epifluorescence microscopes. Using a high-numeric aperture objective, axial resolution (*z*-axis) as low as 0.6–1.2 μm (the corresponding *xy* resolution is usually around 0.2–0.4 μm) are routinely achieved in most confocal setups. The diffraction-limited excitation and probe regions, which have approximately the shape of a 3D cylindrically-symmetric Gaussian whose dimensions correspond with the optical *xy* and *z* resolutions, is called Point Spread Function (PSF). The PSF represents the volume that is actually sampled by the microscope; during a measurement, the PSF is rapidly scanned across a much larger *xy* region to provide a convolution image of each pixel of the sample at the focal plane, which represents the "fluorescence map" of the sample at the objective resolution. Confocal microscopy allows for pH_i detection in focal volumes of less than 1 fl yielding, in principle, high-resolution pH maps in living organisms such as cultured cells. The sequential acquisition process is usually accomplished by fast-moving galvanometric mirrors that deflect the excitation light before entering the objective, thereby scanning line by line the desired region of the sample. Thus, the scanning speed (and pixel dwell time) is a relevant parameter of the measurement because it determines the illumination time of each pixel as well as the time required to collect one image. Modern confocal allows line-scanning speed ranging from a few Hz to 1–2 kHz.

Besides allowing for high-resolution xyz measurements, a confocal microscopy system must possess other properties to maximize the performances of any fluorescent indicator; at the same time, the indicator should be selected and engineered to take full advantage of all the features of the imaging system. The most important rule is to maximize the Signal to Noise Ratio, because it determines the accuracy of the indicator measurement. This can be obtained by using: (1) stable and powerful illumination sources, typically lasers, tuned on the excitation bands of the indicator; (2) very sensitive detectors to reveal weak signals (e.g., highly sensitive photomultipliers, SPADs); (3) highly-selective excitation/emission selection optics (e.g., steep dichroics and filters, Acousto-Optical Beam Splitter). Properties (2) and (3) are particularly critical, as the increasing of excitation light above a certain level typically does not lead to any fluorescence signal increase owing to fluorophore saturation and photobleaching.

Measurements by ratiometric indicators need additional considerations. The microscopy system must be flexible in selecting, and fast in switching, the excitation sources and/or the collection intervals, in dependence of the ratiometric nature of the indicator. Conventional confocal microscopes are usually supplied with multiple excitation laser sources that can be activated sequentially in less than 100 ms, being therefore well-suited for ratiometry by the excitation. We should note again that E^2GFP is an optimized pH indicator for confocal imaging, as it performs best when excited by the 458 and 488 nm lines of Ar laser, which is one of the typical laser source for confocal setups [106]. The large diffusion of solid-state 403/405-nm lasers make also easily accessible ratiometry by emission making use of both E^1GFP and E^2GFP. Some wide-field microscopes come supplied with motorized filter wheels that allow the fast switching of the excitation wavelengths from a lamp source. In such cases excitation ratiometry is possible, although the high axial resolution of the confocal setups is lost (and can be recovered only by post-processing deconvolution).

Confocal microscopes are usually equipped with more than one detector, permitting the concomitant collection of fluorescence in two or more wavelength intervals, which is the basic requirement of emission ratiometry. In the most advanced setups, spectral detector are supplied, i.e., each detector is software-adjusted to collect emission photons of selected wavelength by means of a complex system of motorized gratings and slits within the microscope (bandwidths can be as low as 5 nm). This instrumental feature is particularly interesting in view of modulating both the dynamic range and the apparent pK_a' of ratiometric response [Sect. 2.4, (10)].

Minimization of the detector noise is fundamental for ratiometric measurements. Indeed, the ratio between fluorescence values amplifies the "electronic" error that is present in each image. The nonlinear (sigmoidal) shape of the calibration curve further amplifies this error when the ratio is back-converted to pH. Thus, high amplification of the initial photocurrent should be avoided because it adds a strong electronic noise that degrades the ratiometric measurement. The easiest, and in many cases the only strategy to reduce noise effects in ratiometric measurements is the use of the maximum excitation intensity compatible with minimal probe photobleaching; then, adequate thresholding of images (by means of some reproducible criteria, e.g., the subtraction of background plus a fixed multiple of its standard deviation) is quintessential to obtain a meaningful pH map. pH_i maps collected on cells clamped at a given pH by means of ionophores can help evaluating the intrinsic accuracy of the ratiometric detection method. Notably, the use of photon-counting detectors may represents an interesting option for improving accuracy of ratiometric measurements, as they are characterized by predictable and low-level Poissonian noise [118]. Unfortunately, standard instruments are usually not supplied with photon-counting detectors, which typically find use in complex microscopy setups devoted to single molecule biophysical measurements.

4.2 Two-Photon Microscopy Setups

In recent years, two-photon excitation (TPE) fluorescence microscopy emerged as a valuable alternative of confocal laser microscopy [117, 119]. TPE relies upon the nonlinear (quadratic) sample excitation upon near simultaneous absorption of two photons each bearing half of the energy required to carry out the electronic transition between the ground and the excited state. The nonlinear dependence of TPE on illumination intensity and the high photon flux densities required to yield a significant number of TPE events (~10^{31} photon/cm^2 s), confer to this imaging technique intrinsic high 3D resolution, a property not associated with single-photon excitation microscopy [120]. Indeed, TPE is usually confined in a small volume around the focal point, whose dimensions are critically dependent on the numerical aperture of the objective lens and the illumination wavelength. The calculated TPE lateral and axial resolutions for a 1.2 NA objective at 900 nm are about 0.18 and 0.45 μm, respectively, yielding a focal excitation volume of 0.113 fl [121]. Nonetheless, these high resolutions will degrade steadily with illumination power if fluorophore excitation saturation becomes significant, i.e., for highly two-photon absorbing fluorophores; also, optical aberrations, refraction index mismatch of the sample and other optical factors may contribute to slightly worse resolutions than those predicted by theory.

TPE fluorescence microscopy shows some relevant advantages over both widefield and confocal microscopy. In the latter two techniques, the linear dependence of excitation on illumination intensity produces probe stimulation along an elongated axial region. Since photon absorption is the prerequisite also for photodegradation processes (e.g., photobleaching), the spatially restricted two-photon excitation volume should significantly reduce sample photobleaching and phototoxic effects. Because out-of-focus fluorescence is never generated, there is no need to introduce a pinhole in the detection path of the microscope, as in confocal microscopy [120]. This latter observation implies that in TPE microscopy the "spatially informative" signal coming from the sample is entirely collected by the detector. Finally, the longer wavelength of TPE allows for imaging visible–excitable probes in thick biological samples when IR illumination is adopted [122]; analogously, TPE microscopy with visible light is used to image UV-excitable species such as aromatic residues in proteins and cofactors. It is also worth mentioning that many commercially available confocal setups can be easily upgraded to multiphoton imaging by interfacing a femto- or pico-second mode-locked infrared laser source [121].

As predictable, the development of TPE microscopy methods and setups prompted the engineering of new fluorescent intracellular indicators that could efficiently respond to nonlinear excitation [123]. FPs constitute a valuable class of indicators also in TPE microscopy, owing to their genetically encoded expression that is nonrelated to their optical properties. The TPE fluorescence yielded by a fluorophore in a microscopy setup can be expressed as [124]:

$$\langle \mathrm{F(t)} \rangle = \frac{0.5 \cdot \varphi(\lambda_{\mathrm{em}}) \cdot \delta(\lambda_{\mathrm{ex}}) \cdot \langle \mathrm{P(t)} \rangle^2 \cdot \eta}{\lambda_{\mathrm{ex}}} \cdot \mathrm{C} \qquad (11)$$

where the time averaging of both emitted fluorescence $\langle \mathrm{F(t)} \rangle$ and excitation power $\langle \mathrm{P(t)} \rangle$ reflects the common use of pulsed laser illumination sources; $\varphi(\lambda_{\mathrm{em}})$ and $\delta(\lambda_{\mathrm{ex}})$ are the quantum yield at emission wavelength and the *TPE cross-section* (usually expressed in GM units, $1\mathrm{GM} = 10^{-50}$ cm^4 s/photon) at the excitation wavelength, respectively; η is a multicomponent parameter that takes into account the properties of the optical setup and of the illumination source [125]; C is the concentration of the probe. In ratiometric configuration (11) becomes:

$$\mathrm{R}[1,2] = \frac{\delta(\lambda_{\mathrm{ex1}})}{\delta(\lambda_{\mathrm{ex2}})} \cdot \left(\frac{\langle \mathrm{P_1(t)} \rangle}{\langle \mathrm{P_2(t)} \rangle} \right)^2 \cdot \frac{\eta_1}{\eta_2} \cdot \frac{\lambda_{\mathrm{ex2}}}{\lambda_{\mathrm{ex1}}} \qquad (12)$$

by excitation and

$$\mathrm{R}[1,2] = \frac{\varphi(\lambda_{\mathrm{ex1}})}{\varphi(\lambda_{\mathrm{ex2}})} \cdot \frac{\eta_1}{\eta_2} \qquad (13)$$

by emission. Thus, TPE ratiometric indicators require changes in their cross-section and/or in the emission spectrum to monitor any species, as the other parameters in (12)–(13) would remain constant during the measurement.

Fluorescent proteins display TPE cross sections in the interval 10÷100 GM at wavelengths ranging between 650 and 1,200 nm [126, 127], allowing for good SNR in typical imaging conditions. In GFPs, the neutral and anionic forms of *Chro* have rather distinguishable two-photon excitation wavelengths: the neutral states are usually excited around 800 nm, which corresponds to twice the excitation wavelength at one photon, whereas the anionic state peaks at values (900–1,000 nm) [128]. Often the TPE wavelength falls significantly lower than twice the corresponding one-photon wavelength. Apparently, the latter discrepancy is attributable to a hidden excited state in the vicinity of the lowest excited singlet (S1) of the anionic form of *Chro* [129]. These properties make in principle feasible pH-indicators ratiometric by excitation starting from carefully selected mutants. Compared to single photon excitation, however, the strong overlap between the two excitation bands related to the different selection rules, together with the usually much longer time required to change the excitation wavelength in TPE setups (seconds) make this option not very convenient for physiological measurements. pH-indicator ratiometric by emission is much more attractive. Indeed, TPE usually does not affect the properties of fluorescence emission (a single-photon process), because TPE does not intervene on the nature of the excited electronic–vibrational state that radioactively decays down to ground state. Thus, ESPT and emission shifts identical to the single-photon case are commonly observed in the TPE of pH-sensitive GFP variants [77]. As anticipated in Sect. 2.4, Remington and coworkers described some deGFPs as two-photon ratiometric pH indicators by emission [77]. TPE was demonstrated to give a superior S/N ratio compared with single-photon excitation. This finding was attributed to less autofluorescence under TPE and/or more efficient excitation of the protonated state of deGFP4.

Recently, we demonstrated that also E^2GFP is suitable for emission ratiometry in TPE configuration to monitor external pH. We showed that the emission band obtained by excitation at 780 nm, and attributable principally to **A**′ and **A** states, undergoes a significant red shift upon pH increase [26]. The fluorescence ratio between the 515–555 nm and 470–503 nm intervals was found to increase by a factor of more than 3 when pH was shifted from 4.9 up to 9.5.

Finally, it is worth noting that synapto pHlourin (a fusion construct of ecGFP) was visualized in the olfactory neuron of *Drosophila melanogaster* by TPE microscopy [130]. The nonratiometric pH reporter, however, was just used to monitor real-time exocytic events upon stimulation of the fly by means of the strong fluorescence increase associated to the pH rise upon exocytosis.

4.3 Kinetic Vs. Optical Resolution

In Sect. 1, we noticed that pH is seldom spatially uniform at intracellular level on account of the multicompartmentalized cell biochemistry. High-resolution intracellular pH maps are of great value to understand the biology underlying the pH regulation at *spatial level*, or, in other terms, the scientific relevance of any measurement is directly related to the maximum resolution achievable. Apparently, this would pose a strict requirement on the microscopy imaging setup, asking for the maximum possible resolution. Yet, the photophysical characteristics of the pH indicator are of comparable importance.

A good pH indicator must be capable of reporting an accurate pH estimate in every subcellular region it comes in, i.e., to provide a spatially unbiased pH map. So far we considered the effect of pH on the *equilibrium* optical properties of the target GFP (2), (5) but we have not taken into account the time (τ_c) required to reach the equilibrium state. This value, however, has great importance in view of the translational diffusion of the protein (or, more generally, of the indicator) in the sample under study and the spatial resolution of the imaging setup. In fact, the spatial resolution of the measurement and the diffusive property of the indicator determine together its *permanence time* (τ_d) in each "observation volume" (actually the PSF in a confocal setup): if $\tau_d > \tau_c$, the pH in the "observation volume" is correctly measured, since the indicator ionization has time to get to equilibrium; otherwise, a "smeared" pH map is observed, unless the pH is spatially homogeneous. Note that these observations are rather general and apply to any couple of indicator/monitored species, in any environment, *whenever high spatial resolution maps are the goal.*

Under the assumption of free isotropic diffusion, in the time τ_c the indicator spans a linear distance Δw in the sample given by [131]:

$$\Delta w = \sqrt{4D\tau_c} \tag{14}$$

where D is the diffusion coefficient of the indicator in the sample medium. Δw is called the *kinetic resolution* of the indicator.

Let us now introduce the *optical resolution* of the imaging system Δd. The isotropic diffusion hypothesis now yields:

$$\Delta d = \sqrt{4D\tau_d} \tag{15}$$

Thus, the previously stated $\tau_d > \tau_c$ condition can be recast as $\Delta d > \Delta w$, i.e., the optical resolution *must be larger* than the kinetic resolution to ensure meaningful spatial measurements of pH or any other species.

The kinetic theory of chemical reactions allows for the calculation of the time τ_c, and thereby of the kinetic resolution, once the rate limiting step(s) of the indicator's reaction are identified. For simplicity, we shall consider only the [$\mathbf{A}' \leftrightarrow \mathbf{B} + H^+$] ionization of *Chro* in the present discussion on GFP-based pH indicators. The mathematical treatment, however, can be easily extended to the full 2S model. τ_c can be expressed by [132]:

$$\frac{1}{\tau_c} = k_{\text{on}} \cdot \left([\mathbf{B}]_{\text{eq}} + [H^+]_{\text{eq}}\right) + k_{\text{off}} \tag{16}$$

where $[\mathbf{B}]_{\text{eq}}$ and $[H^+]_{\text{eq}}$ are the concentrations of $\mathbf{B}$ and H^+ at equilibrium, and k_{on}, k_{off} are the protonation and deprotonation rate constants, respectively ($K = k_{\text{off}}/k_{\text{on}}$). The rate constants are implicitly assumed to account for all processes involved in the proton transfer from outside and inside the protein along suitable proton wires. This observation suggests lower protonation kinetics of the *Chro* compared to either ionizable residues on the protein surface or to synthetic acid–base indicators, owing to the β-barrel-restricted proton access to the GFP chromophore. This prediction has been experimentally verified in several GFP mutants [72, 76]. For concentrations similar to those found in vivo, GFPs are typically associated with protonation τ_c values between 100 μs and 1 ms, which are 1–2 orders of magnitude larger than for organic acid–base compounds. However, the nonnegligible size of GFPs is associated with a diffusion coefficient that is 2–4 times smaller than for small organic pH indicators molecules. Also, the cell components offer viscous environments that further reduce the mobility of the GFP; typically, D values between 0.1 and 20 $\mu m^2/s$ are observed [133–135]. GFPs linked to membrane receptors can show D as low as $10^{-3} \div 10^{-2}$ $\mu m^2/s$. Assuming the most unfavorable case of a GFP indicator characterized by $\tau_c = 1$ ms and freely diffusing in the cell cytoplasm ($D = 20$ $\mu m^2/s$), (14) yields an upper limit to the kinetic resolution of 0.28 μm.

In a diffraction-limited optical system, Δd is theoretically given by the Abbe's criteria [136]:

$$\Delta d = \frac{\lambda}{2 \cdot \mathrm{NA}} \tag{17}$$

where NA is the numerical aperture of the imaging lens (e.g., the microscope objective). Considering NA = 1.25 and $\lambda = 488$ nm (typical imaging conditions of green pH indicators), we have: $\Delta d \approx 200$ nm, a value slightly lower than the poorest kinetic resolution. In real cases, however, Δd is usually significantly larger than predicted by 17 owing to the nonideal imaging conditions (e.g., optical aberrations, strong emission spatial gradients, refractive index mismatch, and variation throughout the sample). These observations suggest that in most conditions (and particularly when the protein diffusion is hampered by subcellular targeting compared to cytoplasm), GFP indicators provide really meaningful pH maps of intracellular regions. This conclusion should be carefully considered, as in most cases intracellular indicators are seldom able to afford reliable spatial maps at the high resolutions implied in microscopy imaging. For instance, pH-dependent organic probes display $\tau_d > \tau_c$ only up to pH 6 when monitored by high-resolution microscopy setups [137, 138].

5 Conclusions and Perspectives

To date, FPs are the most valuable probes to monitor biochemical processes in vivo, owing to the genetic encoding of fluorescence in the protein primary sequence. Among FPs, a special role is played by GFPs, the variants sharing the common origin from wtGFP, the fluorescence emitter of *Aequorea victoria*; in fact, GFPs constitute the fluorescent marker of thousands of genetic construct for fluorescence imaging and are characterized by versatile engineering of their properties.

The presence of a phenol-ionizable group on the GFP chromophore leads to two protonation ground states (neutral and anionic) associated with considerably different absorption and fluorescence excitation (and sometime, emission) spectra. We demonstrated that a second ionizable site on a nearby residue, Glu^{222}, can help modulate the chromophore protonation. These properties makes GFPs ideally suited to report on intracellular pH, and some variants were shown to be effective probes of the proton concentration in vivo, either nonratiometrically or ratiometrically. Ratiometric indicators represent the most useful class of fluorescent sensors as they allow a universal calibration within a series of experiments.

In this chapter we reviewed the proton-dependent optical properties and we reported the principal applications of the principal GFP-based pH indicators developed so far. None of these indicators proved satisfactory for all in vivo uses. This leaves room for further improvement of pH-dependent optical properties starting from the present knowledge of GFP photophysics. We should stress that the selection of any indicator for a certain application cannot be made without any knowledge of the photophysical features it possesses, as well as of the characteristics of the imaging setup. In the final section, we discussed the most relevant properties of microscopy systems that are used in conjugation with GFP-based indicators.

The engineering of new pH indicators tailored to specific applications and imaging represents a stimulating task, as it involves the manipulation – through rational amino acid replacements in the primary sequence – of the protonation reactions of the chromophore as well as its nearby residues. The "optical" pK_a (that is the pK_a associated with the optical response) is often determined by a subtle thermodynamic balance between the ionization reactions of the chromophore and some nearby residues. Hence, the introduction/removal of amino acids that stabilize the neutral/anionic forms via H-bonding (e.g., Thr^{203} to stabilize the anionic chromophore) is the best way to tune the protonation thermodynamics. Predictably, new mutants with unusual pK_a values will be reported in the next years, thus opening the way to new applications. For instance, there is a lack of GFP indicator tailored to high pH as displayed by certain intracellular regions that would require higher pK_a values: likely, a few substitutions around the chromophore could decrease further the stability of the anionic state without affecting its optical properties.

The manipulation of the spectral characteristics of the neutral and anionic chromophore forms appears a much more difficult task, as no general photophysical guidelines are available. As previously mentioned, the excitation/emission of the anionic state can be susceptible to modifications of polarity/electronic density associated with the chromophore cavity. The neutral chromophore is characterized by both direct emission from its excited state ($\lambda_{em} \approx 460$ nm) and/or ESPT ($\lambda_{em} > 500$ nm). Since ESPT relies on one or more efficient molecular wires that conveys the proton to one/more final acceptor(s), the rate of ESPT is in principle modulable by modifying the proton relay residues, as elegantly shown by Remington and coworkers [70]. Changes in ESPT rate can affect both the wavelength and the yield of emission, thus allowing for a palette of emission ratiometric pH indicators. Emission ratiometric pH-indicators are very relevant also from another perspective: they can be easily applied to two-photon excitation microscopy, an increasingly relevant technique for biomolecular imaging.

In conclusion, today GFPs comprise an incredibly powerful arsenal of tools that opened new perspectives in biology; pH indicators make no exception. No doubt that the wonder of the light sculpted in the genetic essence of the matter will be still mesmerizing and putting to the test researchers all around the world.

References

1. Srivastava J, Barber DL, Jacobson MP (2007) Intracellular pH sensors: design principles and functional significance. Physiology (Bethesda, Md) 22:30–39
2. Wang H, Singh D, Fliegel L (1997) The Na+/H+ antiporter potentiates growth and retinoic acid-induced differentiation of P19 embryonal carcinoma cells. J Biol Chem 272: 26545–26549
3. Putney LK, Barber DL (2003) Na-H exchange-dependent increase in intracellular pH times G2/M entry and transition. J Biol Chem 278:44645–44649
4. Fliegel L (2005) The Na+/H + exchanger isoform 1. Int J Biochem Cell Biol 37:33–37
5. Hunte C, Screpanti E, Venturi M et al (2005) Structure of a Na+/H + antiporter and insights into mechanism of action and regulation by pH. Nature 435:1197–1202
6. Liang E, Liu P, Dinh S (2007) Use of a pH-sensitive fluorescent probe for measuring intracellular pH of Caco-2 cells. Int J Pharm 338:104–109
7. Simpson JE, Walker NM, Supuran CT et al (2010) Putative anion transporter-1 (Pat-1, Slc26a6) contributes to intracellular pH regulation during H + -dipeptide transport in duodenal villous epithelium. Am J Physiol Gastrointest Liver Physiol 298:G683–691
8. Gruenberg J, Stenmark H (2004) The biogenesis of multivesicular endosomes. Nat Rev Mol Cell Biol 5:317–323
9. Wakabayashi I, Poteser M, Groschner K (2006) Intracellular pH as a determinant of vascular smooth muscle function. J Vasc Res 43:238–250
10. Abad MFC, Di Benedetto G, Magalhaes PJ et al (2004) Mitochondrial pH monitored by a new engineered green fluorescent protein mutant. J Biol Chem 279:11521–11529
11. Meisenholder GW, Martin SJ, Green DR et al (1996) Events in apoptosis. Acidification is downstream of protease activation and BCL-2 protection. J Biol Chem 271:16260–16262
12. Izumi H, Torigoe T, Ishiguchi H et al (2003) Cellular pH regulators: potentially promising molecular targets for cancer chemotherapy. Cancer Treat Rev 29:541–549
13. Carnell L, Moore HP (1994) Transport via the regulated secretory pathway in semi-intact PC12 cells: role of intra-cisternal calcium and pH in the transport and sorting of secretogranin II. J Cell Biol 127:693–705
14. Chanat E, Huttner WB (1991) Milieu-induced, selective aggregation of regulated secretory proteins in the trans-Golgi network. J Cell Biol 115:1505–1519
15. Puri S, Bachert C, Fimmel CJ et al (2002) Cycling of early Golgi proteins via the cell surface and endosomes upon lumenal pH disruption. Traffic 3:641–653
16. Kellokumpu S, Sormunen R, Kellokumpu I (2002) Abnormal glycosylation and altered Golgi structure in colorectal cancer: dependence on intra-Golgi pH. FEBS Lett 516:217–224
17. Chesler M (2003) Regulation and modulation of pH in the brain. Physiol Rev 83:1183–1221
18. Jahn R, Lang T, Sudhof TC (2003) Membrane fusion. Cell 112:519–533

19. Sun HY, Wang NP, Halkos ME et al (2004) Involvement of Na+/H + exchanger in hypoxia/re-oxygenation-induced neonatal rat cardiomyocyte apoptosis. Eur J Pharmacol 486:121–131
20. Mellman I, Warren G (2000) The road taken: past and future foundations of membrane traffic. Cell 100:99–112
21. Schindler M, Grabski S, Hoff E et al (1996) Defective pH regulation of acidic compartments in human breast cancer cells (MCF-7) is normalized in adriamycin-resistant cells (MCF-7adr). Biochemistry 35:2811–2817
22. Han J, Burgess K (2010) Fluorescent indicators for intracellular pH. Chem Rev 110: 2709–2728
23. Esposito A, Schlachter S, Schierle GS et al (2009) Quantitative fluorescence microscopy techniques. Methods Mol Biol 586:117–142
24. Rink TJ, Tsien RY, Pozzan T (1982) Cytoplasmic pH and free Mg2+ in lymphocytes. J Cell Biol 95:189–196
25. Tsien RY (1989) Fluorescent indicators of ion concentrations. Meth Cell Biol 30:127–156
26. Bizzarri R, Serresi M, Luin S et al (2009) Green fluorescent protein based pH indicators for in vivo use: a review. Anal Bioanal Chem 393:1107–1122
27. Shimomura O, Johnson FH, Saiga Y (1962) Extraction, purification and properties of aequorin, a bioluminescent protein from the luminous hydromedusan, Aequorea. J Cell Comp Physiol 59:223–239
28. Tsien RY (1998) The green fluorescent protein. Annu Rev Biochem 67:509–544
29. Prasher DC, Eckenrode VK, Ward WW et al (1992) Primary structure of the Aequorea victoria green-fluorescent protein. Gene 111:229–233
30. Chalfie M, Tu Y, Euskirchen G et al (1994) Green fluorescent protein as a marker for gene expression. Science 263:802–805
31. Miyawaki A (2005) Innovations in the imaging of brain functions using fluorescent proteins. Neuron 48:189–199
32. Shaner NC, Steinbach PA, Tsien RY (2005) A guide to choosing fluorescent proteins. Nat Methods 2:905–909
33. Chudakov DM, Lukyanov S, Lukyanov KA (2005) Fluorescent proteins as a toolkit for in vivo imaging. Trends Biotechnol 23:605–613
34. Zhang J, Campbell RE, Ting AY et al (2002) Creating new fluorescent probes for cell biology. Nat Rev Mol Cell Biol 3:906–918
35. Stepanenko OV, Verkhusha VV, Kuznetsova IM et al (2008) Fluorescent proteins as biomarkers and biosensors: throwing color lights on molecular and cellular processes. Curr Protein Pept Sci 9:338–369
36. Ibraheem A, Campbell RE (2010) Designs and applications of fluorescent protein-based biosensors. Curr Opin Chem Biol 14:30–36
37. Johnson DE, Ai HW, Wong P et al (2009) Red fluorescent protein pH biosensor to detect concentrative nucleoside transport. J Biol Chem 284:20499–20511
38. Schulte A, Lorenzen I, Bottcher M et al (2006) A novel fluorescent pH probe for expression in plants. Plant Methods 2:7
39. Wachter RM (2007) Chromogenic cross-link formation in green fluorescent protein. Acc Chem Res 40:120–127
40. Zhang L, Patel HN, Lappe JW et al (2006) Reaction progress of chromophore biogenesis in green fluorescent protein. J Am Chem Soc 128:4766–4772
41. Niwa H, Inouye S, Hirano T et al (1996) Chemical nature of the light emitter of the Aequorea green fluorescent protein. Proc Natl Acad Sci 93:13617–13622
42. Bell AF, He X, Wachter RM et al (2000) Probing the ground state structure of the green fluorescent protein chromophore using Raman spectroscopy. Biochemistry 39:4423–4431
43. Dong J, Solntsev KM, Tolbert LM (2006) Solvatochromism of the green fluorescence protein chromophore and its derivatives. J Am Chem Soc 128:12038–12039

44. Voityuk AA, Michel-Beyerle ME, Rosch N (1998) Quantum chemical modeling of structure and absorption spectra of the chromophore in green fluorescent proteins. Chem Phys 231:13–25
45. Follenius-Wund A, Bourotte M, Schmitt M et al (2003) Fluorescent derivatives of the GFP chromophore give a new insight into the GFP fluorescence process. Biophys J 85:1839–1850
46. Litvinenko KL, Webber NM, Meech SR (2001) An ultrafast polarisation spectroscopy study of internal conversion and orientational relaxation of the chromophore of the green fluorescent protein. Chem Phys Lett 346:47–53
47. Litvinenko KL, Webber NM, Meech SR (2003) Internal conversion in the chromophore of the green fluorescent protein: temperature dependence and isoviscosity analysis. J Phys Chem A 107:2616–2623
48. Patterson GH, Knobel SM, Sharif WD et al (1997) Use of the green fluorescent protein and its mutants in quantitative fluorescence microscopy. Biophys J 73:2782–2790
49. Sniegowski JA, Phail ME, Wachter RM (2005) Maturation efficiency, trypsin sensitivity, and optical properties of Arg96, Glu222, and Gly67 variants of green fluorescent protein. Biochem Biophys Res Commun 332:657–663
50. Cody CW, Prasher DC, Westler WM et al (1993) Chemical structure of the hexapeptide chromophore of the Aequorea green-fluorescent protein. Biochemistry 32:1212–1218
51. Palm GJ, Zdanov A, Gaitanaris GA et al (1997) The structural basis for spectral variations in green fluorescent protein. Nat Struct Biol 4:361–365
52. Ormo M, Cubitt AB, Kallio K et al (1996) Crystal structure of the Aequorea victoria green fluorescent protein. Science 273:1392–1395
53. Brejc K, Sixma TK, Kitts PA et al (1997) Structural basis for dual excitation and photoisomerization of the Aequorea victoria green fluorescent protein. Proc Natl Acad Sci 94:2306–2311
54. Kummer AD, Wiehler J, Rehaber H et al (2000) Effects of threonine 203 replacements on excited-state dynamics and fluorescence properties of the green fluorescent protein (GFP). J Phys Chem B 104:4791–4798
55. Wachter RM, Yarbrough D, Kallio K et al (2000) Crystallographic and energetic analysis of binding of selected anions to the yellow variants of green fluorescent protein. J Mol Biol 301:157–171
56. Palm GJ, Wlodawer A (1999) Spectral variants of green fluorescent protein. Meth Enzymol 302:378–394
57. Morise H, Shimomura O, Johnson FH et al (1974) Intermolecular energy transfer in the bioluminescent system of Aequorea. Biochemistry 13:2656–2662
58. Webber NM, Litvinenko KL, Meech SR (2001) Radiationless relaxation in a synthetic analogue of the green fluorescent protein chromophore. J Phys Chem B 105:8036–8039
59. Ward WW, Bokman SH (1982) Reversible denaturation of Aequorea green-fluorescent protein: physical separation and characterization of the renatured protein. Biochemistry 21:4535–4540
60. Chattoraj M, King BA, Bublitz GU et al (1996) Ultra-fast excited state dynamics in green fluorescent protein: multiple states and proton transfer. Proc Natl Acad Sci 93:8362–8367
61. Lossau H, Kummer A, Heinecke R et al (1996) Time-resolved spectroscopy of wild-type and mutant Green Fluorescent Proteins reveals excited state deprotonation consistent with fluorophore-protein interactions. Chem Phys 213:1–16
62. McAnaney TB, Park ES, Hanson GT et al (2002) Green fluorescent protein variants as ratiometric dual emission pH sensors. 2. Excited-state dynamics. Biochemistry 41:15489–15494
63. Bonsma S, Purchase R, Jezowski S et al (2005) Green and red fluorescent proteins: photo- and thermally induced dynamics probed by site-selective spectroscopy and hole burning. Chemphyschem 6:838–849
64. Creemers TM, Lock AJ, Subramaniam V et al (1999) Three photoconvertible forms of green fluorescent protein identified by spectral hole-burning. Nat Struct Biol 6:557–560

65. Wiehler J, Jung G, Seebacher C et al (2003) Mutagenic stabilization of the photocycle intermediate of green fluorescent protein (GFP). Chembiochem 4:1164–1171
66. Kennis JT, Larsen DS, van Stokkum IH et al (2004) Uncovering the hidden ground state of green fluorescent protein. Proc Natl Acad Sci 101:17988–17993
67. Wachter RM, Remington SJ (1999) Sensitivity of the yellow variant of green fluorescent protein to halides and nitrate. Curr Biol 9:R628–629
68. McAnaney TB, Zeng W, Doe CF et al (2005) Protonation, photobleaching, and photoactivation of yellow fluorescent protein (YFP 10 C): a unifying mechanism. Biochemistry 44:5510–5524
69. Biteen JS, Thompson MA, Tselentis NK et al (2008) Super-resolution imaging in live Caulobacter crescentus cells using photoswitchable EYFP. Nat Methods 5:947–949
70. Stoner-Ma D, Jaye AA, Ronayne KL et al (2008) An alternate proton acceptor for excited-state proton transfer in green fluorescent protein: rewiring GFP. J Am Chem Soc 130:1227–1235
71. Kneen M, Farinas J, Li Y et al (1998) Green fluorescent protein as a noninvasive intracellular pH indicator. Biophys J 74:1591–1599
72. Abbruzzetti S, Grandi E, Viappiani C et al (2005) Kinetics of acid-induced spectral changes in the GFPmut2 chromophore. J Am Chem Soc 127:626–635
73. Elsliger MA, Wachter RM, Hanson GT et al (1999) Structural and spectral response of green fluorescent protein variants to changes in pH. Biochemistry 38:5296–5301
74. Shinobu A, Palm GJ, Schierbeek AJ et al (2010) Visualizing proton antenna in a high-resolution green fluorescent protein structure. J Am Chem Soc 132:11093–11102
75. Robey RB, Ruiz O, Santos AV et al (1998) pH-dependent fluorescence of a heterologously expressed Aequorea green fluorescent protein mutant: in situ spectral characteristics and applicability to intracellular pH estimation. Biochemistry 37:9894–9901
76. Bizzarri R, Nifosi R, Abbruzzetti S et al (2007) Green fluorescent protein ground states: the influence of a second protonation site near the chromophore. Biochemistry 46:5494–5504
77. Hanson GT, McAnaney TB, Park ES et al (2002) Green fluorescent protein variants as ratiometric dual emission pH sensors. 1. Structural characterization and preliminary application. Biochemistry 41:15477–15488
78. Ullmann GM (2003) Relations between protonation constants and titration curves in polyprotic acids: a critical view. J Phys Chem B 107:1263–1271
79. Jayaraman S, Haggie P, Wachter RM et al (2000) Mechanism and cellular applications of a green fluorescent protein-based halide sensor. J Biol Chem 275:6047–6050
80. Griesbeck O, Baird GS, Campbell RE et al (2001) Reducing the environmental sensitivity of yellow fluorescent protein. Mechanism and applications. J Biol Chem 276:29188–29194
81. Scharnagl C, Raupp-Kossmann R, Fischer SF (1999) Molecular basis for pH sensitivity and proton transfer in green fluorescent protein: protonation and conformational substates from electrostatic calculations. Biophys J 77:1839–1857
82. Jung G, Mais S, Zumbusch A et al (2000) The role of dark states in the photodynamics of the green fluorescent protein examined with two-color fluorescence excitation spectroscopy. J Phys Chem A 104:873–877
83. Cardarelli F, Bizzarri R, Serresi M et al (2009) Probing nuclear localization signal-importin alpha binding equilibria in living cells. J Biol Chem 284:36638–36646
84. Grynkiewicz G, Poenie M, Tsien RY (1985) A new generation of Ca2+ indicators with greatly improved fluorescence properties. J Biol Chem 260:3440–3450
85. Llopis J, McCaffery JM, Miyawaki A et al (1998) Measurement of cytosolic, mitochondrial, and Golgi pH in single living cells with green fluorescent proteins. Proc Natl Acad Sci 95:6803–6808
86. Miesenbock G, De Angelis DA, Rothman JE (1998) Visualizing secretion and synaptic transmission with pH-sensitive green fluorescent proteins. Nature 394:192–195
87. Prosser DC, Whitworth K, Wendland B (2008) Quantitative analysis of endocytosis with cytoplasmic pHluorin chimeras. Traffic 11:1141–1150

88. Jankowski A, Kim JH, Collins RF et al (2001) In situ measurements of the pH of mammalian peroxisomes using the fluorescent protein pHluorin. J Biol Chem 276:48748–48753
89. Haupts U, Maiti S, Schwille P et al (1998) Dynamics of fluorescence fluctuations in green fluorescent protein observed by fluorescence correlation spectroscopy. Proc Natl Acad Sci 95:13573–13578
90. Sankaranarayanan S, De Angelis D, Rothman JE et al (2000) The use of pHluorins for optical measurements of presynaptic activity. Biophys J 79:2199–2208
91. Hess ST, Heikal AA, Webb WW (2004) Fluorescence photoconversion kinetics in novel green fluorescent protein pH sensors (pHluorins). J Phys Chem B 108:10138–10148
92. Nakabayashi T, Wang HP, Kinjo M et al (2008) Application of fluorescence lifetime imaging of enhanced green fluorescent protein to intracellular pH measurements. Photochem Photobiol Sci 7:668–670
93. Ashby MC, Ibaraki K, Henley JM (2004) It's green outside: tracking cell surface proteins with pH-sensitive GFP. Trends Neurosci 27:257–261
94. Sankaranarayanan S, Ryan TA (2001) Calcium accelerates endocytosis of vSNAREs at hippocampal synapses. Nat Neurosci 4:129–136
95. Wachter RM, Elsliger MA, Kallio K et al (1998) Structural basis of spectral shifts in the yellow-emission variants of green fluorescent protein. Structure 6:1267–1277
96. Patterson G, Day RN, Piston D (2001) Fluorescent protein spectra. J Cell Sci 114:837–838
97. Karagiannis J, Young PG (2001) Intracellular pH homeostasis during cell-cycle progression and growth state transition in Schizosaccharomyces pombe. J Cell Sci 114:2929–2941
98. Orij R, Postmus J, Ter Beek A et al (2009) In vivo measurement of cytosolic and mitochondrial pH using a pH-sensitive GFP derivative in Saccharomyces cerevisiae reveals a relation between intracellular pH and growth. Microbiology 155:268–278
99. Bohnert S, Schiavo G (2005) Tetanus toxin is transported in a novel neuronal compartment characterized by a specialized pH regulation. J Biol Chem 280:42336–42344
100. Morimoto YV, Che YS, Minamino T et al (2010) Proton-conductivity assay of plugged and unplugged MotA/B proton channel by cytoplasmic pHluorin expressed in Salmonella. FEBS Lett 584:1268–1272
101. Machen TE, Leigh MJ, Taylor C et al (2003) pH of TGN and recycling endosomes of H+/K + -ATPase-transfected HEK-293 cells: implications for pH regulation in the secretory pathway. Am J Physiol 285:C205–214
102. Shu X, Leiderman P, Gepshtein R et al (2007) An alternative excited-state proton transfer pathway in green fluorescent protein variant S205V. Protein Sci 16:2703–2710
103. Shu X, Kallio K, Shi X et al (2007) Ultrafast excited-state dynamics in the green fluorescent protein variant S65T/H148D. 1. Mutagenesis and structural studies. Biochemistry 46:12005–12013
104. Cody SH, Dubbin PN, Beischer AD et al (1993) Intracellular pH mapping with Snarf-1 and confocal microscopy.1. A quantitative technique for living tissues and isolated cells. Micron 24:573–580
105. Dubbin PN, Cody SH, Williams DA (1993) Intracellular pH mapping with Snarf-1 and confocal microscopy.2. Ph gradients within single cultured-cells. Micron 24:581–586
106. Bizzarri R, Arcangeli C, Arosio D et al (2006) Development of a novel GFP-based ratiometric excitation and emission pH indicator for intracellular studies. Biophys J 90: 3300–3314
107. Serresi M, Bizzarri R, Cardarelli F et al (2009) Real-time measurement of endosomal acidification by a novel genetically encoded biosensor. Anal Bioanal Chem 393:1123–1133
108. Marcello A, Lusic M, Pegoraro G et al (2004) Nuclear organization and the control of HIV-1 transcription. Gene 326:1–11
109. Arosio D, Garau G, Ricci F et al (2007) Spectroscopic and structural study of proton and halide ion cooperative binding to gfp. Biophys J 93:232–244
110. Arosio D, Ricci F, Marchetti L et al (2010) Simultaneous intracellular chloride and pH measurements using a GFP-based sensor. Nat Methods 7:516–518

111. Awaji T, Hirasawa A, Shirakawa H et al (2001) Novel green fluorescent protein-based ratiometric indicators for monitoring pH in defined intracellular microdomains. Biochem Biophys Res Commun 289:457–462
112. Chan FK, Siegel RM, Zacharias D et al (2001) Fluorescence resonance energy transfer analysis of cell surface receptor interactions and signaling using spectral variants of the green fluorescent protein. Cytometry 44:361–368
113. Wlodarczyk J, Woehler A, Kobe F et al (2008) Analysis of FRET signals in the presence of free donors and acceptors. Biophys J 94:986–1000
114. Urra J, Sandoval M, Cornejo I et al (2008) A genetically encoded ratiometric sensor to measure extracellular pH in microdomains bounded by basolateral membranes of epithelial cells. Pflugers Arch 457:233–242
115. Tsien RY (2003) Imagining imaging's future. Nat Rev Mol Cell Biol (Suppl): SS16–SS21
116. Paddock S (2008) Over the rainbow: 25 years of confocal imaging. BioTechniques 44:643–644, 646, 648
117. Diaspro A, Bianchini P, Vicidomini G et al (2006) Multi-photon excitation microscopy. Biomed Eng Online 5:36
118. Digman MA, Dalal R, Horwitz AF et al (2008) Mapping the number of molecules and brightness in the laser scanning microscope. Biophys J 94:2320–2332
119. Petty HR (2007) Fluorescence microscopy: established and emerging methods, experimental strategies, and applications in immunology. Microsc Res Tech 70:687–709
120. Diaspro A, Robello M (2000) Two-photon excitation of fluorescence for three-dimensional optical imaging of biological structures. J Photochem Photobiol B Biol 55:1–8
121. Zipfel WR, Williams RM, Webb WW (2003) Nonlinear magic: multiphoton microscopy in the biosciences. Nat Biotechnol 21:1369–1377
122. Svoboda K, Yasuda R (2006) Principles of two-photon excitation microscopy and its applications to neuroscience. Neuron 50:823–839
123. Albota M, Beljonne D, Bredas JL et al (1998) Design of organic molecules with large two-photon absorption cross sections. Science 281:1653–1656
124. Albota MA, Xu C, Webb WW (1998) Two-photon fluorescence excitation cross sections of biomolecular probes from 690 to 960 nm. Appl Opt 37:7352–7356
125. Xu C, Webb WW (1996) Measurement of two-photon excitation cross sections of molecular fluorophores with data from 690 to 1050 nm. J Opt Soc Am B: Opt Phys 13:481–491
126. Drobizhev M, Tillo S, Makarov NS et al (2009) Absolute two-photon absorption spectra and two-photon brightness of orange and red fluorescent proteins. J Phys Chem B 113:855–859
127. Bestvater F, Spiess E, Stobrawa G et al (2002) Two-photon fluorescence absorption and emission spectra of dyes relevant for cell imaging. Journal Microsc-Oxford 208:108–115
128. Chirico G, Cannone F, Diaspro A et al (2004) Multiphoton switching dynamics of single green fluorescent proteins. Phys Rev E 70:030901
129. Hosoi H, Yamaguchi S, Mizuno H et al (2008) Hidden electronic excited state of enhanced green fluorescent protein. J Phys Chem B 112:2761–2763
130. Roorda RD, Hohl TM, Toledo-Crow R et al (2004) Video-rate nonlinear microscopy of neuronal membrane dynamics with genetically encoded probes. J Neurophysiol 92:609–621
131. Sonnleitner A, Schutz GJ, Schmidt T (1999) Free brownian motion of individual lipid molecules in biomembranes. Biophys J 77:2638–2642
132. Jovin TM, Striker G (1977) Chemical relaxation kinetic studies of E. coli RNA polymerase binding to poly [d(A–T)] using ethidium bromide as a fluorescence probe. Mol Biol Biochem Biophys 24:245–281
133. Digman MA, Brown CM, Sengupta P et al (2005) Measuring fast dynamics in solutions and cells with a laser scanning microscope. Biophys J 89:1317–1327
134. Cardarelli F, Serresi M, Bizzarri R et al (2008) Tuning the transport properties of HIV-1 Tat arginine-rich motif in living cells. Traffic 9:528–539
135. Cardarelli F, Serresi M, Bizzarri R et al (2007) In vivo study of HIV-1 Tat arginine-rich motif unveils its transport properties. Mol Ther 15:1313–1322

136. Hell SW (2003) Toward fluorescence nanoscopy. Nat Biotechnol 21:1347–1355
137. Charier S, Meglio A, Alcor D et al (2005) Reactant concentrations from fluorescence correlation spectroscopy with tailored fluorescent probes. An example of local calibration-free pH measurement. J Am Chem Soc 127:15491–15505
138. Widengren J, Terry B, Rigler R (1999) Protonation kinetics of GFP and FITC investigated by FCS – aspects of the use of fluorescent indicators for measuring pH. Chem Phys 249:259–271

Green Fluorescent Protein-Based Chloride Ion Sensors for In Vivo Imaging

Piotr Bregestovski and Daniele Arosio

Abstract Fluorescence quenching-based optical sensors offer a powerful tool for noninvasive monitoring of halide transmembrane fluxes and intracellular distribution. The observation that the native fluorescence of yellow fluorescent protein (YFP) is quenched by many anions led to the development of intracellular halide sensors genetically encoded and thus able to target specific subcellular compartments. Cyan (C)FP–YFP-based sensors have opened the way for ratiometric estimation of intracellular Cl^- concentrations. Moreover, the complete loss of fluorescence induced by halide binding via static quenching in specific GFP variants led to the possibility of development of combined chloride–pH sensors. In this chapter, the thermodynamics, binding-pocket structure, and spectroscopic properties of halide-sensing GFP variants are reviewed in terms of the detailed mechanism governing the dependence of fluorescence on local ions. Engineering of GFP-based probes and their advantages in a range of applications will be discussed, with emphasis on the measurement in vivo of chloride ion fluxes, which regulate many physiological functions and cellular parameters such as membrane excitability, cell volume, charge balance, and resting potential.

Keywords Chloride imaging · Genetically encoded probes · Green fluorescent protein-based biosensors · Ratiometric imaging

P. Bregestovski
INSERM U751, Universite' de la Mediterranee, 27 Bd Jean Molin, 13385 Marseille, France
e-mail: piotr.bregestovski@univmed.fr

D. Arosio (✉)
Istituto di Biofisica, Consiglio Nazionale delle Ricerche & Fondazione Bruno Kessler, via alla Cascata 56/C, 38123 Trento, Italy
and
Laboratory of Molecular Imaging, Centre for Integrative Biology (CIBIO), University of Trento, via delle Regole 101, 38123 Trento, Italy
e-mail: daniele.arosio@cnr.it

G. Jung (ed.), *Fluorescent Proteins II*, Springer Ser Fluoresc (2012) 12: 99–124,
DOI 10.1007/4243_2011_27
Published online: 18 June 2011

Contents

1 Introduction

Chloride (Cl^-), the most abundant anion in living organisms, is necessary for all forms of biological life. The homeostasis of intracellular Cl^- concentration ($[Cl^-]_i$) is a critical determinant of several physiological functions and cellular processes, including synaptic transmission [1], epithelial fluid absorption and secretion, and regulation of charge balance [2, 3] and cell volume [4, 5]. Furthermore, $[Cl^-]_i$ determines whether GABA or glycine excites or inhibits postsynaptic target neurons [6]. In cells, $[Cl^-]_i$ is highly regulated by various Cl^--selective channels and Cl^- transporters [7–9]. Alterations in these protein functions lead to disturbance of Cl^- homeostasis and occur in hereditary or acquired human diseases as diverse as cystic fibrosis, myotonia congenita, epilepsy, hyperekplexia, lysosomal storage disease, Bartter syndrome, deafness, renal salt loss, kidney stones, and osteopetrosis [10]. Therefore, there is a great need to both clarify the molecular mechanisms underlying these diseases and find new drugs that specifically modulate the function of Cl^--selective ion channels or Cl^- transporters [11].

Direct measurement of $[Cl^-]_i$ in cells is a challenging task owing to the low Cl^- transmembrane ratio of 10:1 (cf. Ca^{2+}, 10000:1) and the small Cl^- driving force, as the resting potential of cells is usually very close to the Cl^- equilibrium potential (E_{Cl}). Nevertheless, several methods have been proposed for the analysis of $[Cl^-]_i$ distribution and its functional variations in living cells. This chapter considers genetically encoded Cl^- sensors, as does a recent review [12]. Two excellent sources of information about alternative methods of measuring Cl^- homeostasis are [8] and [13].

Visualization of molecular and cellular phenomena by means of the green fluorescent protein (GFP) from the jellyfish *Aequorea victoria* is widely recognized as a revolution in cell biological imaging [14–17]. The superior attractiveness of

GFPs as biological tags resides in the fact that no cofactors other than O_2 are needed for its heterologous expression (in organisms other than jellyfish) [18, 19]. The spectroscopy characteristics of GFP are substantially improved by the single-point mutation S65T [20], which, together with the 37°C folding efficiency F64L mutation, yield enhanced GFP (EGFP) [21]. EGFP is, de facto, a benchmark for using GFPs in mammalian cells. Much effort has been made to modify and advance GFP properties. A variety of GFP derivatives with different spectral characteristics have been developed and some of them can be used as sensors of specific cellular processes. For instance, semi-rational mutagenesis of residues in the chromophore environment led to pH-sensing mutants known as pHluorins and super-ecliptic pHluorins. By exploiting the rapid change in pH upon synaptic vesicle fusion, pHluorins tagged to synaptobrevin can be used to visualize synaptic activity in neurons [22].

GFP-based biosensors have become a powerful tool for the functional imaging of Ca^{2+}, cAMP, and other cellular constituents [23, 24], and they also represent the most promising tools for effective analysis of Cl^- homeostasis in living cells [12]. In general, genetically encoded biosensors address many of the shortcomings of small-molecule organic dyes: (1) they can be delivered in a straightforward manner to specific cell types or subcellular compartments; (2) they are retained in the target cell compartment and thus are compatible with chronic repeated measurements from the same cells in vivo; and (3) the Cl^--binding kinetic and equilibrium constants are independent of the sensor concentration.

2 Halide Ion-Binding Site in GFP Variants

In the initial attempts to extend the usefulness of GFP tags in cell biology, new colors have been sought by random and semi-rational mutagenesis [25, 26]. The quadruple mutant T203Y/S65G/V68L/S72A Yellow FP (YFP) [27] has been of particular interest, especially in fluorescence resonance energy transfer (FRET) measurements, because the YFP spectrum is significantly shifted toward the red spectral region and adequately matched with that of the Cyan FP (CFP). Straightaway, it was recognized that YFP is sensitive to the bulk pH and anion concentrations; so various attempts were made to reduce the sensitivity of YFP to its environment [28, 29]. Conversely, YFP mutants with improved halide sensitivity, such as YFP-H148Q, have been proposed as a tool to report halide concentrations in living cells [30]. Several other mutations of YFP were developed for measuring Cl^- concentration in subcellular organelles [31] (see more in Sect. 3). It was also found that the single mutation T203Y in EGFP (E^2GFP; wtGFP-S65T/F64L/H231L) is sufficient to introduce a halide-binding site with higher Cl^- affinity than YFP-H148Q; moreover, the E^2GFP binding site is in a different location in the protein structure [32]. The GFP structure is characterized by 11 β-strands that protect the chromophore from the environment, surrounding it and preserving its fluorescence.

The chromophore, which is generated by spontaneous cyclization and oxidation of the three amino acids S65–Y66–G67, consists of a conjugated planar π-system, connecting the autocatalytically formed heterocycle to the phenol ring of Y66. Variants containing the T203Y substitution are furthermore characterized by a nearly coplanar π–π stacking between the phenol ring of Y203 and the chromophore, although YFP and E^2GFP structures exhibit other substantial differences [32]. Above all, superimposing the structures of iodide-containing complexes for E^2GFP and YFP-H148Q (Fig. 1), the different position of the binding sites can be seen. In YFP-H148, iodide is involved in a charge interaction with R96 NE2 (4.1 Å), two hydrogen bonds with Y203 OH (3.3 Å) and Q69 NE2 (3.2 Å), and hydrophobic contacts with the aromatic rings of the chromophore (CD2 4.5 Å) and Y203 (CE1 4.1 Å). In E^2GFP, the binding site is still delimited by the same aromatic rings of chromophore (3.5 Å), and Y203 and hydrogen bond with Y203

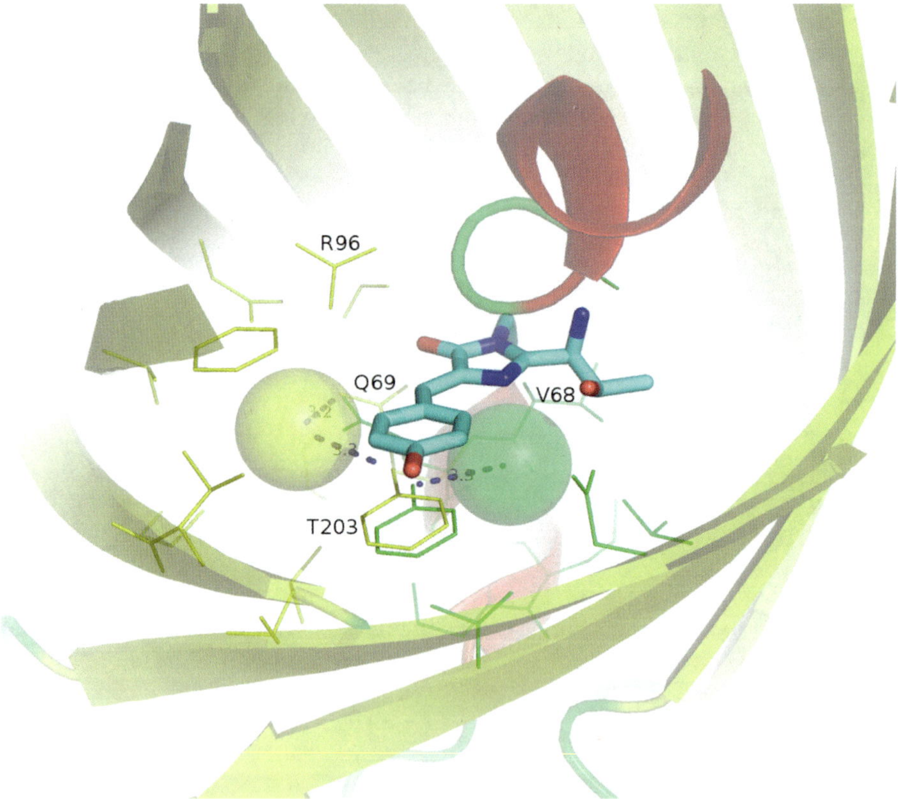

Fig. 1 Ribbon representation of YFP-H148Q•I^- complex X-ray structure (pdb code 1f09). The chromophore is shown in stick representation and only residues that define the binding site are explicitly shown. The iodide-binding site of EGFP-T203Y (E^2GFP) (pdb code 2o2b) is colored in *green* and superimposed for comparison. The bonded iodide anion has been superimposed and is represented as *yellow* (YFP-H148Q) or *green* (E^2GFP) spheres

OH (3.3 Å); nevertheless, it is located on the opposite side of Y203 OH with a second hydrogen bond with V68 N (3.4 Å) (Fig. 1). In contrast with YFP, the E^2GFP-binding pocket contains three water molecules that are displaced upon halide binding. Most importantly, the anion is located in proximity to the chromophore heterocyclic carbonyl oxygen atom (~ 4.3 Å) in both structures, leading to the linkage between anion binding and chromophore protonation, a relevant thermodynamic property of all halide-binding GFP variants.

2.1 Thermodynamics of GFP Halide Binding and pH Linkage

Two chromophore protonation states, neutral and anionic, are relevant in intact GFP, as revealed by accurate structure and spectroscopic analysis [33–35]. The ionization constant pK_a ranges approximately from 4.5 to 7.5 depending on the GFP variant. Moreover, the chromophore pK_a of GFPs carrying the T203Y substitution is strongly dependent on the concentration of small anions such as chloride [32, 36, 37]. The simplest model accounting for a molecule of YFP (or GFP) reacting with protons and chloride ions entails a macromolecule M with two distinct binding sites for ligands X (H^+) and Y (Cl^-). In 1948, Wyman highlighted for the first time the effects on each other elicited by two ligands binding to a macromolecule; he developed the linked functions theory [38, 39], which is best illustrated by a reference cycle [40] depicted for the present case as follows:

$$\begin{array}{ccccc} & M_X & \overset{{}^0pK_a}{\leftrightarrows} & M & \\ {}^1K_Y & \uparrow\downarrow & & \uparrow\downarrow & {}^0K_Y \\ & M_{XY} & \underset{{}^1pK_a}{\leftrightarrows} & M_Y & \end{array} \tag{1}$$

Thermodynamic principles allow us to derive the properties of either ligand from the binding of the other. In particular, let ${}^0K_Y$ and ${}^1K_Y$ be the binding affinities of Y (Cl^-) to the unligated (i.e., anionic YFP) and X-ligated (neutral YFP), and let ^{0}pK_a be ^{1}pK_a analogous quantities for protonation equilibriums. Then, by definition:

$$\begin{aligned} {}^0K_Y &= \frac{[M_Y]}{[M][Y]} & \qquad {}^1K_Y &= \frac{[M_{XY}]}{[M_X][Y]} \\ {}^0\mathrm{p}K_\mathrm{a} &= \log\frac{[M_X]}{[M][X]} & \qquad {}^1\mathrm{p}K_\mathrm{a} &= \log\frac{[M_{XY}]}{[M_Y][X]}. \end{aligned} \tag{2}$$

Of the four reactions depicted, only three are independent; in fact

$$^{0}K_Y \cdot 10^{^{1}\mathrm{p}K_\mathrm{a}} = \frac{[M_Y]}{[M][Y]}\frac{[M_{XY}]}{[M_Y][X]} = \frac{[M_{XY}]}{[M][X][Y]} = {}^{1}K_Y \cdot 10^{^{0}\mathrm{p}K_\mathrm{a}}. \tag{3}$$

An interaction constant (C_{XY}) between the two sites is introduced as a measure of the free energy of coupling (ΔG_c) for the cycle as introduced by Weber [41]:

$$C_{XY} = \frac{^{1}K_Y}{^{0}K_Y} = \frac{10^{^{1}\mathrm{p}K_\mathrm{a}}}{10^{^{0}\mathrm{p}K_\mathrm{a}}} = \exp\left(-\frac{\Delta G_\mathrm{c}}{k_\mathrm{B}T}\right). \tag{4}$$

Thus, C_{XY} is the equilibrium constant governing the $[M_X] + [M_Y] \leftrightarrow [M] + [M_{XY}]$ equilibrium. Taking the unligated form of the macromolecule as a reference, the system partition function P is given by:

$$\begin{aligned} P &= 1 + 10^{^{0}\mathrm{p}K_\mathrm{a}}[X] + {}^{0}K_Y[Y] + 10^{^{0}\mathrm{p}K_\mathrm{a}} \cdot {}^{1}K_Y[X][Y], \\ &= 1 + A_{10}x + A_{01}y + A_{11}xy, \end{aligned} \tag{5}$$

and the fraction of each intermediate is simply derived:

$$\alpha_{ij} = \frac{A_{ij}x^i y^j}{P}. \tag{6}$$

If the binding of Cl^- (Y) is examined as a function of pH (X), an obvious choice for Cl^--sensing applications, the apparent dissociation constant for Cl^- ($1/K_\mathrm{d}$) conforms to the following equation (see Fig. 2):

$$\frac{1}{K_\mathrm{d}} = \frac{^{0}K_Y + {}^{1}K_Y 10^{(^{1}\mathrm{p}K_\mathrm{a}-\mathrm{pH})}}{1 + 10^{(^{1}\mathrm{p}K_\mathrm{a}-\mathrm{pH})}}. \tag{7}$$

Unfortunately, it is only possible to determine accurate experimental estimates of the Cl^- dissociation constant $1/^{0}K_Y$ in the absence of H+ (high pH values); likewise, the ionization constant $^{1}\mathrm{p}K_\mathrm{a}$ of the Y-ligated form is poorly determined by experiments. However, data impose high values for $1/^{0}K_Y$ and $^{1}\mathrm{p}K_\mathrm{a}$ parameters, suggesting a large free energy of coupling (4). Indeed, as previously observed, the anionic chromophore leads to unfavorable charge interaction with the bound halide ion so that the anion-bound anionic chromophore intermediate (M_Y) is highly unlikely. In other words, at high pH values Cl^- binding is concomitant with the binding of one proton (to the chromophore), and so an infinite-cooperativity model (8) fits experimental data as well [32].

$$K_\mathrm{d} = \frac{1}{^{1}K_Y}\frac{1 + 10^{(^{0}\mathrm{p}K_\mathrm{a}-\mathrm{pH})}}{10^{(^{0}\mathrm{p}K_\mathrm{a}-\mathrm{pH})}}. \tag{8}$$

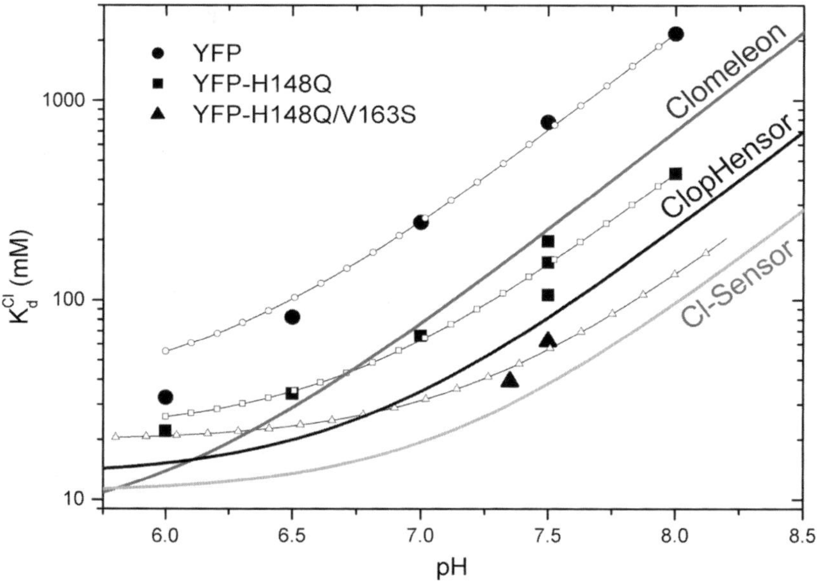

Fig. 2 Cl^- dissociation constant as a function of pH for the indicated YFP and GFP variants and biosensors. Fitting curves were drawn using (8)

The dissociation constant is a fundamental property to consider when selecting a sensor. GFP-based Cl^- sensors exhibit apparent Cl^- dissociation constants that are strongly dependent on pH. Here, we introduce a theoretical framework allowing correct comparison among different sensors (Fig. 2).

2.2 *Halide-Induced Fluorescence Quenching*

Fluorescence quenching is fundamental for Cl^- sensors, where the decrease in emission intensity or lifetime can be measured and related to the halide concentration via different quenching models [42–44]. Quantification of halides by fluorescence quenching is a popular technique because of the high sensitivity typical of fluorescence assays.

Fluorescence quenching of quinine by chloride ions was first (in 1869) discovered by George Stokes [45]. In this example, Cl^- acts as a collisional quencher: within the lifetime of the excited state of the fluorophore (quinine), the fluorophore is returned to the ground state by a diffusive encounter with a quencher (Cl^-). For halides, *dynamic quenching* is the result of the fluorophore intersystem crossing to an excited tripled state, promoted by spin–orbit coupling between the fluorophore (singlet) excited state and the halide. Another relevant quenching model is *static quenching*, where the fluorophore forms a nonfluorescent complex with the quencher. As *dynamic quenching* only affects the excited state of the fluorophore, there are typically no changes in the fluorophore absorption spectra. Conversely, *static quenching* usually causes alterations in the chromophore absorption spectra via the formation of a ground-state complex. Measuring lifetime dependence on quencher concentration allows distinguishing between static and dynamic quenching models rigorously. The excited state lifetime of fluorophore is shortened when

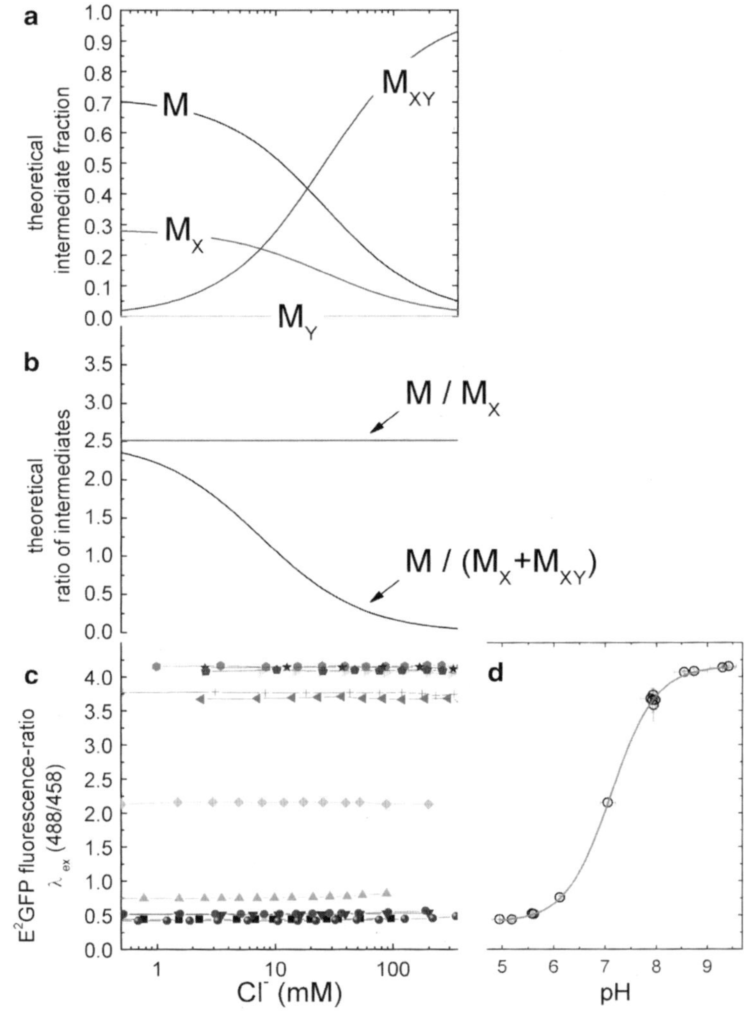

Fig. 3 (**a**) Simulated fractions (6) of bound intermediates for a theoretical GFP-based sensor with $pK_a = 7.0$, $K_d = 20$ mM and $C_{XY} = 10^4$ as a function of [Cl] at pH = 7.4. (**b**) Corresponding intermediate ratios M/M_X and $M/(M_X + M_{XY})$. (**c**) Experimental fluorescence excitation ratio (488/458 nm) as a function of [Cl] at various pH values, from 4.5 to 9.5. Fluorescence spectrophotometer data were obtained with purified E^2GFP at 37°C and emission set at 523 nm. (**d**) Average of ratio values obtained at various [Cl], reported in (C), as a function of pH. Sigmoid curve fitting provides the thermodynamic parameter ^0pK_a

quencher concentration is increased; in fact, the presence of quenchers in solution makes available additional pathways from the excited to the ground state.

Fluorescence of GFP-based Cl^- sensors is thought to be a result of static quenching processes [31, 32, 37], because halide-dependent alterations of the chromophore absorption spectra are usually observed.

Distinguishing between static and dynamic quenching bears some important consequences for reading pH via fluorescence ratio. As *static quenching* implies the formation of a ground-state nonfluorescent complex M_{XY} [see the thermodynamic cycle depicted in (1)], the fluorescence ratios measured for $M/(M_X + M_{XY})$ and M/M_X are identical. Therefore, pH calibrations derived from fluorescence ratios are independent of $[Cl^-]$ (Fig. 3a, b), and determine the value of ${}^0pK_a$ [see (1)] univocally. While GFP apparent pK_a depending on $[Cl^-]$ is expected when using spectroscopic signal (e.g., absorbance) proportional to the ratio $M/(M_X + M_{XY})$, only the combination of static and dynamic quenching or the presence of heterogeneously emitting mixtures of fluorophores [42] might justify apparent pK_a, and hence pH calibration curves, depending on the $[Cl^-]$ for data derived from fluorescence titration.

In conclusion, if halide-dependent fluorescence changes of a GFP-based Cl^- sensor are totally governed by static quenching, then Cl^--independent pH reading is attainable (Fig. 3c, d).

3 Chloride Ion Sensors

The discovery, cloning, and heterologous expression of the GFP from the jellyfish *A. victoria* opened tremendous opportunities for visualization of proteins and nucleic acid-binding domains, analysis of protein–protein interactions [46], monitoring of Ca^{2+} [23], H^+ [22, 47, 48], high-throughput screening of drugs, and analysis of various physiological processes in biological organisms [49].

In the last decade, several GFP-based Cl^- sensors have become available, with a variety of characteristics: spectral properties, Cl^--binding affinities, and pH dependence of excitation and emission wavelengths.

3.1 YFP-Based Sensors

As we mentioned earlier, among the widely used derivatives of GFP, the subfamily of yellow fluorescent proteins (YFPs) carries a key T203Y substitution resulting in a yellowish shift (~20 nm) of excitation and emission spectra [50]. In 1999, Wachter and Remington presented the important observation that the fluorescence of YFP, which, in addition to T203Y, contains substitutions S65G/V68L/S72A (to improve brightness in living cells), is modulated by halide or nitrate ions, of which Cl^- is the most physiologically relevant. It was found that YFP fluorescence was

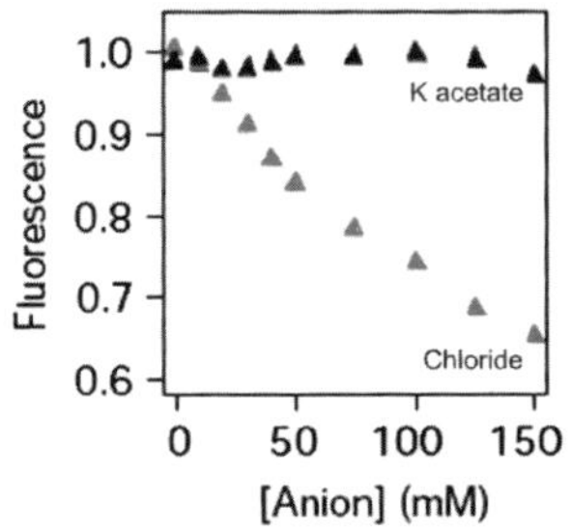

Fig. 4 Halide sensitivity of YFP. Normalized fluorescence emission of YFP as a function of anion concentration. Only dependences for chloride (*red*) and potassium acetate (*black*) are shown (modified from [30])

sensitive to anions with relative potencies I > NO3 > Cl > Br ≫ formate > acetate (Fig. 4; [30]).

YFP sensitivity to these small anions results from ground-state binding near the chromophore [36, 51], which apparently alters the chromophore ionization constant and hence the fluorescence emission. The halides behave like ligands that affect the chromophore charge state [30, 32, 37]. To improve spectral characteristics and brightness, YFP was subjected to further mutagenesis and the most successful variants were selected. A mutant with modified polar groups was created: YFP-H148G, where substitution of His with Gly would open up a solvent channel to the chromophore without perturbing its structure [36]. YFP-H148G was found to be highly fluorescent. It exhibits higher sensitivity to Cl^-, and this sensitivity strongly depends on H^+ concentrations (see Fig. 2). At pH 6.0, the apparent dissociation constant (K_{app}) for Cl^- is 32mM for YFP and 22 mM for YFP-H148Q, whereas at pH 7.5, K_{app} values are 777 mM and 154 mM, respectively [37].

As in the majority of cell types the intracellular Cl^- concentration varies in the range 5–60 mM (Fig. 6c; see also [12]); at physiological pH, the apparent dissociation constants for YFP and even YFP-H148Q are far from ideal values. This seriously limited the use of these proteins as Cl^- probes. To enhance the sensitivity of YFP-H148Q to Cl, libraries of mutants were generated in which pairs of residues in the vicinity of the halide-binding site were randomly mutated [31]. Analysis of over a 1,000 clones revealed improved anion sensitivity with K_{app} down to 40 mM for YFP-H148Q/V163S. Interestingly, another double mutant, YFP-H148Q/I152L, was selected for the relatively rapid-binding kinetics (association time constant about 50 ms [31]). These mutants were used for a high-throughput screening and as the basis for development of a ratiometric probe, called Cl-sensor (see below).

3.2 *Ratiometric Sensors*

Serious problems in quantifying optical signals from living cells arise because of variations in optical path length, local probe concentration, light scattering, illumination intensity, and photobleaching. These variations, which are present in different cells and different regions of the same cell, also change with time.

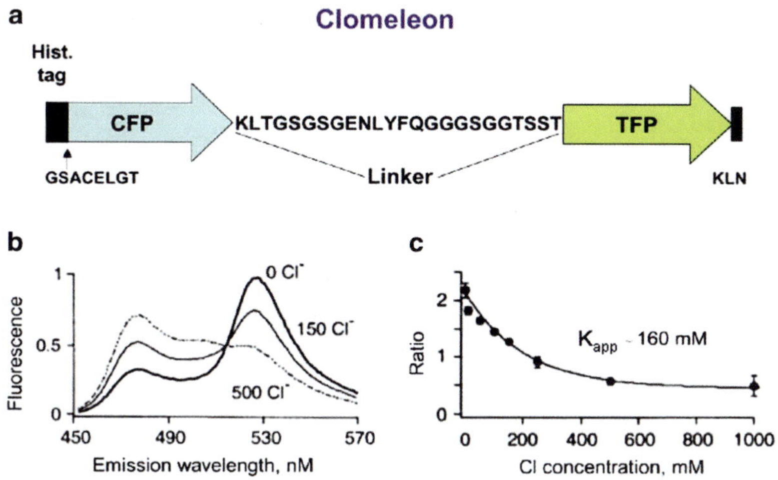

Fig. 5 Design and fluorescence properties of Clomeleon. (**a**) Scheme of Clomeleon construct with indication of the 24-amino acid flexible linker. (**b**) Emission spectra of Clomeleon in the presence of different Cl^- concentrations. Excitation was at 434 nm and the emission spectra were normalized to their peaks at 527 nm. (**c**) The relationship between fluorescence emission ratio (527/485 nm) and Cl^- concentrations. The line is an exponential function fit to the data (modified from [52])

Measuring the ratio between two wavelengths (for either excitation or emission fluorescence) is an effective way to overcome most of these experimental obstacles. The important development of genetically encoded Cl^- probes was achieved by constructing ratiometric YFP-based Cl^- sensors. Three main constructs were developed: Clomeleon [52], Cl-sensor [53], and ClopHensor [54].

3.2.1 Clomeleon

This construct consists of two fluorescent proteins, CFP and a variant of YFP, topaz fluorescence protein (TFP; GFP/S65G/S72A/K79R/T203Y/H231L), connected with a polypeptide linker of 24 amino acids (Fig. 5a).

Clomeleon has two important features. First, Clomeleon is a so-called FRET-based sensor because halide binding, altering the chemical and spectroscopic properties of YFP, affects FRET between CFP and TFP. FRET is a widely used approach for monitoring protein–protein interactions and development of sensors for different ions [55–57]. This phenomenon represents interaction between the two closely spaced fluorophores, when the excited fluorophore operating at shorter excitation–emission range of wavelengths directly transfer its excitation energy to the acceptor fluorophore operating at longer wavelengths. This process was called FRET by Förster [58]. Its main principles and features are described in various reviews [55–57, 59–62]. In Clomeleon, excited at 430–435 nm, CFP transfers its energy to closely adjacent TFP. Binding of a Cl^- to this anion-sensitive protein reduces its emission, leading to a decrease in FRET efficiency.

The second main advantage of Clomeleon is the ability to perform ratiometric measurements based on the Cl^- sensitivity of TFP and Cl^- insensitivity of CFP. Analysis of normalized emission spectra of this construct at different Cl^- concentrations revealed the existence of the isosbestic point (Fig. 5b), assuring reliable ratiometric analysis and the possibility of estimating intracellular Cl^- concentrations. The ratiometric capabilities of Clomeleon allow noninvasive optical measurements that are minimally influenced by the thickness of the specimen, intensity of the excitation light, and concentration of the sensor. This, in turn, makes it possible to accurately determine Cl^- values even in cells with complicated geometry, such as neurons.

Clomeleon possesses several other valuable features: excitation at visible wavelengths, good signal-to-noise ratio, absence of leakage from cells, and the possibility of targeting the probe to different cell types using specific promoters. Moreover, the construct exhibits high fluorescence and proteolitical stability: the absence of Clomeleon bleaching over 2 h of recording [63] and its presence in mice in vivo for up to 9 months [64].

The important disadvantage of Clomeleon is that at physiological pH, it has a rather low sensitivity to Cl. The K_{app} of Clomeleon (the concentration of Cl which induces a 50% change in the fluorescence ratio) is more than 160 mM (Fig. 5c), which is far from the physiological $[Cl^-]_i$ (of 3–60 mM). For this reason, the development of ratiometric probes with higher sensitivity to Cl^- was required. This was achieved by the development of the probe called "Cl-sensor," whose sensitivity to Cl^- at physiological concentrations is about fivefold higher than that of Clomeleon.

3.2.2 Cl-Sensor

Similar to Clomeleon, Cl-sensor is based on the coupling of CFP with YFP via a peptide linker of 20 amino acids (Fig. 6a). In order to enhance the sensitivity of YFP to Cl^- and accelerate its binding kinetics, three mutations (H148Q, I152L, and V163S) were introduced.

This construct exhibits two major advantages. First, it was found that normalized spectra obtained at different Cl^- concentrations have a common point near 465 nm (Fig. 6b). For every spectrum, the intensity at this point is Cl^- independent. Thus, Cl-sensor with triple-mutated YFP can be used as a ratiometric excitation wavelength probe. This feature is particularly useful for recordings of Cl^- transients using conventional setups with polychromators or other devices for a rapid change of excitation wavelength.

Second, the calibration obtained in cell lines showed that the K_d of this construct is about 30 mM. This indicates that Cl-sensor possesses a sensitivity corresponding

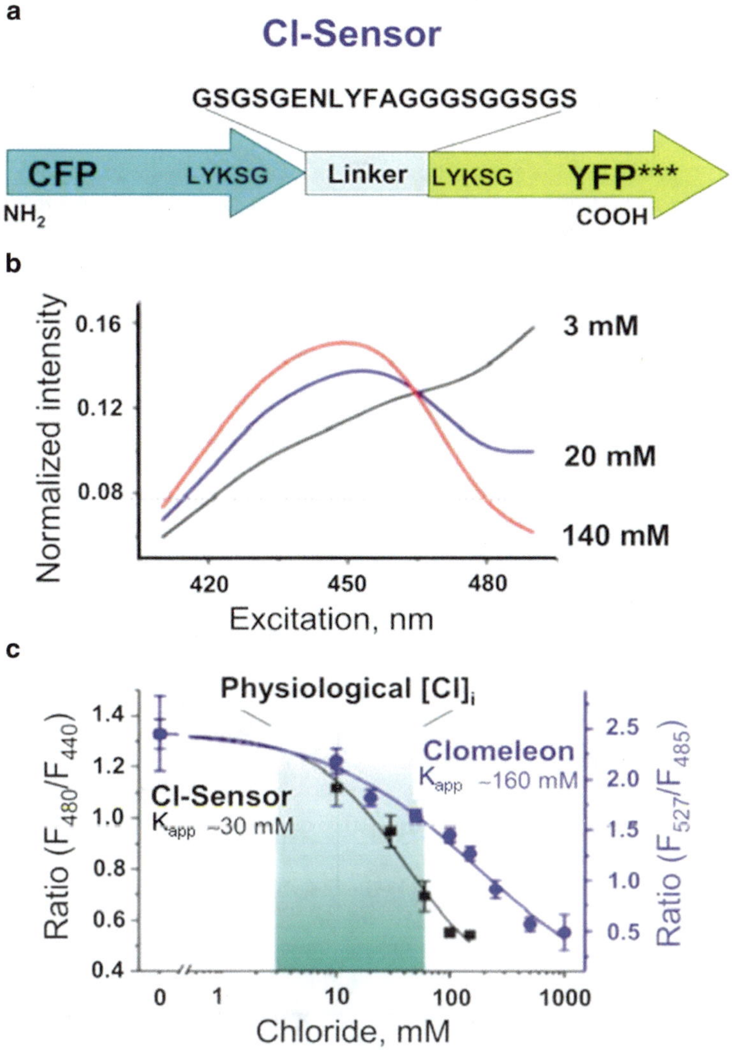

Fig. 6 Design and fluorescence properties of Cl-sensor. (**a**) Scheme of Cl-sensor construct with indication of the 20-amino acid flexible linker. *Three asterisks* indicate three mutations: YFP-H148Q, -I152L and -V163S in the YFP sequence. (**b**) Normalized excitation spectra of Cl-sensor. Whole-cell recordings from CHO cells with pipettes containing different Cl^- concentrations (shown in the graph). Note that spectra have a common point at 465 nm. (**c**) Comparison of Cl^- sensitivities of Cl-sensor and Clomeleon. Cl-sensor (*black squares* and *line*): the relationship between fluorescence excitation ratio (480/440 nm) and $[Cl]_i$ obtained from whole-cell recordings with pipettes containing solutions with different Cl concentrations (from [53]). Clomeleon (*blue circles* and *line*): the relationship between fluorescence emission ratio (527/485 nm) and $[Cl]_i$ plotted from [52]. Note that the calibration curve for Clomeleon is mainly out of the physiological range of $[Cl]_i$ (from [12])

to the physiological range of intracellular Cl^- concentrations. Indeed, comparison of concentration dependencies for Clomeleon and Cl-sensor (Fig. 6c) shows that the main part of the curve for Clomeleon is out of physiological range of intracellular Cl^- detected in different cell types (Fig. 6c). In contrast, the main changes in fluorescence ratio for Cl-sensor are in this area.

As many other GFP-derived fluorophores, Clomeleon and Cl-sensor exhibit relatively high sensitivity to pH (Fig. 2). Although at low $[Cl^-]_i$ (below 25 mM) the errors introduced by pH variations are relatively small (less than 10–20%), at $[Cl^-]_i$ approaching 150 mM errors may be about 50% [52]. This disadvantage has been elegantly overcome by the development of a probe allowing simultaneous recording of pH and Cl^- from the same point in space and time.

3.2.3 ClopHensor Allows Simultaneous pH and Chloride Measurements

Because of the strong pH dependence of the affinity of Cl^- for YFP-derived sensors, the concomitant assessment of intracellular pH is a requirement of precise $[Cl^-]_i$ measurements. Furthermore, pH-induced modulation of the fluorescence signal of most GFP derivatives may complicate the interpretation of FRET signals. Therefore, the development of ClopHensor was not based on FRET, allowing concomitant pH and $[Cl^-]_i$ ratiometric readout.

ClopHensor is obtained by fusion of a red fluorescent protein (DsRed-monomer) to the E^2GFP variant that contains a specific chloride-binding site (Fig. 7a). DsRed-monomer fluorescence is insensitive to variation in either $[Cl^-]$ or pH and is spectrally distinct from that of E^2GFP, so that it allows straightforward normalization of sensor signal, e.g., to eliminate the dependence of the signal on sensor concentration.

Key features of ClopHensor are: (1) the static quenching mechanism governing halide binding (Fig. 7b and [32]), allowing for reading of pH values free from the influence of $[Cl^-]_i$ variations; and (2) the presence of a pH isosbestic point at 458 nm (Fig. 7c), which considerably simplifies post-experiment calibrations. Two prominent peaks centered at nearly 515 and 560 nm, stemming from E^2GFP and DsRed excitation, characterize the ClopHensor fluorescence excitation spectrum (Fig. 7c, d). Overall, ClopHensor spectral changes are ideally suited for ratiometric operation at three excitation wavelengths: 488, 458, and 560 nm. $[Cl^-]_i$ measurements and Cl^--independent pH measurements are derived from the ratios 458/560 and 488/458, respectively. ClopHensor is intrinsically characterized by two molecular parameters (1): ${}^0pK_a = 6.81 \pm 0.05$ and $1/{}^1K_Y = 13.1 \pm 0.5$ mM, which correspond to a construct K_d for Cl^- of about 50 mM at physiological pH (7.2–7.3).

ClopHensor characteristics determine the following procedural sequence: (1) estimate pH in a region of interest; (2) calculate [using (8)] the correct affinity for Cl^- corresponding to the estimated pH value; (3) estimate $[Cl^-]_i$ [54]. This sensor promises to be a reliable tool for accurate analyses of Cl^- homeostasis in different cells of biological organisms regardless of intracellular pH variations.

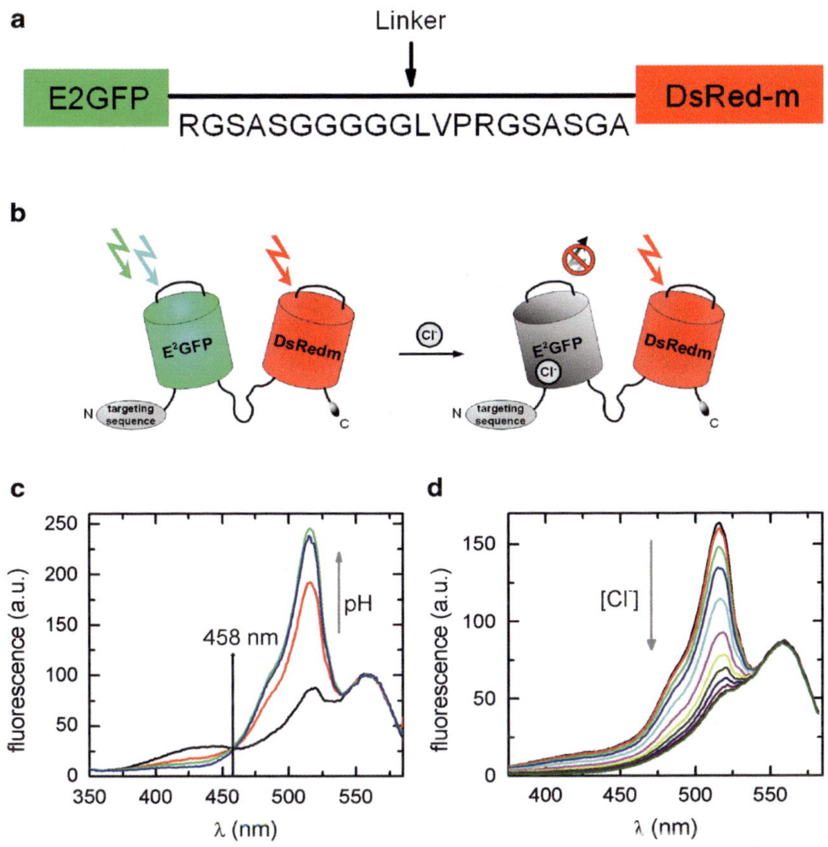

Fig. 7 ClopHensor design and spectral and functional characterization. (**a**) Schematic representation of ClopHensor construct with indication of the 20-amino acid flexible linker. (**b**) Cl^- binding to E^2GFP induces quenching of its fluorescence via a *static quenching* mechanism. (**c**, **d**) Excitation spectra of purified ClopHensor with emission set at 590 nm collected at different pH values (5.9, 6.9, 7.4, and 8.4) in the absence of Cl^- (**c**) and at increasing Cl^- concentration (0–1 M) and constant pH = 6.9 (**d**) (from [54])

4 Imaging Procedure and Data Analysis

Two kinds of measurements are typically made with fluorescent sensors: relative and absolute. Relative measurements would reflect changes in $[Cl^-]_i$ levels without reference to resting levels or the absolute size of $[Cl^-]_i$ changes, and are usually reported as the change in fluorescence normalized by the overall mean fluorescence ($\Delta F/F$). In contrast, absolute measurements would give a quantitative estimate of $[Cl^-]_i$ changes. Absolute measurements typically require post-experiment calibration, which can be achieved in one of three ways: single-wavelength measurements (e.g., single YFP sensors), ratiometric measurements (e.g., linked GFP sensors), or hybrid measurements.

Ratiometric sensors, showing a Cl^--dependent spectral shift in either their excitation or emission spectrum, undoubtedly give the most reliable estimates of $[Cl^-]_i$ changes. In fact, ratiometric measurements eliminate or attenuate distortions caused by variations in probe concentrations, cell thickness, excitation intensity, and detection efficiency. Furthermore, ratiometric methods attenuate artifacts such as photobleaching, nonuniform sensor distribution, and instrumentation instability [65]. When experimental conditions require high recording bandwidth, hybrid protocols can be followed performing single-wavelength and ratiometric measurements at different instant in time [65, 66].

The calibration equation for single-wavelength measurements can be written in terms of a generic fluorescence signal F as:

$$[\mathrm{Cl}^-] = K_\mathrm{d} \frac{F - F_\mathrm{free}}{F_\mathrm{bound} - F}, \tag{9}$$

where F_free and F_bound are fluorescence intensity values in zero and saturating Cl^- conditions, respectively. We note that in the case of a fluorescence static quenching mechanism $F_\mathrm{bound} = 0$, owing to the formation of a ground-state nonfluorescent Cl^--GFP complex.

Ratiometric measurements are commonly made at two wavelengths (λ_N, λ_D) to obtain the ratio and calibration equation, known also as Grynkiewicz equation [67], written as follows:

$$R = \frac{F_{\lambda_N}}{F_{\lambda_D}} \qquad [\mathrm{Cl}^-] = K_\mathrm{d} \left(\frac{R - R_\mathrm{free}}{R_\mathrm{bound} - R} \right) \left(\frac{F_{\lambda_D,\mathrm{free}}}{F_{\lambda_D,\mathrm{bound}}} \right). \tag{10}$$

It is worth noting that if a Cl^--independent signal (i.e., an isosbestic wavelength) is used in the denominator of the ratio R, then the last term in (10) cancels out, giving a simple equation:

$$[\mathrm{Cl}^-] = K_\mathrm{d} \left(\frac{R - R_\mathrm{free}}{R_\mathrm{bound} - R} \right). \tag{10a}$$

The calibration ratiometric equation for dual-wavelengths pH sensors can be written as follows:

$$\mathrm{pH} = \mathrm{p}K_\mathrm{a} + \log \left(\frac{R - R_\mathrm{A}}{R_\mathrm{B} - R} \right) + \log \left(\frac{F_{\lambda_D,\mathrm{A}}}{F_{\lambda_D,\mathrm{B}}} \right), \tag{11}$$

where R_A and R_B are the ratio values in acidic and basic conditions, respectively; again, the last additive term is absent in the case where a pH isosbestic point is used in the denominator of R (e.g., ClopHensor procedure).

Finally, ratiometric measurements include those made at: two-excitation and one-emission wavelengths (excitation ratio imaging); one-excitation and two-emission wavelengths (emission ratio imaging); and two-excitation and two-emission wavelengths (excitation–emission ratio imaging). Cl^- sensors are available for excitation (Cl-sensor), emission (Clomeleon), and excitation–emission (ClopHensor) ratio imaging. Excitation ratio imaging typically requires switching excitation filters, or polychromator devices, and collecting two images sequentially. Emission ratio imaging uses a single excitation source with the concomitant recording of two images in different emission channels. Advantageously, acquisition speed of emission ratio imaging can be very high, and, above all, noise generated by variations in source intensity is effectively removed. In fact, excitation intensity fluctuation is probably the most common noise source in optical recording experiments, especially with laser sources where such intensity variations can be as much as 10% peak-to-peak [68–70]. To remove sources of noise variations in excitation ratio imaging, it is crucial to introduce a reference measurement in the signals used in ratio formation. Unfortunately, measured signals in commercially available confocal microscopes do not provide such correction and so are not suitable for quantitative (spectroscopy) application such as ratio imaging. However, in excitation and excitation–emission ratio imaging, this problem can be remedied, as suggested by Zucker and Price [68, 69], using the transmission-detector signal to update laser-intensity fluctuations dynamically. In addition, it is important to carefully subtract any autofluorescence and other offset background before forming ratio images or calculating $\Delta F/F$.

5 Application of Genetically Encoded Cl^- Sensors

5.1 YFP-Based Sensors

Genetically encoded YFP-based sensors have a number of advantages over other techniques for intracellular Cl^- monitoring [71]. These advantages stimulated rapid expansion of YFP-based sensors for use in different applications. They have been used for extensive screening of large collections of drugs in a search for new ligands for Cl^--selective channels and proteins involved in Cl^- homeostasis.

For instance, several classes of modulators of the cystic fibrosis transmembrane conductance regulator (CFTR) chloride channel were identified: inhibitors [72], activators [73–76], potentiators [77], and correctors [78, 79]. These approaches were particularly useful in clarifying the role of CFTR in submucosal fluid secretion in excised human airways [80].

For identification of inhibitors of a human intestinal Ca-activated Cl^- channel (CaCC), a high-throughput screening was established using stable expression of YFP-H148Q/I152L with lentivirus. It allowed screening of 50,000 compounds and selection of two drugs, which inhibited by >95% iodide influx in HT-29 cells in response to agonists of CaCC [81].

Derivatives of YFP have also been used for pharmacological analysis of Cl^- selective receptor-operated channels activated by GABA and glycine [82]. In flow cytometry experiments, YFP-I152L was successfully used for high-throughput discrimination between functional and nonfunctional mutations of the glycine receptor [83].

YFP can be targeted to specific cell types by cell-specific promoters, or to defined cellular compartments by fusion to short sequence tags or to specific proteins. Transgenic mice expressing EYFP under control of the Kv3.1 K^+ channel promoter (pKv3.1) have been generated [84] and used for analysis of glutamate-induced changes in intracellular Cl^- and pH [84, 85].

5.2 Clomeleon and Cl-Sensor

Development of ratiometric YFP-based Cl^- sensors has greatly expanded their use for monitoring of halides in different preparations. Particularly important is the possibility of noninvasive estimation of intracellular Cl^- concentration in different cell types.

Transient expression of Clomeleon and Cl-sensor has permitted accurate determination of resting $[Cl^-]_i$ in cultured hippocampal neurons [52, 53], in spinal cord neurons [71], and in plant cells [86]. Moreover, it allowed monitoring of changes in $[Cl^-]_i$ elicited by the activation of GABA receptors or changes in extracellular Cl^- concentration or membrane potential.

Using electroporation in vivo, Cl-sensor was expressed in rat retina (Fig. 8a). Quantitative fluorescence analysis of $[Cl^-]_i$ distribution in different cell types, as well as its changes during development and depolarization, has been performed [71, 87]. These observations demonstrated that the efficiency of electroporation in the developing postnatal retina (at P0) is high, and that the expression persists for more than 6 month.

The widest field of application for genetically encoded probes comes from the possibility of targeting them to specific cell types or to cellular compartments and membrane domains by fusion to respective tags, to proteins with known location, or using unique promoters. Several techniques to express Clomeleon in selected populations of neurons have been developed [88]. The most successful were: in vivo viral transfer of the Clomeleon gene to targeted neurons and creation of transgenic mouse lines by insertion of Clomeleon, under control of the thy1 promoter. The high expression of Clomeleon in transgenic mice lines has allowed, for instance, the visualization of the entire neocortical layer 5 pyramidal neurons in living mice and estimation of $[Cl^-]_i$ distribution (Fig. 8b, d). Moreover, it allowed noninvasive monitoring of synaptic inhibition in different brain regions: the hippocampus, the deep cerebellar nuclei, the basolateral nucleus of the amygdala, and the superior colliculus [88].

Using ratiometric two-photon imaging in retina slices from transgenic mice expressing Clomeleon, Duebel et al. performed elegant analysis of $[Cl^-]_i$ values

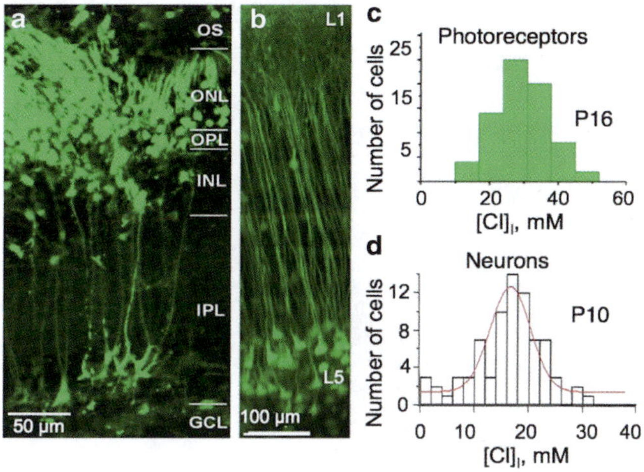

Fig. 8 Expression of Cl-sensor (**a**, **c**) and Clomeleon (**b**, **d**) in different preparations. (**a**) Confocal image of a vertical P20 rat retina slice electroporated in vivo at P0. Retinal layers are indicated: *OS* outer segment, *ONL* outer nuclear layer, *OPL* outer plexiform layer, *INL* inner nuclear layer, *IPL* inner plexiform layer, *GCL* ganglion cell layer (from [71]).(**b**) Expression of Clomeleon in transgenic mice. Two-photon image of neocortical layer 5 (L5) neurons obtained from the brain surface (L1) of living mice (from [88]). (**c**) Histogram of $[Cl^-]_i$ distribution in photoreceptor cells from P15 rat retina slices measured noninvasively with Cl-sensor. Note high range of $[Cl^-]_i$ variations from 10 to 50 mM with mean of 31 ± 9 mM ($n = 73$) (from [87]). (**d**) Distribution of $[Cl^-]_i$ in cortical neurons at P10 estimated with Clomeleon. Mean $[Cl^-]_i = 16.6 \pm 5.2$ mM (from [64])

in different parts of bipolar cells. They found a somatodendritic gradient of $[Cl^-]_i$ indicating variations in the distribution of this anion along bipolar cells [89].

These and previously obtained electrophysiological observations on differential intracellular distribution of Cl^- [90, 91] have been confirmed and extended in a recent study in which Cl-sensor was used for mapping $[Cl^-]_i$ in cellular processes of hippocampal and spinal cord neurons in culture [71].

Mapping of intracellular Cl^- concentration in different neuronal areas and in small cellular compartments such as dendritic spines is a difficult and important task. Using Cl-sensor with improved neuronal promoter, Waseem et al. [71] performed systematic mapping of the $[Cl^-]_i$ distribution in different neuronal compartments of cultured hippocampal and spinal cord neurons. It was shown that the two types of neurons exhibit similar tendencies for a somatodendritic gradient of $[Cl^-]_i$: the maximum $[Cl^-]_i$ was observed in soma and it gradually decreased along dendritic branches, particularly after bifurcations. Cl^- concentration in distal dendrites was frequently more than 50% smaller than in the soma (Fig. 9a, b). Moreover, quantitative estimation of $[Cl^-]_i$ was performed in dendritic spines, which represent small postsynaptic compartments of glutamatergic synapses. Application of glutamate caused a transient elevation of Cl^- of 5–10 mM [71].

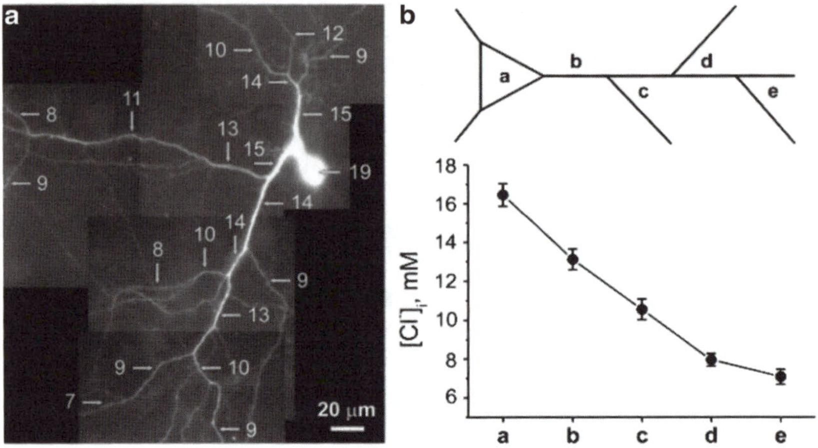

Fig. 9 Distribution of $[Cl^-]_i$ along dendritic trees in cultured hippocampal neurons. (**a**) Hippocampal neuron (15 DIV) transfected with Cl-sensor. Numbers are $[Cl^-]_i$ in different regions of the cell. Note the decrease in $[Cl^-]_i$ at increasing distance from the soma. Scale bar 20 μm. (**b**) Alteration in the $[Cl^-]_i$ levels in different parts of hippocampal neurons (10–15 DIV). *Top*: schematic representation of regions which have been analyzed in hippocampal neurons. *Bottom*: mean values of estimated $[Cl^-]_i$ in soma (1) and different parts of the dendritic tree (data from 13 neurons) (from [71])

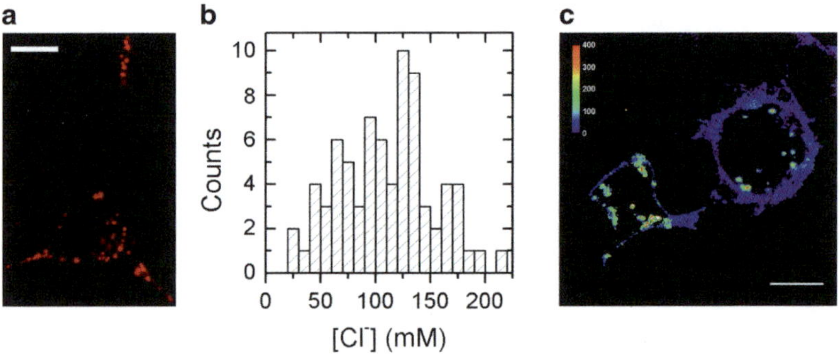

Fig. 10 Measurement of chloride concentration in LDCV. (**a**) Micrograph of PC12 cells transfected with the NPY-ClopHensor plasmid that were differentiated by treatment with nerve growth factor (*red channel*). (**b**) Distribution of the measured LDCV $[Cl^-]$ in PC12 cells ($n = 76$). (**c**) $[Cl^-]_i$ map of WSS-1 cells co-expressing ClopHensor in LDCV and at the plasma membrane. *Scale bars*, 5 μm (from [54])

5.3 *ClopHensor*

Fusion of GFP tags to various portions of prepro-neuropeptide Y (NPY) precursor has been successfully used to investigate the trafficking of regulated dense-core secretory granules (LDCV) [92], the dominant secretory organelles in neuroendocrine cells. Fused to the N-terminal signal sequence of the NPY, ClopHensor has been successfully targeted to LDCV (Fig. 10a), recording an average $pH_i = 5.2$

± 0.4 and revealing a high intravesicular Cl^- concentration of ~110 ± 48 mM (Fig. 10). This finding has been further confirmed by measuring Cl^- concentration in LDCV and cytoplasm (at the plasma membrane), using two constructs concurrently: one with the NPY tag and another with a double palmitoylation-site tag (Fig. 10c).

These studies demonstrate that different constructs of Cl^- sensors represent an effective noninvasive tool allowing systematic analysis of $[Cl^-]_i$ distribution in different cell types or neuronal branches and also quantitative estimation of $[Cl^-]_i$ changes in small cellular compartments such as dendritic spines during excitation.

5.4 *Visualization of Activity of Cl^--Selective Receptor-Operated Channels*

The other important perspective in using genetically encoded Cl^- sensors consists of its insertion into the protein's sequence without changing its functional properties. Recently, this idea has been successfully explored and a new genetic probe, termed BioSensor-GlyR, has been developed [87]. This construct represents a Cl^--selective glycine receptor (GlyR) channel with Cl-sensor incorporated into the long cytoplasmic domain (Fig. 11a). Importantly, after expression in cell lines (HEK-

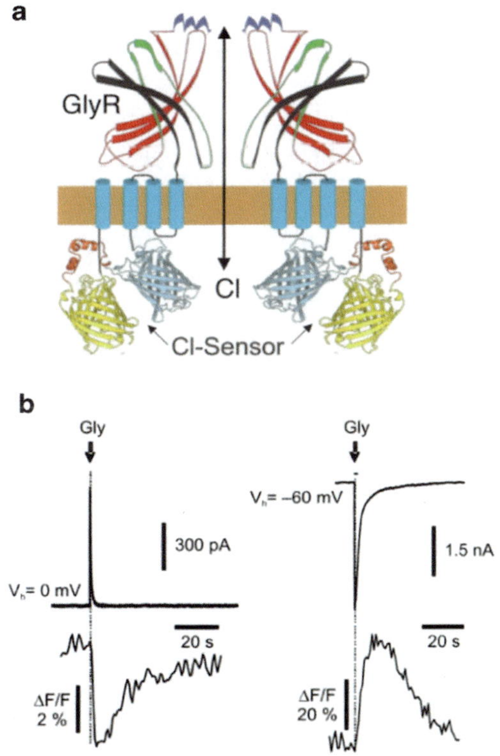

Fig. 11 Structure and fluorescence response of BioSensor-GlyR. (**a**) Scheme of BioSensor-GlyR construct. Only two subunits of this pentameric channel are shown. (**b**) Examples of simultaneous whole-cell and fluorescence recordings from BioSensor-GlyR expressed in BHK cells. The recording pipette contained either nominally 0 (*left traces*) or 150 (*right traces*) mM Cl^-. Glycine (1 mM) was pressure-applied for 10 ms (from [87])

293 and CHO), the main functional properties, i.e., kinetics, agonist sensitivity, and Cl^- selectivity of this modified ionotropic receptor, were not perturbed by the inserted Cl-sensor. Application of glycine to cells expressing this construct induced changes in fluorescence, corresponding to the direction and amplitude of glycone-induced Cl^- currents (Fig. 11b). This demonstrates that BioSensor-GlyR is a good probe for spectroscopic monitoring of GlyR activation in live cells. While the sensitivity of BioSensor-GlyR was high enough to resolve changes in $[Cl^-]_i$ induced by activation of postsynaptic receptors in glycinergic synapses, the decay kinetics of fluorescence responses were slow compared with those of ionic currents (Fig. 11b). This construct also exhibits weak expression in neurons (unpublished observations).

In spite of these limitations, it is important to continue development of improved probes for noninvasive monitoring of Cl^--selective channels activity. This would open additional ways for visualization functional synapses and noninvasive monitoring of network signaling in cell circuits.

6 Conclusions

Development of genetically encoded Cl^- sensors has opened a new perspective for noninvasive monitoring of Cl^- in different cells and tissues of biological organisms. This approach allows measurement of $[Cl^-]_i$ in small cellular compartments, and simultaneous monitoring of Cl^- and pH. Development of transgenic mice provides a highly promising direction for the analysis of Cl^- homeostasis under different experimental conditions, during development of the nervous system and in pathological conditions. The possibility of targeting Cl^- probes to Cl^--selective transporters, allowing the visualization of their functional activity, is intriguing. This task, however, is highly challenging, as it needs the development of new constructs with both improved brightness and intact functionality under a variety of intracellular environments.

Acknowledgments The financial support of Telethon – Italy (Grant no. GGP10138D) for DA and the European FP-7 (Grant no. HEALTH-F2-2008-202088) ("Neurocypres" Project) for PB are gratefully acknowledged.

References

1. Hamill OP, Bormann J, Sakmann B (1983) Activation of multiple-conductance state chloride channels in spinal neurones by glycine and GABA. Nature 305(5937):805–808
2. Faundez V, Hartzell HC (2004) Intracellular chloride channels: determinants of function in the endosomal pathway. Sci STKE 2004(233):re8
3. Graves AR, Curran PK, Smith CL, Mindell JA (2008) The Cl^-/H^+ antiporter ClC-7 is the primary chloride permeation pathway in lysosomes. Nature 453:788–792

4. Pedersen M, Carmosino M, Forbush B (2008) Intramolecular and intermolecular fluorescence resonance energy transfer in fluorescent protein-tagged Na-K-Cl cotransporter (NKCC1): sensitivity to regulatory conformational change and cell volume. J Biol Chem 283:2663–2674
5. Miyazaki H, Shiozaki A, Niisato N, Marunaka Y (2007) Physiological significance of hypotonicity-induced regulatory volume decrease: reduction in intracellular Cl- concentration acting as an intracellular signaling. Am J Physiol Renal Physiol 292:F1411–F1417
6. Marty A, Llano I (2005) Excitatory effects of GABA in established brain networks. Trends Neurosci 28(6):284–289
7. Chen T, Hwang T (2008) CLC-0 and CFTR: chloride channels evolved from transporters. Physiol Rev 88(2):351–387
8. Blaesse P, Airaksinen MS, Rivera C, Kaila K (2009) Cation-chloride cotransporters and neuronal function. Neuron 61(6):820–838
9. Jentsch TJ (2008) CLC chloride channels and transporters: from genes to protein structure, pathology and physiology. Crit Rev Biochem Mol Biol 43(1):3–36
10. Planells-Cases R, Jentsch TJ (2009) Chloride channelopathies. Biochim Biophys Acta 1792 (3):173–189
11. Verkman AS, Galietta LJV (2009) Chloride channels as drug targets. Nat Rev Drug Discov 8 (2):153–171
12. Bregestovski P, Waseem T, Mukhtarov M (2009) Genetically encoded optical sensors for monitoring of intracellular chloride and chloride-selective channel activity. Front Mol Neurosci 2:15
13. Geddes CD, Apperson K, Karolin J, Birch DJ (2001) Chloride-sensitive fluorescent indicators. Anal Biochem 293:60–66
14. Giepmans BNG, Adams SR, Ellisman MH, Tsien RY (2006) The fluorescent toolbox for assessing protein location and function. Science 312(5771):217–224
15. Verkhusha VV, Lukyanov KA (2004) The molecular properties and applications of Anthozoa fluorescent proteins and chromoproteins. Nat Biotechnol 22(3):289–296
16. Nienhaus GU, Wiedenmann J (2009) Structure, dynamics and optical properties of fluorescent proteins: perspectives for marker development. ChemPhysChem 10(9):1369–1379
17. Xie XS, Yu J, Yang WY (2006) Living cells as test tubes. Science 312(5771):228
18. Chalfie M, Tu Y, Euskirchen G, Ward WW, Prasher DC (1994) Green fluorescent protein as a marker for gene expression. Science 263(5148):802–805
19. Inouye S, Tsuji FI (1994) Aequorea green fluorescent protein. Expression of the gene and fluorescence characteristics of the recombinant protein. FEBS Lett 341(2):277–280
20. Tsien RY (1998) The green fluorescent protein. Annu Rev Biochem 67(1):509–544
21. Cormack BP, Valdivia RH, Falkow S (1996) FACS-optimized mutants of the green fluorescent protein (GFP). Gene 173(1):33–38
22. Miesenböck G, De Angelis DA, Rothman JE (1998) Visualizing secretion and synaptic transmission with pH-sensitive green fluorescent proteins. Nature 394(6689):192–195
23. Miyawaki A, Griesbeck O, Heim R, Tsien RY (1999) Dynamic and quantitative $Ca2^{+}$ measurements using improved cameleons. Proc Natl Acad Sci U S A 96(5):2135
24. Griesbeck O (2004) Fluorescent proteins as sensors for cellular functions. Curr Opin Neurobiol 14(5):636–641
25. Heim R, Prasher DC, Tsien RY (1994) Wavelength mutations and posttranslational autoxidation of green fluorescent protein. Proc Natl Acad Sci U S A 91(26):12501–12504
26. Heim R, Tsien RY (1996) Engineering green fluorescent protein for improved brightness, longer wavelengths and fluorescence resonance energy transfer. Curr Biol 6(2):178–182
27. Wachter RM, Elsliger MA, Kallio K, Hanson GT, Remington SJ (1998) Structural basis of spectral shifts in the yellow-emission variants of green fluorescent protein. Structure 6(10):1267–1277
28. Rekas A (2002) Crystal structure of venus, a yellow fluorescent protein with improved maturation and reduced environmental sensitivity. J Biol Chem 277(52):50573–50578
29. Griesbeck O, Baird GS, Campbell RE, Zacharias DA, Tsien RY (2001) Reducing the environmental sensitivity of yellow fluorescent protein. J Biol Chem 276(31):29188–29194

30. Wachter RM, Remington SJ (1999) Sensitivity of the yellow variant of green fluorescent protein to halides and nitrate. Curr Biol 9(17):R628–R629
31. Galietta LJ, Haggie PM, Verkman AS (2001) Green fluorescent protein-based halide indicators with improved chloride and iodide affinities. FEBS Lett 499(3):220–224
32. Arosio D et al (2007) Spectroscopic and structural study of proton and halide ion cooperative binding to gfp. Biophys J 93(1):232–244
33. Bell AF, He X, Wachter RM, Tonge PJ (2000) Probing the ground state structure of the green fluorescent protein chromophore using Raman spectroscopy. Biochemistry 39(15):4423–4431
34. Jung G (2005) The photophysics of green fluorescent protein: influence of the key amino acids at positions 65, 203, and 222. Biophys J 88(3):1932–1947
35. Elsliger MA, Wachter RM, Hanson GT, Kallio K, Remington SJ (1999) Structural and spectral response of green fluorescent protein variants to changes in pH. Biochemistry 38 (17):5296–5301
36. Jayaraman S, Haggie P, Wachter RM, Remington SJ, Verkman AS (2000) Mechanism and cellular applications of a green fluorescent protein-based halide sensor. J Biol Chem 275 (9):6047–6050
37. Wachter RM, Yarbrough D, Kallio K, Remington S (2000) Crystallographic and energetic analysis of binding of selected anions to the yellow variants of green fluorescent protein1. J Mol Biol 301(1):157–171
38. Wyman J (1948) Heme proteins. Adv Protein Chem 4:407–531
39. Wyman J, Gill SJ (1990) Binding and linkage: functional chemistry of biological macromolecules. University Science Books, Mill Valley, CA
40. DiCera E (2005) Thermodynamic theory of site-specific binding processes in biological macromolecules. Cambridge University Press, Cambridge
41. Weber G (1992) Protein Interactions. 1st ed. New York, Chapman and Hall
42. Geddes CD (2001) Optical halide sensing using fluorescence quenching. Meas Sci Technol 12:R53–R88
43. Valeur B (2002) Molecular fluorescence principles and applications. WILEY-VCH, Weinheim
44. Lakowicz JR (1999) Principles of fluorescence spectroscopy. Kulwer Academic/Plenum Publishers, New York
45. Stokes GG (1869) On a certain reaction of quinine. J Chem Soc 22:174
46. Heim R (1999) Green fluorescent protein forms for energy transfer. Methods Enzymol 302:408–423
47. Kneen M, Farinas J, Li Y, Verkman AS (1998) Green fluorescent protein as a noninvasive intracellular pH indicator. Biophys J 74(3):1591–1599
48. Llopis J, McCaffery JM, Miyawaki A, Farquhar MG, Tsien RY (1998) Measurement of cytosolic, mitochondrial, and Golgi pH in single living cells with green fluorescent proteins. Proc Natl Acad Sci U S A 95(12):6803
49. Chudakov DM, Matz MV, Lukyanov S, Lukyanov KA (2010) Fluorescent proteins and their applications in imaging living cells and tissues. Physiol Rev 90(3):1103–1163
50. Ormo M, Cubitt AB, Kallio K, Gross LA, Tsien RY, Remington SJ (1996) Crystal structure of the Aequorea victoria green fluorescent protein. Science 273:1392–1395
51. Jayaraman S, Verkman AS (2000) Quenching mechanism of quinolinium-type chloride-sensitive fluorescent indicators. Biophys Chem 85(1):49–57
52. Kuner T, Augustine GJ (2000) A genetically encoded ratiometric indicator for chloride: capturing chloride transients in cultured hippocampal neurons. Neuron 27(3):447–459
53. Markova O, Mukhtarov M, Real E, Jacob Y, Bregestovski P (2008) Genetically encoded chloride indicator with improved sensitivity. J Neurosci Methods 170:67–76
54. Arosio D, Ricci F, Marchetti L, Gualdani R, Albertazzi L, Beltram F (2010) Simultaneous intracellular chloride and pH measurements using a GFP-based sensor. Nat Methods 7:516–518
55. Hink MA, Bisselin T, Visser AJWG (2002) Imaging protein-protein interactions in living cells. Plant Mol Biol 50(6):871–883

56. Hallworth R, Currall B, Nichols MG, Wu X, Zuo J (2006) Studying inner ear protein-protein interactions using FRET and FLIM. Brain Res 1091(1):122–131
57. Tsien RY (2009) Indicators based on fluorescence resonance energy transfer (FRET). Cold Spring Harb Protoc 2009(7):pdb.top57
58. Förster T (1948) Zwischenmolekulare Energiewanderung und Fluoreszenz. Annalen der Physik 437(1):55–75
59. Pollok BA, Heim R (1999) Using GFP in FRET-based applications. Trends cell biol 9(2):57–60
60. Sekar RB, Periasamy A (2003) Fluorescence resonance energy transfer (FRET) microscopy imaging of live cell protein localizations. J Cell Biol 160(5):629
61. Wallrabe H, Periasamy A (2005) Imaging protein molecules using FRET and FLIM microscopy. Curr Opin Biotechnol 16(1):19–27
62. Jose M, Nair DK, Reissner C, Hartig R, Zuschratter W (2007) Photophysics of Clomeleon by FLIM: discriminating excited state reactions along neuronal development. Biophys J 92 (6):2237–2254
63. Pond BB, Berglund K, Kuner T, Feng G, Augustine GJ, Schwartz-Bloom RD (2006) The chloride transporter Na^{+}–K^{+}–Cl-cotransporter isoform-1 contributes to intracellular chloride increases after in vitro ischemia. J Neurosci 26(5):1396
64. Berglund K et al (2006) Imaging synaptic inhibition in transgenic mice expressing the chloride indicator, Clomeleon. Brain Cell Biol 35(4):207–228
65. Bullen A, Saggau P (1999) Optical Recording from Individual Neurons in Culture. In: Modern Techniques in Neuroscience Research. Berlin, Springer
66. Lev-Ram V, Miyakawa H, Lasser-Ross N, Ross WN (1992) Calcium transients in cerebellar Purkinje neurons evoked by intracellular stimulation. J Neurophysiol 68(4):1167–1177
67. Grynkiewicz G, Poenie M, Tsien RY (1985) A new generation of Ca2+ indicators with greatly improved fluorescence properties. J Biol Chem 260(6):3440
68. Zucker RM, Price OT (2001) Statistical evaluation of confocal microscopy images. Cytometry Part A 44(4):295–308
69. Zucker RM, Price O (2001) Evaluation of confocal microscopy system performance. Cytometry 44(4):273–294
70. Bullen A, Saggau P Optical Recording from Individual Neurons in Culture. In: Modern Techniques in Neuroscience Research. Berlin, Springer 1999
71. Waseem T, Mukhtarov M, Buldakova S, Medina I, Bregestovski P (2010) Genetically encoded Cl-Sensor as a tool for monitoring of Cl-dependent processes in small neuronal compartments. J Neurosci Methods 193:14–23
72. Ma T et al (2002) Thiazolidinone CFTR inhibitor identified by high-throughput screening blocks cholera toxin-induced intestinal fluid secretion. J Clin Invest 110(11):1651–1658
73. Galietta LJV et al (2001) Novel CFTR chloride channel activators identified by screening of combinatorial libraries based on flavone and benzoquinolizinium lead compounds. J Biol Chem 276(23):19723–19728
74. Ma T et al (2002) High-affinity activators of cystic fibrosis transmembrane conductance regulator (CFTR) chloride conductance identified by high-throughput screening. J Biol Chem 277(40):37235–37241
75. Muanprasat C, Sonawane ND, Salinas D, Taddei A, Galietta LJV, Verkman AS (2004) Discovery of glycine hydrazide pore-occluding CFTR inhibitors: mechanism, structure-activity analysis, and in vivo efficacy. J Gen Physiol 124(2):125–137
76. Muanprasat C, Kaewmokul S, Chatsudthipong V (2007) Identification of new small molecule inhibitors of cystic fibrosis transmembrane conductance regulator protein: in vitro and in vivo studies. Biol Pharm Bull 30(3):502–507
77. Yang H et al (2003) Nanomolar affinity small molecule correctors of defective Delta F508-CFTR chloride channel gating. J Biol Chem 278(37):35079–35085
78. Pedemonte N et al (2005) Phenylglycine and sulfonamide correctors of defective delta F508 and G551D cystic fibrosis transmembrane conductance regulator chloride-channel gating. Mol Pharmacol 67(5):1797–1807

79. Pedemonte N et al (2005) Small-molecule correctors of defective DeltaF508-CFTR cellular processing identified by high-throughput screening. J Clin Invest 115(9):2564–2571
80. Verkman AS (2004) Drug discovery in academia. Am J Physiol Cell Physiol 286(3): C465–C474
81. De La Fuente R, Namkung W, Mills A, Verkman AS (2008) Small-molecule screen identifies inhibitors of a human intestinal calcium-activated chloride channel. Mol Pharmacol 73 (3):758–768
82. Kruger W et al (2005) A yellow fluorescent protein-based assay for high-throughput screening of glycine and GABAA receptor chloride channels. Neurosci Lett 380(3):340–345
83. Gilbert DF, Wilson JC, Nink V, Lynch JW, Osborne GW (2009) Multiplexed labeling of viable cells for high-throughput analysis of glycine receptor function using flow cytometry. Cytometry A 75(5):440–449
84. Metzger F et al (2002) Transgenic mice expressing a pH and Cl- sensing yellow-fluorescent protein under the control of a potassium channel promoter. Eur J Neurosci 15(1):40–50
85. Slemmer JE, Matsushita S, De Zeeuw CI, Weber JT, Knöpfel T (2004) Glutamate-induced elevations in intracellular chloride concentration in hippocampal cell cultures derived from EYFP-expressing mice. Eur J Neurosci 19(11):2915–2922
86. Lorenzen I, Aberle T, Plieth C (2004) Salt stress-induced chloride flux: a study using transgenic Arabidopsis expressing a fluorescent anion probe. Plant J 38(3):539–544
87. Mukhtarov M, Markova O, Real E, Jacob Y, Buldakova S, Bregestovski P (2008) Monitoring of chloride and activity of glycine receptor channels using genetically encoded fluorescent sensors. Philos Transact A Math Phys Eng Sci 366(1880):3445–3462
88. Berglund K et al (2008) Imaging synaptic inhibition throughout the brain via genetically targeted Clomeleon. Brain Cell Biol 36(1):101–118
89. Duebel J et al (2006) Two-photon imaging reveals somatodendritic chloride gradient in retinal ON-type bipolar cells expressing the biosensor Clomeleon. Neuron 49(1):81–94
90. Szabadics J, Varga C, Molnár G, Oláh S, Barzó P, Tamás G (2006) Excitatory effect of GABAergic axo-axonic cells in cortical microcircuits. Science 311(5758):233–235
91. Khirug S, Yamada J, Afzalov R, Voipio J, Khiroug L, Kaila K (2008) GABAergic depolarization of the axon initial segment in cortical principal neurons is caused by the Na-K-2Cl cotransporter NKCC1. J Neurosci 28(18):4635–4639
92. El Meskini R et al (2001) A signal sequence is sufficient for green fluorescent protein to be routed to regulated secretory granules. Endocrinology 142:864–873

Fluorescent Genetically Encoded Calcium Indicators and Their In Vivo Application

Thomas Gensch and Dagmar Kaschuba

Abstract Calcium ions have a pivotal role in all living organisms. Monitoring and quantifying changes of the calcium concentration in living cells, tissues or organisms is a challenging task. To achieve this goal, specific calcium sensor proteins were designed and artificially assembled. An important group of these proteins is commonly referred to as fluorescent genetically encoded calcium indicators (fGECIs). In fGECIs, fluorescent proteins are fused to calcium-binding protein motifs and additionally harbour modules that interact with the calcium-binding domains. Since the first publications in 1997, a multitude of fGECIs has been generated, characterized and applied in many studies. This chapter gives an overview on the usage of fGECIs in the recent past, considering 236 publications from January 2008 to August 2010. Outstanding studies are presented that investigate intracellular calcium changes with respect to physiological functions at the level of cellular organelles, cells, as well as cellular networks and whole animals such as worms, flies and mice.

Keywords Calcium in vivo measurements · GECI · Mitochondria · Microdomains · Neuronal activity · Chronic imaging · Genetically encoded sensors · Fluorescent protin

Contents

T. Gensch (✉) and D. Kaschuba
Institute of Complex Systems 4 (ICS-4), Cellular Biophysics, Forschungszentrum Jülich, 52425 Jülich, Germany
e-mail: t.gensch@fz-juelich.de

G. Jung (ed.), *Fluorescent Proteins II*, Springer Ser Fluoresc (2012) 12: 125–162,
DOI 10.1007/4243_2011_29
Published online: 18 June 2011

1 Fluorescent Genetically Encoded Calcium Indicators

In all kingdoms of life, calcium ions serve as ubiquitous second messengers [1]. Within a cell, calcium participates in the regulation, modulation and control of vital functions. In addition, calcium constitutes an important intercellular communication signal. The presence of intracellular calcium stores allows calcium signals to be generated independently of the extracellular calcium concentration. However, for calcium changes to be an effective cellular signal, the free cytosolic calcium concentration is strictly controlled at a rather low level (about 100 nM) in eukaryotic cells. Notably, unphysiologically elevated concentrations of calcium can lead to the aggregation of proteins and nucleic acids, as well as precipitation of calcium phosphate [2] and consequently, they may induce apoptotic cell death as well.

Fluorescence microscopy emanated as the method of choice to observe and determine calcium concentrations as well as its changes in living cells. Calcium imaging via fluorescence is based on calcium-sensitive fluorescent dyes – most of them are low molecular weight (0.3–2 kDa) organic molecules that allow detection of calcium-dependent changes through fluorescence. The first of these dyes have been available already for more than 30 years. They have a modular design with a fluorescent moiety, e.g. fluorescein or rhodamine, and a calcium-binding moiety, e.g. BAPTA or EGTA. The binding of calcium ions to the calcium-binding moiety of the calcium indicator dye (CID) affects the absorption and fluorescence properties of the fluorescent moiety allowing the readout of the change in free calcium. CIDs are commercially available and come also as modified forms that can easily cross cellular membranes and get enriched in the cell. After more than a decade of development and optimization, an arsenal of CIDs is now available, varying in absorption maximum, emission maximum, absorption coefficient, fluorescence quantum yield and calcium sensitivity. This allows the experimenter to choose the "right" excitation and emission wavelength from the UV to orange and violet to red spectrum, as well as between calcium sensitivities ranging from nanomolar to micromolar values [3]. CIDs can have either a pure fluorescence intensity change upon calcium binding as seen, e.g. in Fluo-4 (increase) or Fura Red (decrease), or more sophisticated properties like a fluorescence lifetime change, e.g. in Calcium Green-1, or a ratio change in, e.g. Fura-2 (excitation) or Indo-1 (emission).

However, CIDs suffer from several experimental limitations:

1. While loading of cultured cells with CIDs works well, tissue slices or whole organs are often insufficiently accessible to CIDs.
2. It is not possible to direct or target CIDs to specific subcellular structures, e.g. nucleus, mitochondria, endoplasmic or sarcoplasmic reticulum and plasma membrane.
3. Similarly, cell-type specific loading with CIDs in tissue, e.g. neurons vs. glia cells, is impossible.

4. The CIDs can only be used for a limited time period after loading, i.e. several hours (at most). Chronic imaging, aimed to observe one and the same cell(s) and/or its neighbours over days or weeks, is not possible.
5. Loading cells with CIDs is an acute intrusion that may cause detrimental stress to cells or tissues.

These shortcomings of CIDs limit the application of calcium imaging in a lot of modern biological research areas, where not the isolated cell, but a tissue, an organ or the whole organism is the focus of research. Therefore, genetically encoded calcium sensors (GECIs) were introduced to overcome the mentioned limitations of CIDs. GECIs can be integrated into the genome of cells, so that they are constitutively expressed and transferred to daughter cells. This allows chronic imaging over long time periods. Furthermore, using suitable promoters, expression can be limited to a specific cell type or a tissue, e.g. an organ (see Fig. 1). There are moreover genetic and pharmacological tools available to induce expression at a defined time

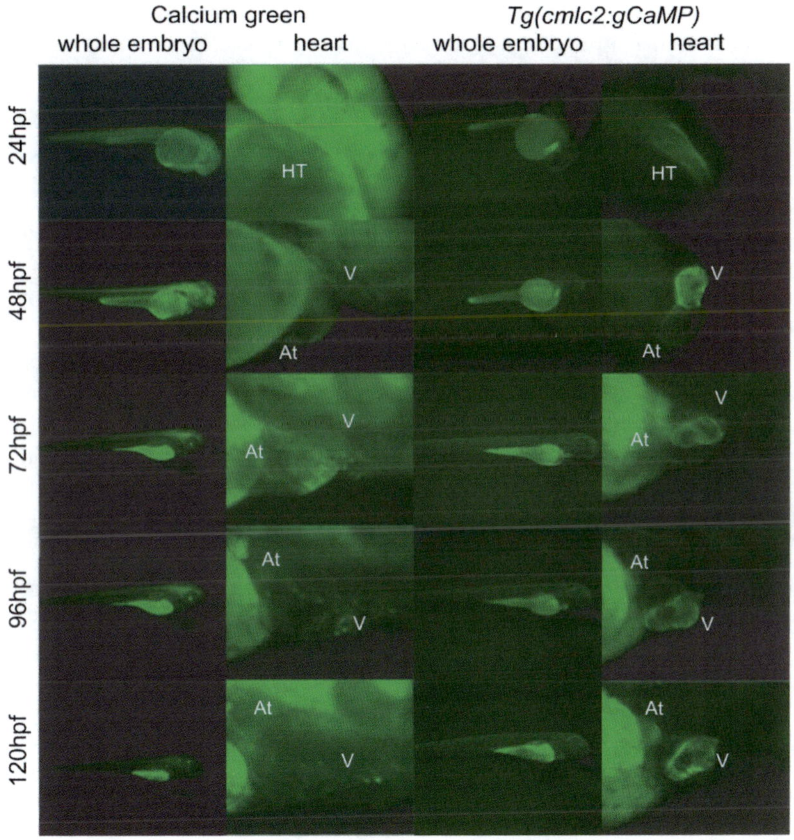

Fig. 1 Imaging heart function in a living zebrafish embryo (0–120 h post-fertilization): Comparison of images using a CID [*left*, calcium green (injected)] and a transgenic zebrafish line that specifically expresses GCaMP at all developmental stages in the heart using a cardiac-specific promoter, adapted from [4] (*right*; *HT* heart tube, *V* ventricle, *At* atrium)

point. In addition, by attaching small amino acid sequences, so-called target sequences, to GECIs – derived from proteins that are highly localized to specific parts of the cell – GECIs can be targeted, e.g. to the nucleus, mitochondria, endoplasmic or sarcoplasmic reticulum, peroxisomes, Golgi apparatus or the plasma membrane.

2 Bioluminescent GECIs

The topic of this book is fluorescent proteins (FPs), and this chapter deals with GECIs based on FPs (fGECI). Nevertheless, a few words should be spent on another type of GECI, namely those that are based on bioluminescence. The best known representative is aequorin (for review, see [5, 6]). Apo-aequorin is a luciferase that forms a blue-emitting photoprotein by binding covalently the chromophore coelenterazine. Binding of calcium ions to aequorin leads to conformational changes of the protein and finally coelenterazine is oxidized to coelenteramine. The latter is in its excited state and a blue photon is emitted upon relaxation to the ground state. In this way, calcium changes in a concentration range from 100 nM to 1 mM can be registered [7]. Aequorin has been cloned and expressed in 1985–1987 and has been used as a GECI since then. One of its most remarkable features is the high signal-to-noise ratio. This property can be explained by the fact that aequorin does not require excitation by light, which eliminates illumination-induced background signals such as autofluorescence or scattered excitation light. However, aequorin also has some drawbacks. The most obvious ones are the irreversible reaction from coelenterazine to coelenteramine, the long formation time of aequorin from apo-aequorin, the necessity of supplying cells or tissue with the cofactor coelenterazine and a relatively low spatial resolution (due to the low intensities of the emitted bioluminescence). In a recent review, the application of bioluminescent vs. fluorescent GECIs to image mitochondrial calcium changes in living animals has been compared [8]. The spatiotemporal resolution obtained in measurements with a mitochondria-targeted cameleon (fGECI) was superior to that obtained with a mitochondria-targeted aequorin. When considering the whole organism, non-invasive imaging methods can be realized with aequorin. Calcium signals were observed in living, unrestrained and non-anaesthetized animals. On the other hand, fGECIs are superior when imaging neuronal activity in brain or intact slices (see, e.g. [9, 10]). The best aequorin-based GECIs are 5–15 times less sensitive in detection of action potentials.

3 fGECIs: A Short History

Despite the experimental advantages of fluorescent GECIs, it is interesting to know that the original idea to construct such sensor proteins was born out of the failure to produce a functional, FRET-based inositol 1,4,5-triphosphate (InsP3) sensor. Originally, two differently coloured GFPs and the cytosolic part of the InsP3 receptor were fused, but no FRET changes were observed when the InsP3 concentration was varied. Nobel laureate Dr. R. Tsien suggested to his postdoctoral fellow Dr. A. Miyawaki to practice on a different sensor, similarly engineered but with a sensing unit of known structure. This protein consisted of a fusion of calmodulin (CaM) with its target peptide M13 from skeletal muscle myosin light chain kinase [11]. Two fGECIs were produced named cameleons that differed by the GFP-FRET-pairs, i.e. blue-green and cyan-yellow [12]. Upon calcium binding, calmodulin wraps around the M13 peptide and FRET increases. Shortly before this hallmark publication, the group of Dr. Persechini presented another type of fGECI also relying on FRET between two GFP mutants [13]. The construct lacked calmodulin in the linker; instead it used binding of endogenous calmodulin to the M13 peptide on the sensor when the calcium concentration is elevated. Interestingly, the FRET change was opposite compared with cameleon: at high calcium concentration, FRET was almost completely abolished. In the same year, the group introduced another fGECI, which contained calmodulin at the C-terminal end of the sensor protein [14]. While the first design did not find follow-up sensors, the second architecture is still used and can be found in some of the single GFP fGECIs, which use a circular permuted GFP flanked by the M13-peptide and calmodulin [15, 16].

Since then, almost every year further improved versions of fGECIs are published that have higher brightness, enhanced dynamic range, improved photostability, etc. Unfortunately, a comparison to older versions or to other fGECI families is often incomplete and not standardized. Therefore, it would be desirable that the laboratories developing fGECIs would agree on a standardized protocol that can be reliably performed in every laboratory, e.g. calcium responses to histamine treatment of HeLa cells and to signals that can be quantified by a ratiometric dye, like Fura-2. There exist some "heroic" and very detailed studies in the literature that tried to compare the performance of fGECIs shortly after their introduction [17–22], but they were outdated rather quickly. In addition, test expression systems were chosen that are not easily accessible for non-experts.

At this point, we would like to highlight a problem for scientists who aim to use fGECIs without having a background in fGECI development. Several parameters are important for choosing one over the other fGECI: calcium sensitivity (K_{Ca2+}), dynamic range, visibility of the calcium-bound and calcium-free form, pH insensitivity, magnesium insensitivity, calcium on- and off-rates (speed), toxicity and successful targeting. But even well-defined parameters like the first two suffer from the fact that the reproducibility of protein purity and functionality are not always given. Often, the temperature is different, which is also known to affect the performance of some fGECI. In addition, the use of calcium buffers for defined calcium concentrations is error prone, especially in the presence of other ions, like Mg^{2+}. For example, a laboratory may prefer one fGECI over the other because of its higher sensitivity, like YC3.6 over D3cpv, because YC3.6 is 2.5 times more sensitive according to the originally published value ([23, 24], see Table 1). Re-examination of both sensors in two later publications yielded similar K_{Ca2+} values for D3cpv but much higher values for YC3.6 [20, 31]. Along with these results, the advantage of YC3.6 over D3cpv no longer exists! Also for GCaMP2, a single FP

Table 1 Overview on fGECI used in January 2008 to August 2010

fGECI	Signal change[a]	K_{Ca2+}	Original publication	Improvement[b]	Articles[c]
Single GFP					
GCaMP[d]	3.5-fold	0.24 μM	Nakai et al. [16]		35
GCaMP1.3[d]	n.d.	n.d., likely similar to GCaMP	Ohkura et al. [25]	Elimination of restriction sites	6
GCaMP1.6[d]	4-fold	0.15 μM	Ohkura et al. [25]	Ca^{2+}-free form with higher fluorescence, chromophore maturation	7
		1.23 μM	Hendel et al. [20]		
GCaMP2	4-fold	0.15 μM	Tallini et al. [26]	Ca^{2+}-free form with higher fluorescence, stability at 37°C	34
	6.4-fold	0.44 μM (0.70 μM at 2 mM Mg^{2+})	Hires et al. [27]		
	4-fold	0.84 μM	Tian et al. [22]		
GCaMP3	12-fold	0.66 μM	Tian et al. [22]	Dynamic range, Ca^{2+}-free form with higher fluorescence	4
Ratiometric pericam	10-fold	1.7 μM	Nagai et al. [15]		31
Inverse pericam	6.6-fold	0.20 μM	Nagai et al. [15]		6
Case 12	12-fold	1.1 μM	Souslova et al. [28]		6
GFP pair (FRET)					
YC2	1.5-fold	0.07 μM (35%)/ 11 μM (65%)	Miyawaki et al. [12]		5
YC2.1	1.9-fold	0.10 μM (70%)/ 4.3 μM (30%)	Miyawaki et al. [29]	pH-sensitivity	18
YC2.12	n.d.	n.d., likely similar to YC2.1	Nagai et al. [30]	Maturation at 37°C	8
YC3.6	5.6-fold	0.25 μM	Nagai et al. [23]	Dynamic range	37
	3.5-fold	0.63 μM	Hendel et al. [20]		
	14.5-fold	0.22 μM (33%)/ 0.78 μM (67%)	Horikawa et al. [31]		
PremoCameleon		(Most probably like YC3.60)	Beacham et al. [32] Hanson et al. [33]		5
D1 (D1ER)	2-fold	0.58 μM (28%)/ 56 μM (72%)	Palmer et al. [34]	Reduced crosstalk with endogenous calmodulin	27
D3cpv	4.1-fold	0.60 μM	Palmer et al. [24]	Reduced interaction with endogenous calmodulin	16
	1.9-fold	0.66 μM	Hendel et al. [20]		

(*continued*)

Table 1 (continued)

fGECI	Signal change[a]	K_{Ca2+}	Original publication	Improvement[b]	Articles[c]
TN-L15	1.4-fold	1.2 μM	Heim et al. [35]		8
	0.8-fold	0.72 μM	Hendel et al. [20]		
TN-XL	4-fold	2.5 μM	Mank et al. [36]	Mg^{2+} insensitive	7
TN-XXL	2.3-fold	0.80 μM	Mank et al. [37]	Mg^{2+} insensitive, more sensitive	6
YC-Nano50	12.5-fold	0.05 μM (82%)/ 0.4 μM (18%)	Horikawa et al. [31]	More sensitive	1
YC-Nano15	14.5-fold	0.016 μM (72%)/ 0.32 μM (28%)	Horikawa et al. [31]	More sensitive	1

[a]All signal changes are increases except for inverse pericam
[b]In case the sensor is an off-spring of another fGECI
[c]Published between January 2008 and August 2010
[d]In the original papers by Nakai et al. [16] and Ohkura et al. [25], the sensors are written with a dash between G and C (G-CaMP, G-CaMP1.3 and G-CaMP1.6). In Tallini et al. [26] – still with Dr. Ohkura and Dr. Nakai as co-authors – the dash is "lost" (although still reminiscent in Table 1 of that article). From here on, both spelling ways can be found in the literature. The option without dash predominates (i.e. GCaMP)

sensor, various K_{Ca2+} values have been reported differing by 300–500% [22, 26, 27]. Even more puzzling though, in the original publication GCaMP2 is reported to be insensitive to Mg^{2+} [26], while Hires et al. found a significant influence of Mg^{2+} [27]. The differences in the maximal signal change (dynamic range) are also large (on the order of 50–200% !) for many fGECIs (see GCaMP2, YC3.6, D3cpv or TN-L15, Table 2). In addition, data obtained from intracellular or in vivo experiments introduce further sources for variations in the experimental results and will be even more imprecise or non-reproducible. For these reasons, it is highly recommended to compare fGECIs in one and the same experimental setup and conditions as near as possible to the system that will be investigated. This is much more conclusive for choosing the best-suited fGECI than to rely on published absolute values determined under different experimental conditions.

4 Utilization of fGECIs from January 2008 to August 2010

The focus of this chapter is, however, not the detailed description of existing fGECIs, but a comprehensive investigation in which research areas the sensors have been used and which improvements in in vivo calcium measurements have been achieved. At the same time, we are going to analyze to what extent fGECIs have entered laboratories pursuing research in physiology and cell biology, and also to show that fGECIs are on the way to become a standard tool for calcium measurements just like the commercially available CIDs currently are.

Table 2 Overview on applications of fGECIs (January 2008 to August 2010)

Application	References	Articles
Photophysics + structure + sensor development	[22, 27, 31, 37–48, 220]	16
C. elegans	[22, 49–75]	28
Drosophila	[17, 20, 22, 37, 76–112]	41
Transgenic mice	[113–130]	18
Mice/rat (virus/electroporation)	[10, 22, 27, 31, 37, 38, 129, 131–145]	22
Zebrafish	[4, 31, 146–157]	14
Plants	[158–173]	17
Culture cells	[22, 45, 59, 88, 135, 140, 144, 145, 152, 174–219, 221–262]	98
Xenopus	[263]	1
Dictyostelium	[31]	1
Yeast	[264, 265]	2
Targets		
Mitochondria	[17, 88, 105, 174, 176, 179–181, 185, 186, 188–191, 199–201, 205–207, 209, 213, 216, 221, 224, 226, 230, 233, 234, 241, 243, 247, 249, 250, 253, 254, 257–262]	42
SR/ER	[139, 145, 175, 177, 181, 182, 188, 192, 194, 197, 202–204, 210–212, 216–219, 224, 226, 229, 232–234, 248, 250, 252, 253, 255, 257, 258, 261]	34
Nucleus	[167, 185, 191, 206, 252]	5
Peroxisome	[187]	1
Golgi	[214, 223]	2
Plasma membrane	[141, 237, 242, 251, 262]	5
Fusion protein	[175] (bestrophin), [185] (FKBP), [148] (synaptophysin), [149] (synaptophysin), [141] (actin), [152] (P2X receptor), [240] (TRP channel), [251] (adenylyl cyclases)	8

For the period from January 2008 to August 2010, we found 236 articles that made use of at least one fGECI. In total, 36 fGECIs have been explored in these articles reviewed in this chapter. The building principles and biophysical properties of fGECIs are covered in three excellent and informative recent reviews (Mank and Griesbeck [266], McCombs and Palmer [267], Garaschuk and Griesbeck [268]). In Table 1, we listed all fGECIs that have been used in more than four articles in the time period from January 2008 to August 2010, together with the two most recently published sensors GCaMP3 [22] and the YC-Nanos [31] for comparison reasons. It is noted here that the group of Dr. J. Nakai has announced two improved GCaMP2 versions (GCaMP4s), one with enhanced fluorescence signal changes upon calcium binding [46] and the other with reduced cross-interaction with calcineurin [44]. In another article, a GCaMP4.1 is mentioned and used [263], but no details about the mutations or type of improvements are given. Also listed in Table 2 are the in vitro calcium-binding constant K_{Ca2+} and the signal change upon calcium binding (dynamic range) for each sensor protein.

There are clearly some "major players" that have been used in more than 25 studies, i.e. GCaMP, GCaMP2, ratiometric pericam, YC3.6 and D1ER. Two of them, D1ER and ratiometric pericam, were established for calcium imaging in the sarco-/endoplasmic reticulum and mitochondria, respectively. They were used in more than 50% of the calcium imaging studies in these two cell compartments. D1ER has a tailored low calcium affinity to detect calcium changes above 100 μM in the endoplasmic reticulum. Ratiometric pericam is well targeted to mitochondria. It has a very high dynamic range and its somewhat lower calcium affinity might be also of advantage. The original GCaMP is – despite some drawbacks including the very low fluorescence of its calcium-free form – very popular for in vivo studies in model organisms such as *Caenorhabditis elegans* and *Drosophila melanogaster*. In many of the surveyed studies, calcium changes are large and cell-specific expression is straightforward in these animals. Therefore – it seems – the urgency of using the best fGECI available is relatively low in these animals. YC3.6 and GCaMP2 – "on the market" for 6 and 4 years, respectively – have obviously proven to work fine for a number of more sophisticated calcium imaging issues. The latest generation of fGECIs (TN-XXL, GCaMP3, YC-Nano, Case12) have some real advantages compared with the older ones. Nevertheless, it is not clear whether they will replace those soon.

fGECIs have been used in several different application fields. From a total of 236 articles, 53% of the studies have been performed in awake or anaesthetized animals or living tissue samples. The remaining 47% of studies – besides a small fraction dealing with photophysical and structural properties of fGECIs – were performed in cell lines. In the upper part of Table 2, publications are listed according to the organisms studied – experiments in flies and mice obviously leading this group. The lower part of the table refers to publications in which fGECIs have been specifically targeted. In 33% of the articles, the fGECI was targeted to a subcellular structure or fused to a protein. Coming as no surprise, the most frequently targeted cellular structures are mitochondria and the endoplasmic reticulum.

4.1 fGECIs in Cell Organelles

Looking back on the first 13 years of research with fGECIs, it appears that the most successful application certainly was to elucidate the calcium homoeostasis in mitochondria and the endoplasmic reticulum (ER). Together with targeted aequorins, fGECIs allowed to specifically measure calcium concentrations in the cytosol, mitochondria and the endoplasmic reticulum manifesting in an accumulation of articles dealing with organelle-targeted fGECIs (Table 2). There are not only original publications but also several very instructive reviews available [269, 270]. Also specific insights into an organelles calcium homoeostasis and the role of calcium in apoptosis [271, 272] or in heart physiology [273] were provided. While most of the targeting signals place an fGECI into the lumen of the organelle, in some recent studies, fGECIs were directed to the outer membrane of the ER [175] or mitochondria [191]. The latter study could nicely demonstrate the existence of microdomains with high local calcium concentrations near the outer mitochondrial membrane after agonist-evoked calcium release from the ER (see Fig. 2). Such microdomains [274] have been proposed since 15 years. The microdomains were only faintly recognizable in the original raw data

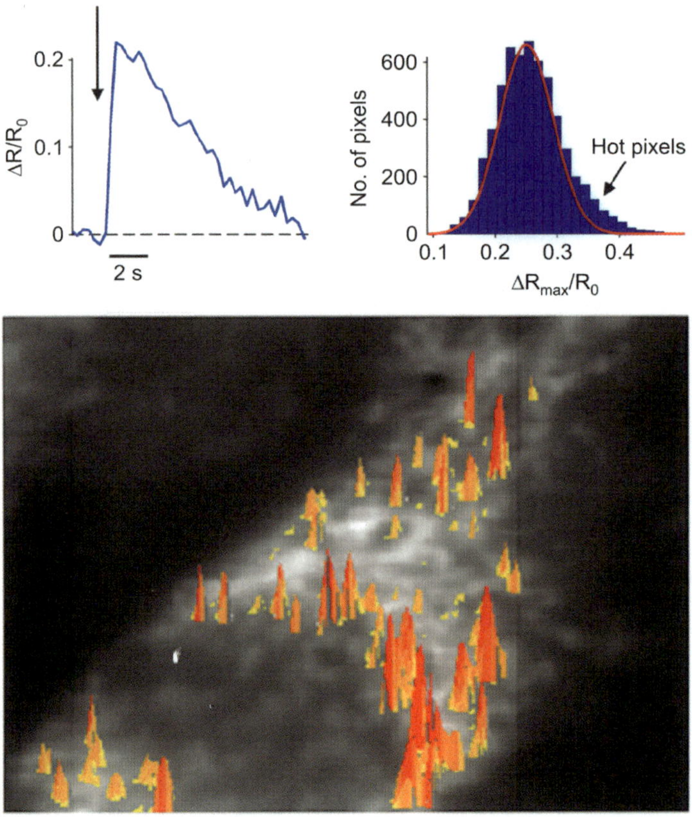

Fig. 2 Microdomains of high calcium elevations near mitochondria in HeLa cells after stimulation with histamine. Changes of the calcium concentration were measured with a cameleon-like fGECI (D1cpv [34]) targeted to the outer mitochondrial membrane. The microdomains can be interpreted as ER–mitochondria connections, adapted from [191]

and needed sophisticated data evaluation and a number of control experiments to convincingly visualize the places of high local calcium changes on the surface of the mitochondria. Giacomello and co-workers not only identified the microdomains but also provided functional and morphological information. Mitochondria are excluded from the place of capacitative calcium entry (CCE), but they are involved in calcium uptake after voltage-operated calcium entry [275]. Very recently, the possibility of precise and exclusive localization of cameleon-like fGECIs into the lumen of mitochondria and ER as well as the cytosol has helped to identify and characterize a first member (MICU1) of a long-searched protein family of mitochondrial calcium uniporters [233]. Upon downregulation of MICU1 by RNA silencing, the mitochondrial calcium concentration was reduced, while that of the cytosol and ER was not affected.

Recent studies on calcium changes in the Golgi apparatus [214, 223] broadened the perspective of future fGECI applications in single cell studies: It may be soon possible to elucidate the interplay between different cell organelles during cellular signalling.

4.2 fGECIs: Incorporation into the Genome of Mammals

Several groups have used fGECI-encoding genes to generate transgenic animals that express the sensors in pre-defined cell types and allow non-invasive imaging of "freely behaving" animals. Dr. Tsiens group mentioned this potential strategy already in their original publication on cameleons in 1997: "Cameleons should be well suited to genetically tractable organisms such as bacteria, yeast, nematodes, flies, plants and transgenic mice" [12]. Many years later, however, they (and others) "encountered problems in using CaM-based genetic sensors" [23] in mammals mainly due to much lower in situ/in vivo signals compared to the in vitro results. This statement from 2004 is still valid. While fGECIs are commonly used in some genetically manipulatable organisms such as worms and flies as well as in cell culture cells, only a few transgenic mouse lines expressing fGECIs have been actively exploited so far for physiological experiments.

In "conventionally" generated transgenic animals, the majority of cells express the sensor protein. On the other hand, it is also possible to limit the expression to a certain tissue or cell type to minimize unwanted side effects. A restricted expression pattern depends on the insertion site of the transgene and its connection to gene regulatory elements like promoters or ribosomal entry sites. An appropriate promoter would be, e.g. connexin40 if intracellular calcium changes in endothelial cells have to be investigated as shown by the group of Kotlikoff [276]. They monitored calcium waves with GCaMP2 in arterioles of transgenic mice in vivo.

Generally, it is laborious to generate transgenic, homozygous mouse lines. Facilities that already offer established mouse lines (e.g. JAX® Mice Database of the Jackson Laboratory in Maine, USA) or founder lines that can be crossed to achieve cell type specific expression via the CreloxP system may improve this situation in the future. In the last years, other fast-paced developing methods emerged, which allow a rapid onset of transgene expression. These methods have become convenient either to induce spatially and temporally controlled transgene expression or to limit expression to adult or developing animals. This development is also manifested in the statistics of fGECI use in living mice and mice tissues (see Table 2). Only 45% of all studies with fGECIs in mice have been performed with transgenic, homozygous animals expressing the fGECI in a certain cell type. The majority of the investigations (55%) work with mice, where the fGECI has been introduced in a different way. At this point, we are going to highlight two of these new methods: electroporation and viral-mediated gene transfer.

DNA is a negatively charged molecule and therefore moves in an electric field towards the anode. If an electric field is applied to a living cell, DNA can be dragged into the cell by a process referred to as electroporation [277, 278]. The transfer efficiency might reach up to 80%. This makes electroporation a versatile method to deliver genetic material into bacteria or cells in culture. Due to variations of electrode design, it is also possible to electroporate slice preparations of tissue and even tissue in a nearly natural environment. Another sophisticated method is electroporation in utero. This approach allows introducing DNA to mice pups starting at embryonic day 12. To drive transgene expression, a CAG promoter, a combination of cytomegalovirus (CMV) early enhancer element and chicken beta actin promoter, is often used. It is active during early development. The injected embryos continue their development normally. The tissue can be imaged either a few days after electroporation or post-natally.

Tian and colleagues introduced GCaMP2 and GCaMP3 sensors into mice via in utero electroporation. They compared the sensor properties in neurons under the same experimental conditions. GCaMP3 could resolve calcium increases evoked by single APs up to a frequency of 6 Hz, whereas GCaMP2 was less sensitive [22]. Figure 3 shows the normalized fluorescence change vs. the number of APs used for stimulation.

Another outstanding study not only introduced an improved version of the troponin- and FRET-based fGECI family (TN-L15 [35, 279], TN-XL [36] and here TN-XXL [37]) but also demonstrated the power of fGECIs for functional chronic imaging of neuronal activity. After in utero electroporation, the TN-XXL sensor was expressed specifically in the upper layer of the visual cortex. By anatomical means, pyramidal neurons were selected for analysis. Visual stimulation was performed on anaesthetized mice and the calcium level of the neurons was measured (see Fig. 4). Individual neurons displayed specific patterns of activity or, depending on the stimulus given, were silenced. Photobleaching was negligible during measurements. Most notably, identified neurons could be investigated for weeks. Over this time period, they showed a rather similar behaviour, although some changes could be detected as well.

Chronic calcium imaging has been performed in a number of studies ranging from a few up to 120 days [22, 37, 131, 137, 225]. The possibility of chronic imaging with fGECIs may assist to answer new scientific questions ranging from learning to cell differentiation and cell division.

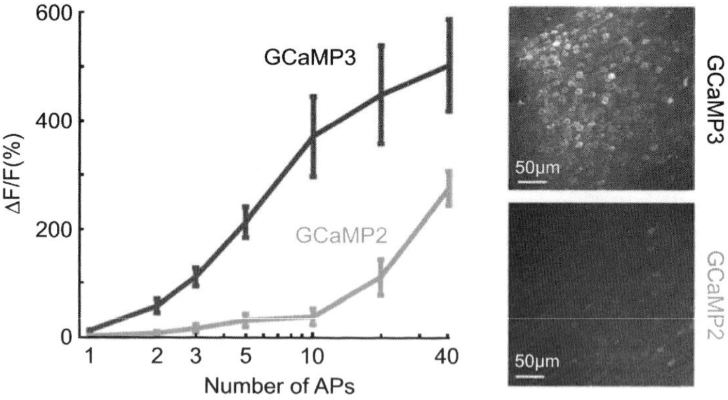

Fig. 3 Comparison of GCaMP2 and GCaMP3 in layer 2/3 cortical neurons (*right*) and hippocampal pyramidal neurons (*left*) of in utero electroporated mice, adapted from [22]. GCaMP3 shows a fourfold higher fluorescence intensity (*right*) and a significantly improved detection of action potentials compared with GCaMP2 (*left*)

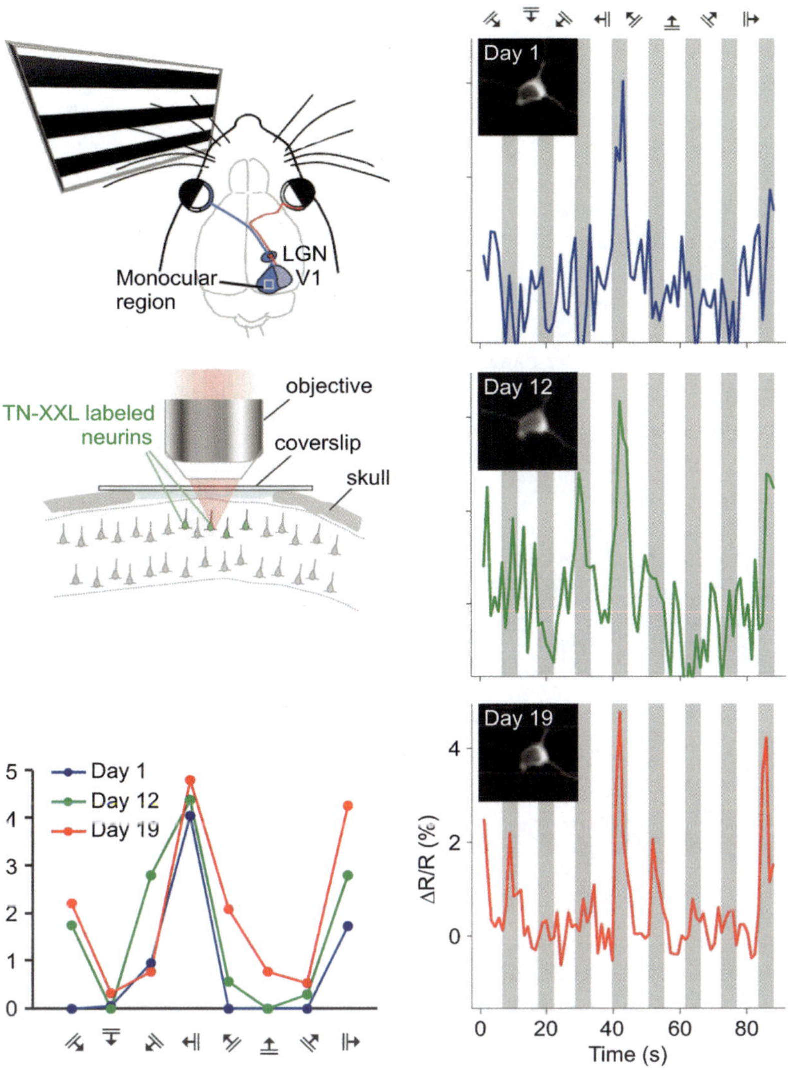

Fig. 4 *Left*: visual stimuli are presented to mice expressing TN-XXL in visual cortex neurons (*top*). Scheme of two-photon fluorescence microscopy performed through a cranial window (*middle*). *Right*: changes of TN-XXL FRET ratios in a single pyramidal neuron upon stimulation with stripe bar presentations in varying orientation. Measurements were performed on days 1, 12 and 19. *Left* (*bottom*): summary of the three measurement days, adapted from [37]

Another alternative to generate transgenic mice is to introduce foreign DNA by viral-mediated gene transfer. This strategy allows a fast onset of transgene expression paired with high transduction efficiency. The generation of virus-based expression constructs is very flexible. The assembly of recombinant viruses often relies on the expression of gene products from different plasmids, from which only one plasmid contains the transgene expression cassette. This cassette can be exchanged on demand thus, allowing to generate viruses containing different promoters, sensors (here fGECIs) or sensor mutants.

A variety of viruses can be used to deliver a transgene to certain cells or tissues. The decision on which viral system should be used depends on a variety of factors like the size of DNA to be transferred, the cell type that should be transduced or for how long the sensor should be expressed. For further reading and a comprehensive overview on viral-mediated transgene expression, see for instance the review of Davidson and Breakfield [280]. Lentivirus-, adenovirus- and adeno-associated virus-based approaches are the most convenient virus-based expression systems to date. For some standard applications, they are even commercially available. Adeno-associated viruses (AAVs) offer a variety of possible cell tropisms based on their capsid composition; i.e. they can be tailored such that they selectively infect and transduce specifically certain cell types.

Lütcke and colleagues expressed the ratiometric sensor YC3.6 driven by the human synapsin promoter in neocortical cells of mice with recombinant AAVs [10]. They stereotactically injected 6–8-week-old mice and detected the neuron-specific YC3.6 fluorescence after 3 weeks. More than 80% of the neurons were transduced. The successful measurement of intracellular calcium changes in vivo, i.e. in neurons of freely moving animals, e.g. during behavioural experiments, is a challenging task (Fig. 5). The authors performed slice and in vivo imaging via one- and two-photon fluorescence microscopy, respectivily. They simultaneously measured – in some cases – local field potentials or recorded changes of the membrane voltage in the cell-attached mode. They also analysed normalized fluorescence changes of the acceptor fluorescence of YC3.6 in relation to the animals' activity in a cage considering different motion events (walking, rest, wall touching). By doing so, they opened the way for future studies, where calcium changes in neurons and behaviour (here motion) can be correlated.

At present, however, only the relative change of the acceptor fluorescence can be monitored. Fibre-optic recordings in particular will allow a multitude of opportunities for functional studies on the neural basis of animal behaviour, from dendrites up to the level of local and large-scale neuronal populations.

4.3 fGECIs in Worms

As discussed above, the use of fGECIs in mammals is not yet a standard technique. The situation is completely different for the nematode *C. elegans*. Its less than 100 developmentally and functionally well-defined neurons make it a textbook example for the use of fGECIs. Many key experiments with fGECIs have been performed already years ago. The observed calcium signals were robust and often large, which might be the reason, why more than 50% of the studies were using "old" fGECIs that have been published before 2002. With both, single GFP or FRET-based fGECIs, a great variety of phenomena was investigated since January 2008: regenerative

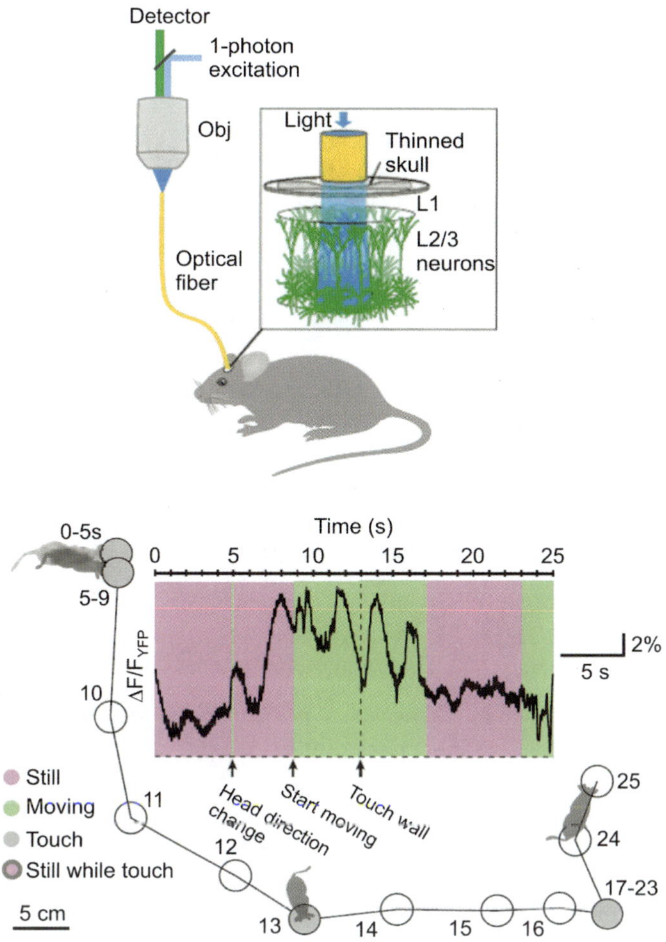

Fig. 5 *Top*: schematic picture of the experimental setup. *Bottom*: motion analysis of a freely moving mouse in a cage with simultaneous detection of neuronal activity (layer 2/3 neocortical neurons in the mouse brain). Plotted is not the YC3.6 ratio change but the normalized fluorescence change of the acceptor of YC3.6 (circular permuted Venus, an YFP variant), adapted from [10]

growth, behaviour in temperature gradients, mating activity, swimming, pheromone or chemoattraction, odour reception, molecular oxygen concentration, locomotion and egg-laying.

So far, imaging experiments were mostly conducted either in head-fixed or in agar-based environments to immobilize the animals. Ben Arous and colleagues generated *C. elegans* strains expressing D3cpv and YC3.6 in different neurons. One strain expresses the sensor in a subset of neurons (AVA neurons), which regulate the backward motion [51]. They performed bright field imaging, while worms moved in a natural environment and with normal speed. Signals with up to 30% FRET change in AVA neurons could be directly correlated to the reversal of motion (see Fig. 6).

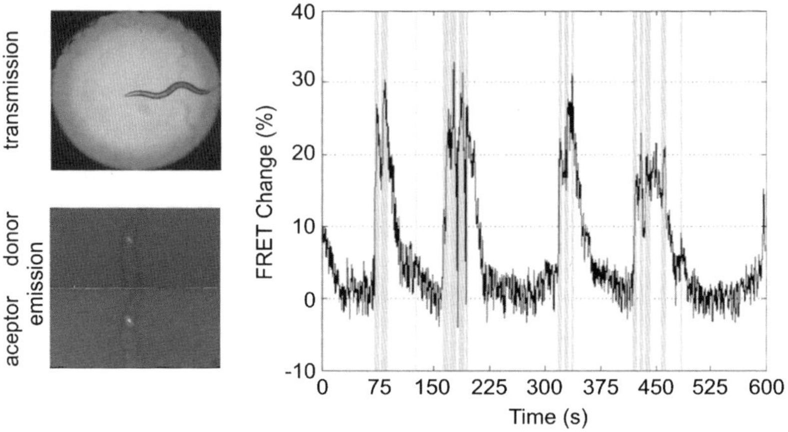

Fig. 6 Motion analysis of freely moving *Caenorhabditis elegans* with simultaneous detection of calcium changes in AVA neurons (*left top*: transmission image; *left bottom*: two-colour fluorescence image). The fGECI D3cpv is specifically expressed in AVA neurons. Robust calcium increases were measured when animals crawled backward (*grey lines*), adapted from [51]

4.4 fGECIs in Flies

Another well-investigated model organism is the fly *D. melanogaster*. More than 40 studies with fGECIs in flies have been published since 2008. As for *C. elegans*, a large variety of physiological research topics have been investigated: e.g. odour and visual perception, stress response, air flow, walking, gravity, hearing, temperature sensing and the neuron-motor junction.

A recent application was the determination of a fly's behaviour simultaneous with calcium imaging in those neurons that are responsible for a sensory signal. Seelig and colleagues analysed flies expressing GCaMP3 in brain neurons that are responsible for a compensatory turning response upon visual stimulation [100]. A single dendrite was imaged, while the flies were able to move freely on a treadmill ball. The normalized fluorescence showed about 200% intensity change when a preferred stimulus was given. The observed fluorescence signal correlated strongly with the behavioural response of the fly to motion stimuli (see Fig. 7).

One crucial point is the reduction of motion artefacts. Seelig and colleagues did so by investigating the awake flies in a fixed, open head configuration through removal of brain-moving muscles [100]. A multitude of control experiments had to be performed to show that the behavioural responses are not affected significantly.

Unwanted emerging motion artefacts that interfere with calcium measurements can be omitted when ratiometric sensors such as YC3.6, D3cpv, ratiometric pericam or TN-XXL are used. When GCaMPs, Case12, Camgaroo or inverse pericam are used, quasi-ratiometric measurement can be performed by applying additionally a fluorescent dye with a different colour that is insensitive for calcium changes.

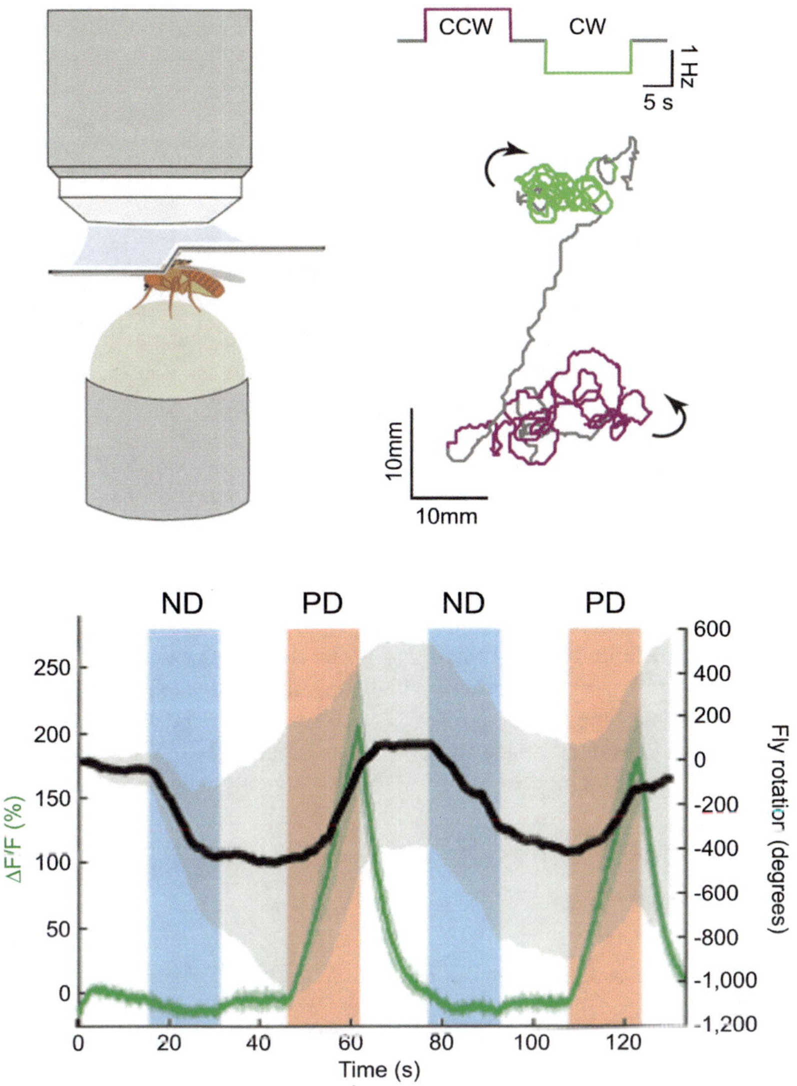

Fig. 7 *Top*: apparatus (*left*) for simultaneous measurement of tethered walking and calcium response in horizontal system (HS) neurons of alive, head-fixed flies and motion pattern (*right*) obtained for a fly in a sequence of visual pattern stimulation for anticlockwise (CCW) and clockwise rotation (CW) and *z*-projection of two-photon fluorescence image of HS neurons expressing the fGECI GCaMP3. *Bottom*: average time course of behavioural response (fly's turning response) and calcium elevations in HS neurons during null-direction (ND) and preferred-direction (PD) response, adapted from [100]

These examples of the use of fGECIs in different model organisms (mice, flies, worms) show that fGECIs can be faithfully applied to determine cellular calcium changes in freely behaving animals. In future, it will be interesting to compare results obtained from freely behaving animals with data from in vitro experiments, the latter being the vast majority of the physiological experiments and therefore the basis of the existing knowledge.

4.5 *fGECIs in Zebrafish*

The zebrafish is an animal model that has a lot of promises. Unlike the invertebrate model organisms (worms, flies), the zebrafish is – being a vertebrate – a much closer relative to higher vertebrates by morphological and physiological means. The generation of transgenic lines is, however, much more easy to compare with higher vertebrates like rodents. Zebrafishes – in particular the embryo – are also much more suitable for full animal or organ imaging. It comes therefore as a surprise that the zebrafish is the one with the lowest number of publications with fGECI applications (see Table 2) among the popular animal species used for studies with genetically changed animals. In addition, more than one-third of the studies uses transgenic zebrafish lines with fGECI for other reasons than calcium imaging (e.g. as cell marker). About 30% of all studies are based on a zebrafish line that expresses GCaMP in cardiac-specific manner allowing impressive organ-specific calcium imaging in the living, functional heart (see Fig. 1). There are only a few studies that utilize fGECIs expressed in neurons for functional calcium imaging. One reason might be, that the use of CIDs with their better imaging properties compared to the fGECIs and their tailored calcium sensitivities is advantageous in zebrafish, although the CID distribution is unspecific (see Fig. 1). Unlike mice, zebra fish embryos can be loaded sufficiently with CIDs.

4.6 *fGECIs in Plants*

Plants can also be genetically modified rather easily. Therefore plants were supposed to be one of the major application fields for fGECIs [12]. A surprisingly low number (17) of applications in plants, however, have been published from 2008 to 2010. One obvious problem is that plant cells – compared to animal cells – contain a large variety of chromophores absorbing and emitting in the whole visible spectrum. Therefore, autofluorescence is a severe problem in plants. This favours bioluminescent aequorins over fGECIs, especially where high spatial resolution is not an issue. Probably for the same reason (i.e. strong autofluorescence), only cameleon-like fGECIs have been used in plants (with one exception [173]). Most studies (70%) have been performed with *Arabidopsis thaliana*, where calcium elevations after external stimulation with ATP or mechanical treatments have been studied. Other nice examples include approaches, where intracellular calcium changes have been measured simultaneously with other cell parameters such as intracellular pH, stomatal closure or cell growth rate. Monshausen et al. investigated in parallel the intracellular calcium concentration and cell growth [164] in

A. thaliana root hair cells. Cell growth occurs in a periodic manner with growth periods between 15 and 30 s. A calcium gradient exists showing a maximum near the root hair tip. The calcium concentration oscillates almost in phase with cellular growth (see Fig. 8). Yet, cellular growth is preceding the highest calcium concentration by about 5 s. The peaks in calcium concentrations are assumed to either prime the root hair cell for the next growth surge or to regulate processes that stabilize the cell wall after a growth period.

Nevertheless, the application of fGECIs in plants is far behind the original expectations. Approaches to target fGECIs to different cell types or subcellular compartments are rarely used compared with the technical possibilities. Nevertheless, the use of fGECIs in plants may intensify in the future for two reasons:

1. Modern fluorescence microscopy techniques, e.g. spectral unmixing, may allow to observe the signature of fGECIs even in pigment-rich plant cells.
2. The study of calcium homoeostasis in different plant cell types may become as important as in animals to clarify the intracellular and intercellular signalling processes.

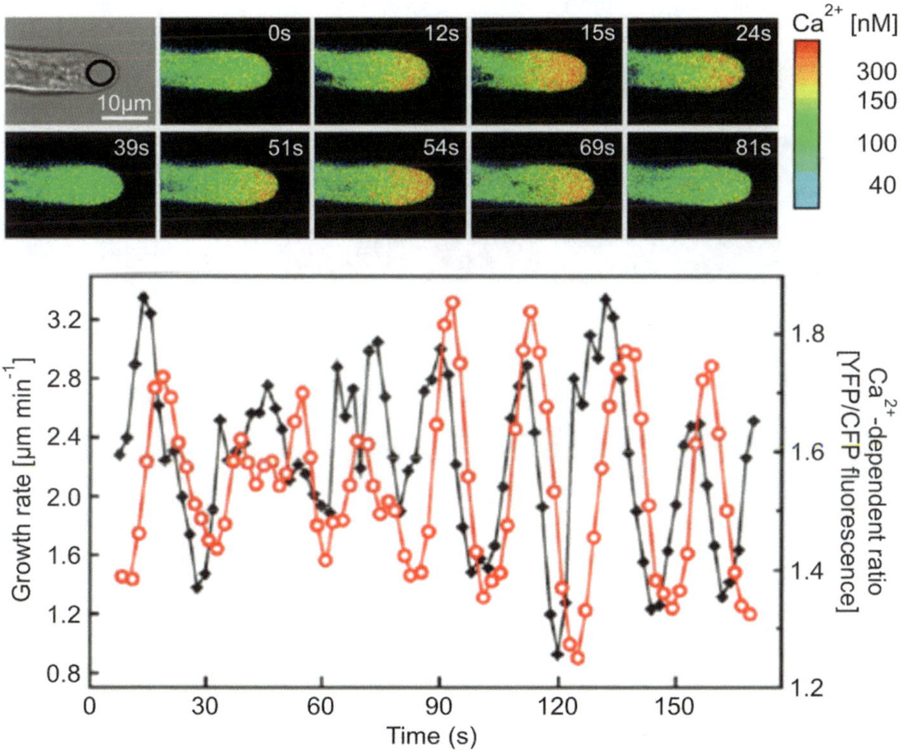

Fig. 8 Arabidopsis root hairs show an oscillating, tip-located calcium gradient that peaks after maximal growth. *Top*: root hairs of Arabidopsis undergoing tip growth expressing the cytosolic fGECI YC3.6. *Bottom*: quantitative analysis of root hair growth rates (*black*, *full circles*) and cytosolic calcium levels (*red*, *open circles*) at the root hair apex, adapted from [164]

4.7 fGECIs Fused to Proteins

Until now, the application of fGECIs targeted to cell organelles or specific cell types in different organisms has been reviewed. Another way to place the "calcium spy" inside the cell is to fuse an fGECI directly to a key-player protein of the process under investigation. A nice example can be found in a recent publication on the calcium regulation of adenylyl cyclase isoforms fused to GCaMP [251].

Adenylyl cyclases (ACs) are enzymes that synthesize cAMP from ATP. From a total of ten isoforms, nine are integral membrane proteins and react to intracellular calcium changes. AC1 and AC8 are stimulated, whereas AC5 and AC6 are inhibited by calcium. Willoughby et al. expressed GCaMP2 in the cytosol. Furthermore, they generated modified GCaMP2 sensors. The N-terminal end of GCaMP2 was extended with consensus motifs for myristoylation and palmytoylation to allow membrane association of the sensor to AC2 and AC8, respectively (Fig. 9). The measurements were performed in HEK293 cells. All modified proteins were

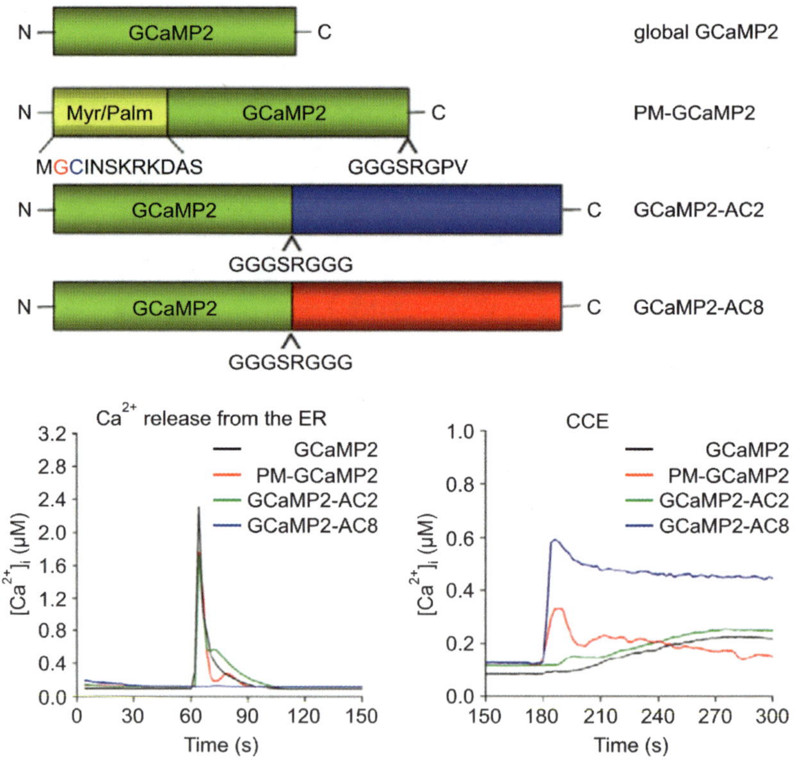

Fig. 9 *Top*: GCaMP2 and three differently modified GCaMP2 versions that all three show strong localization in the plasma membrane. *Bottom*: different local calcium changes detected by the four GCaMP2 sensors for two important routes of intracellular calcium elevation – calcium release from the endoplasmatic reticulum (ER) and the capacitative calcium entry (CCE), adapted from [251]

positively checked for functionality, resulting in four very similar fGECIs that differed only by their local distribution. The three "extended" GCaMP2 versions were all localized on the plasma membrane.

Transient intracellular calcium elevations were induced in two different ways: Ins(1,4,5)P_3-mediated calcium release from the endoplasmic reticulum and CCE through calcium channels from the extracellular side. Both scenarios were measured with the four GCaMP2 proteins. The GCaMP2-AC2 fusion protein detected signals very similar to those measured with the cytosolic GCaMP2 construct. On the contrary, the GCaMP-AC8 fusion protein responded much stronger to CCE, while calcium release from the ER was barely visible. This could be explained by AC8 residing in a special microdomain that promotes for calcium signals arising from CCE. The calcium-insensitive AC2 may also be localized in special microdomains, but the calcium changes were indistinguishable compared with those in the cytosol. These results clearly demonstrate that AC2 and AC8 are located in different microdomains. The plasma membrane-targeted GCaMP2 is facing many different microdomains and shows therefore an intermediate behaviour. Willoughby and colleagues demonstrate in their study directly discrete compartmentalized calcium signals associated with specific signalling proteins. The fusion of fGECI to proteins specific for a certain process is a very promising approach, however, surprisingly rarely realized (only eight publications in total). Especially ion transporters and ion channels seem to be good candidates as fusion partners with fGECIs. (see also [152, 175, 240]).

5 Outlook

The fGECIs that we discussed in this chapter have absorption and emission maxima between 420–520 nm and 470–540 nm, respectively. In general cyan, green and yellow fluorescent proteins are incorporated in these fusion proteins.

Observing the calcium concentration in more than one organelle simultaneously in a single cell asks, however, for fGECIs with an expanded range of excitation and emission wavelengths. Not much progress to solve this issue has been achieved until now, although the first cameleon publications actually used blue/green GFP-FRET pairs [12, 14]. Furthermore, soon after the cloning of the first orange/red-emitting fluorescent proteins (e.g. DsRed [281]), cameleons with orange/red-emitting FRET acceptors have been published [282]. Sapphire/DsRed, which showed so far the best performance is a FRET sensor based on the yellow cameleon YC2.1 [29]. But since then, no real improvement and no successful application of such spectrally distinct fGECIs have been achieved, although recently found new fluorescent proteins extended the spectral range of FPs from ultramarine [245] to the far red [283, 284]. Several laboratories – including those of Dr. Miyawaki, Dr. Griesbeck and Dr. Campbell – tried to generate fGECIs with more red-shifted excitation and emission, but the new fGECIs performed worse (small dynamic range!) than the established "mother fGECIs" from the cyan-yellow spectral range (H. Mizuno, personal communication

2010, [285]). Different dimerisation tendencies as well as different geometries of transition dipole moments of the different fluorescent proteins are partly responsible for the lower FRET efficiencies and FRET changes in these constructs using non-Aequorea fluorescent proteins. Circular permuted green and yellow GFP mutants were the basis for the GCaMP [16] and the pericam family [15] of fGECIs, as well as for major improvements of the cameleons [23] and the troponin-based sensors [36]. The low performance of red-shifted fGECIs can be partly attributed to the bad fluorescent properties of the circular permuted variants of orange/red fluorescent proteins. Very recently, promising variants have been proposed [286, 287]. Nonetheless, just one group has used a "red cameleon" (Sapphire–DsRed) since 2008. The purpose was to demonstrate the possibility to observe the time course of two physiological parameters in one cells using FPs [245]. The laboratory of Dr. French has used YC3.6 for calcium imaging and an orange/red FRET pair to monitor the activation of a GTPase [193] together with intracellular calcium changes. In both studies, sophisticated imaging methods were used like spectral unmixing [245] and combination of ratiometric and fluorescence lifetime imaging [193]. Although, in future, the combination of specifically subcellular targeted, more sensitive, faster responding and spectrally different fGECIs as well as other genetically encoded sensors with the application of the fast increasing arsenal of modern imaging methods will certainly promote the usage of fGECIs to new levels. This chapter, reviewing the applications of fGECIs since 2008, shows an impressive diversity of physiologically relevant questions in biology that have been tackled with the help of fGECIs. This collection shall motivate and give some ideas to life science researchers from various fields to consider, start or continue using fGECIs in their experiments.

Acknowledgements The authors would like to thank the following members of the Institute: Dr. A. Baumann, Dr. J. Hendriks and A. Meisenberg for careful proofreading, and H.D. Grammig for helping in generating the figures. In addition, we would like to thank Dr. S. Dangwal (Institute of Pharmacology and Clinical Pharmacology, University Hospital, Heinrich Heine University Düsseldorf, Germany), A. Geiger (Max Planck Institute of Neurobiology, Munich, Germany), Dr. M. Zeidler and Dr. J. Hughes (Institute of Plant Physiology, Justus Liebig University Giessen, Germany) for helpful discussions.

References

1. Shemarova IV, Nesterov VP (2005) Evolution of mechanisms of Ca2+ -signaling: role of calcium ions in signal transduction in prokaryotes. J Evol Biochem Physiol 41(1):12–19
2. Jaiswal JK (2001) Calcium – how and why? J Biosci 26(3):357–363
3. Paredes RM, Etzler JC, Watts LT, Zheng W, Lechleiter JD (2008) Chemical calcium indicators. Methods 46(3):143–151
4. Chi NC, Shaw RM, Jungblut B, Huisken J, Ferrer T, Arnaout R, Scott I, Beis D, Xiao T, Baier H, Jan LY, Tristani-Firouzi M, Stainier DYR (2008) Genetic and physiologic dissection of the vertebrate cardiac conduction system. PLoS Biol 6(5):1006–1019
5. Brini M (2008) Calcium-sensitive photoproteins. Methods 46(3):160–166
6. Shimomura O (2005) The discovery of aequorin and green fluorescent protein. J Microsc Oxf 217:3–15

7. Montero M, Alonso MT, Carnicero E, Cuchillo-Ibanez I, Albillos A, Garcia AG, Garcia-Sancho J, Alvarez J (2000) Chromaffin-cell stimulation triggers fast millimolar mitochondrial Ca2+ transients that modulate secretion. Nat Cell Biol 2(2):57–61
8. Pozzan T, Rudolf R (2009) Measurements of mitochondrial calcium in vivo. Biochim Biophys Acta Bioenerg 1787(11):1317–1323
9. Drobac E, Tricoire L, Chaffotte AF, Guiot E, Lambolez B (2010) Calcium imaging in single neurons from brain slices using bioluminescent reporters. J Neurosci Res 88 (4):695–711
10. Lütcke H, Murayama M, Hahn T, Margolis DJ, Astori S, Meyer zum Alten Borgloh S, Goebel W, Yang Y, Tang W, Kuegler S, Sprengel R, Nagai T, Miyawaki A, Larkum ME, Helmchen F, Hasan MT (2010) Optical recording of neuronal activity with a genetically encoded calcium indicator in anestetized and freely moving mice. Front Neural Circuits 4 (article 9):1–12
11. Tsien RY (2009) Constructing and exploiting the fluorescent protein paintbox (Nobel lecture). Angew Chem Int Ed 48(31):5612–5626
12. Miyawaki A, Llopis J, Heim R, McCaffery JM, Adams JA, Ikura M, Tsien RY (1997) Fluorescent indicators for Ca2+ based on green fluorescent proteins and calmodulin. Nature 388(6645):882–887
13. Romoser VA, Hinkle PM, Persechini A (1997) Detection in living cells of Ca2+ -dependent changes in the fluorescence emission of an indicator composed of two green fluorescent protein variants linked by a calmodulin-binding sequence – a new class of fluorescent indicators. J Biol Chem 272(20):13270–13274
14. Persechini A, Lynch JA, Romoser VA (1997) Novel fluorescent indicator proteins for monitoring free intracellular Ca2+. Cell Calcium 22(3):209–216
15. Nagai T, Sawano A, Park ES, Miyawaki A (2001) Circularly permuted green fluorescent proteins engineered to sense Ca2+. Proc Natl Acad Sci USA 98(6):3197–3202
16. Nakai J, Ohkura M, Imoto K (2001) A high signal-to-noise Ca2+ probe composed of a single green fluorescent protein. Nat Biotechnol 19(2):137–141
17. Chouhan AK, Zhang JH, Zinsmaier KE, Macleod GT (2010) Presynaptic mitochondria in functionally different motor neurons exhibit similar affinities for Ca2+ but exert little influence as Ca2+ buffers at nerve firing rates in situ. J Neurosci 30(5):1869–1881
18. Filippin L, Abad MC, Gastaldello S, Magalhaes PJ, Sandona D, Pozzan T (2005) Improved strategies for the delivery of GFP-based Ca2+ sensors into the mitochondrial matrix. Cell Calcium 37(2):129–136
19. Hasan MT, Friedrich RW, Euler T, Larkum ME, Giese G, Both M, Duebel J, Waters J, Bujard H, Griesbeck O, Tsien RY, Nagai T, Miyawaki A, Denk W (2004) Functional fluorescent Ca2+ indicator proteins in transgenic mice under TET control. PLoS Biol 2(6):763–775
20. Hendel T, Mank M, Schnell B, Griesbeck O, Borst A, Reiff DF (2008) Fluorescence changes of genetic calcium indicators and OGB-1 correlated with neural activity and calcium in vivo and in vitro. J Neurosci 28(29):7399–7411
21. Reiff DF, Ihring A, Guerrero G, Isacoff EY, Joesch M, Nakai J, Borst A (2005) In vivo performance of genetically encoded indicators of neural activity in flies. J Neurosci 25 (19):4766–4778
22. Tian L, Hires SA, Mao T, Huber D, Chiappe ME, Chalasani SH, Petreanu L, Akerboom J, McKinney SA, Schreiter ER, Bargmann CI, Jayaraman V, Svoboda K, Looger LL (2009) Imaging neural activity in worms, flies and mice with improved GCaMP calcium indicators. Nat Methods 6(12):875–881
23. Nagai T, Yamada S, Tominaga T, Ichikawa M, Miyawaki A (2004) Expanded dynamic range of fluorescent indicators for Ca2+ by circularly permuted yellow fluorescent proteins. Proc Natl Acad Sci USA 101(29):10554–10559
24. Palmer AE, Giacomello M, Kortemme T, Hires SA, Lev-Ram V, Baker D, Tsien RY (2006) Ca2+ indicators based on computationally redesigned calmodulin-peptide pairs. Chem Biol 13(5):521–530

25. Ohkura M, Matsuzaki M, Kasai H, Imoto K, Nakai J (2005) Genetically encoded bright Ca-2+ probe applicable for dynamic Ca-2+ imaging of dendritic spines. Anal Chem 77 (18):5861–5869
26. Tallini YN, Ohkura M, Choi BR, Ji GJ, Imoto K, Doran R, Lee J, Plan P, Wilson J, Xin HB, Sanbe A, Gulick J, Mathai J, Robbins J, Salama G, Nakai J, Kotlikoff MI (2006) Imaging cellular signals in the heart in vivo: cardiac expression of the high-signal Ca2+ indicator GCaMP2. Proc Natl Acad Sci USA 103(12):4753–4758
27. Hires SA, Tian L, Looger LL (2008) Reporting neural activity with genetically encoded calcium indicators. Brain Cell Biol 36(1–4):69–86
28. Souslova EA, Belousov VV, Lock JG, Stromblad S, Kasparov S, Bolshakov AP, Pinelis VG, Labas YA, Lukyanov S, Mayr LM, Chudakov DM (2007) Single fluorescent protein-based Ca2+ sensors with increased dynamic range. BMC Biotechnol 7 (article 37):1–10
29. Miyawaki A, Griesbeck O, Heim R, Tsien RY (1999) Dynamic and quantitative Ca2+ measurements using improved cameleons. Proc Natl Acad Sci USA 96(5):2135–2140
30. Nagai T, Ibata K, Park ES, Kubota M, Mikoshiba K, Miyawaki A (2002) A variant of yellow fluorescent protein with fast and efficient maturation for cell-biological applications. Nat Biotechnol 20(1):87–90
31. Horikawa K, Yamada Y, Matsuda T, Kobayashi K, Hashimoto M, Matsu-ura T, Miyawaki A, Michikawa T, Mikoshiba K, Nagai T (2010) Spontaneous network activity visualized by ultrasensitive Ca2+ indicators, yellow cameleon-nano. Nat Methods 7(9):729–732
32. Beacham D, Hanson G, Batchelor R, Dzubay J, Clements I, Thompson D, Bazar A, Rutten M, Gregory K, O'Grady M, Premo TM (2006) Biosensors and organelle lights: content and delivery for cellular analysis. Invitrogen. http:tools.invitrogen.com/content/sfs/brochures/625.pdf
33. Hanson GT, Cassutt KJ, O'Grady M (2007) H1 histamine GPCR screen using PremoTM Cameleon, a fluorescent protein based calcium indicator delivered with BacMam technology. Invitrogen. http://tools.invitrogen.com/content/sfs/brochures/626.pdf
34. Palmer AE, Jin C, Reed JC, Tsien RY (2004) Bcl-2-mediated alterations in endoplasmic reticulum Ca2+ analyzed with an improved genetically encoded fluorescent sensor. Proc Natl Acad Sci USA 101(50):17404–17409
35. Heim N, Griesbeck O (2004) Genetically encoded indicators of cellular calcium dynamics based on troponin C and green fluorescent protein. J Biol Chem 279(14):14280–14286
36. Mank M, Reiff DF, Heim N, Friedrich MW, Borst A, Griesbeck O (2006) A FRET-based calcium biosensor with fast signal kinetics and high fluorescence change. Biophys J 90 (5):1790–1796
37. Mank M, Santos AF, Direnberger S, Mrsic-Flogel TD, Hofer SB, Stein V, Hendel T, Reiff DF, Levelt C, Borst A, Bonhoeffer T, Hubener M, Griesbeck O (2008) A genetically encoded calcium indicator for chronic in vivo two-photon imaging. Nat Methods 5 (9):805–811
38. Akerboom J, Rivera JDV, Guilbe MMR, Malave ECA, Hernandez HH, Tian L, Hires SA, Marvin JS, Looger LL, Schreiter ER (2009) Crystal structures of the GCaMP calcium sensor reveal the mechanism of fluorescence signal change and Aid rational design. J Biol Chem 284(10):6455–6464
39. Borst JW, Laptenok SP, Westphal AH, Kuhnemuth R, Hornen H, Visser NV, Kalinin S, Aker J, van Hoek A, Seidel CAM, Visser AJWG (2008) Structural changes of yellow cameleon domains observed by quantitative FRET analysis and polarized fluorescence correlation spectroscopy. Biophys J 95(11):5399–5411
40. Guilbe MMR, Malave ECA, Akerboom J, Marvin JS, Looger LL, Schreiter ER (2008) Crystallization and preliminary X-ray characterization of the genetically encoded fluorescent calcium indicator protein GCaMP2. Acta Crystallogr Sect F Struct Biol Cryst Commun 64:629–631
41. Holder AN, Ellis AL, Zou J, Chen N, Yang JJ (2009) Facilitating chromophore formation of engineered Ca2+ binding green fluorescent proteins. Arch Biochem Biophys 486 (1):27–34

42. Kotera I, Iwasaki T, Imamura H, Noji H, Nagai T (2010) Reversible dimerization of *Aequorea victoria* fluorescent proteins increases the dynamic range of FRET-based indicators. ACS Chem Biol 5(2):215–222
43. Leder L, Stark W, Freuler F, Marsh M, Meyerhoffer M, Stettler T, Mayr L, Britanova OV, Strukova LA, Chudakov DM, Souslova EA (2010) The structure of Ca2+ sensor Case16 reveals the mechanism of reaction to low Ca2+ concentrations. Sensors 10(9):8143–8160
44. Ohkura M, Kitahara T, Norota I, Nakai J, Kubota I, Ishii K (2009) An improved genetically-encoded fluorescent Ca2+ probe that less affects calcineurin signaling. J Physiol Sci 59:528
45. Ouyang MX, Sun J, Chien S, Wang YX (2008) Determination of hierarchical relationship of Src and Rac at subcellular locations with FRET biosensors. Proc Natl Acad Sci USA 105 (38):14353–14358
46. Usami A, Ikegaya Y, Matsuki N, Nakai J (2009) Improved genetically encoded fluorescent calcium sensor G-CaMP4. Neurosci Res 65:S225
47. Visser AJWG, Laptenok SP, Visser NV, van Hoek A, Birch DJS, Brochon JC, Borst JW (2010) Time-resolved FRET fluorescence spectroscopy of visible fluorescent protein pairs. Eur Biophys J Biophys Lett 39(2):241–253
48. Wang Q, Shui B, Kotlikoff MI, Sondermann H (2008) Structural basis for calcium sensing by GCaMP2. Structure 16(12):1817–1827
49. Allman E, Johnson D, Nehrke K (2009) Loss of the apical V-ATPase a-subunit VHA-6 prevents acidification of the intestinal lumen during a rhythmic behavior in *C. elegans*. Am J Physiol Cell Physiol 297(5):C1071–C1081
50. Bacaj T, Tevlin M, Lu Y, Shaham S (2008) Glia are essential for sensory organ function in *C. elegans*. Science 322(5902):744–747
51. Ben Arous J, Tanizawa Y, Rabinowitch I, Chatenay D, Schafer WR (2010) Automated imaging of neuronal activity in freely behaving *Caenorhabditis elegans*. J Neurosci Methods 187(2):229–234
52. Biron D, Wasserman S, Thomas JH, Samuel ADT, Sengupta P (2008) An olfactory neuron responds stochastically to temperature and modulates *Caenorhabditis elegans* thermotactic behavior. Proc Natl Acad Sci USA 105(31):11002–11007
53. Chalasani SH, Kato S, Albrecht DR, Nakagawa T, Abbott LF, Bargmann CI (2010) Neuropeptide feedback modifies odor-evoked dynamics in *Caenorhabditis elegans* olfactory neurons. Nat Neurosci 13(5):615–621
54. Chatzigeorgiou M, Yoo S, Watson JD, Lee WH, Spencer WC, Kindt KS, Hwang SW, Miller DM, Treinin M, Driscoll M, Schafer WR (2010) Specific roles for DEG ENaC and TRP channels in touch and thermosensation in *C. elegans* nociceptors. Nat Neurosci 13 (7):861–868
55. Ghosh-Roy A, Wu ZL, Goncharov A, Jin YS, Chisholm AD (2010) Calcium and cyclic AMP promote axonal regeneration in *Caenorhabditis elegans* and require DLK-1 kinase. J Neurosci 30(9):3175–3183
56. Gruninger TR, Gualberto DG, Garcia LR (2008) Sensory perception of food and insulin-like signals influence seizure susceptibility. PLoS Genet 4(7):e1000117
57. Guo ZV, Hart AC, Ramanathan S (2009) Optical interrogation of neural circuits in *Caenorhabditis elegans*. Nat Methods 6(12):891–896
58. Haspel G, O'Donovan MJ, Hart AC (2010) Motoneurons dedicated to either forward or backward locomotion in the nematode *Caenorhabditis elegans*. J Neurosci 30(33): 11151–11156
59. Huang H, Delikanli S, Zeng H, Ferkey DM, Pralle A (2010) Remote control of ion channels and neurons through magnetic-field heating of nanoparticles. Nat Nanotechnol 5(8): 602–606
60. Kuhara A, Okumura M, Kimata T, Tanizawa Y, Takano R, Kimura KD, Inada H, Matsumoto K, Mori I (2008) Temperature sensing by an olfactory neuron in a circuit controlling behavior of *C. elegans*. Science 320(5877):803–807

61. Lockery SR, Lawton KJ, Doll JC, Faumont S, Coulthard SM, Thiele TR, Chronis N, McCormick KE, Goodman MB, Pruitt BL (2008) Artificial dirt: microfluidic substrates for nematode neurobiology and behavior. J Neurophysiol 99(6):3136–3143
62. Macosko EZ, Pokala N, Feinberg EH, Chalasani SH, Butcher RA, Clardy J, Bargmann CI (2009) A hub-and-spoke circuit drives pheromone attraction and social behaviour in *C. elegans*. Nature 458(7242):1171–1175
63. Nehrke K, Denton J, Mowrey W (2008) Intestinal Ca2+ wave dynamics in freely moving *C. elegans* coordinate execution of a rhythmic motor program. Am J Physiol Cell Physiol 294(1):C333–C344
64. Ortiz CO, Faumont S, Takayama J, Ahmed HK, Goldsmith AD, Pocock R, McCormick KE, Kunimoto H, Iino Y, Lockery S, Hobert O (2009) Lateralized gustatory behavior of *C. elegans* is controlled by specific receptor-type guanylyl cyclases. Curr Biol 19(12):996–1004
65. Persson A, Gross E, Laurent P, Busch KE, Bretes H, de Bono M (2009) Natural variation in a neural globin tunes oxygen sensing in wild *Caenorhabditis elegans*. Nature 458(7241):1030–1033
66. Pierce-Shimomura JT, Chen BL, Mun JJ, Ho R, Sarkis R, McIntire SL (2008) Genetic analysis of crawling and swimming locomotory patterns in *C. elegans*. Proc Natl Acad Sci USA 105(52):20982–20987
67. Saheki Y, Bargmann CI (2009) Presynaptic CaV2 calcium channel traffic requires CALF-1 and the alpha(2)delta subunit UNC-36. Nat Neurosci 12(10):1257–1265
68. Senti G, Ezcurra M, Lobner J, Schafer WR, Swoboda P (2009) Worms with a single functional sensory cilium generate proper neuron-specific behavioral output. Genetics 183(2):595–605
69. Suzuki H, Thiele TR, Faumont S, Ezcurra M, Lockery SR, Schafer WR (2008) Functional asymmetry in *Caenorhabditis elegans* taste neurons and its computational role in chemotaxis. Nature 454(7200):114–117
70. Thiele TR, Faumont S, Lockery SR (2009) The neural network for chemotaxis to tastants in *Caenorhabditis elegans* is specialized for temporal differentiation. J Neurosci 29 (38):11904–11911
71. Tsunozaki M, Chalasani SH, Bargmann CIA (2008) Behavioral switch: cGMP and PKC signaling in olfactory neurons reverses odor preference in *C. elegans*. Neuron 59(6):959–971
72. Wakabayashi T, Kimura Y, Ohba Y, Adachi R, Satoh Y, Shingai R (2009) In vivo calcium imaging of OFF-responding ASK chemosensory neurons in *C. elegans*. Biochim Biophys Acta Gen Subj 1790(8):765–769
73. Wang Y, Wang J, Du W, Feng X, Liu B-F (2010) Identification of the neuronal effects of ethanol on *C. elegans* by in vivo fluorescence imaging on a microfluidic chip. Anal Bioanal Chem 1–7
74. Zhang M, Chung SH, Fang-Yen C, Craig C, Kerr RA, Suzuki H, Samuel ADT, Mazur E, Schafer WR (2008) A self-regulating feed-forward circuit controlling *C. elegans* egg-laying behavior. Curr Biol 18(19):1445–1455
75. Zimmer M, Gray JM, Pokala N, Chang AJ, Karow DS, Marletta MA, Hudson ML, Morton DB, Chronis N, Bargmann CI (2009) Neurons detect increases and decreases in oxygen levels using distinct guanylate cyclases. Neuron 61(6):865–879
76. Asahina K, Louis M, Piccinotti S, Vosshall L (2009) A circuit supporting concentration-invariant odor perception in Drosophila. J Biol 8(1):9
77. Cameron P, Hiroi M, Ngai J, Scott K (2010) The molecular basis for water taste in Drosophila. Nature 465(7294):91–95
78. Cheng LE, Song W, Looger LL, Jan LY, Jan YN (2010) The role of the TRP channel NompC in Drosophila larval and adult locomotion. Neuron 67(3):373–380
79. Chiappe ME, Seelig JD, Reiser MB, Jayaraman V (2010) Walking modulates speed sensitivity in Drosophila motion vision. Curr Biol 20(16):1470–1475
80. Dacks AM, Green DS, Root CM, Nighorn AJ, Wang JW (2009) Serotonin modulates olfactory processing in the antennal lobe of Drosophila. J Neurogenet 23(4):366–377
81. Datta SR, Vasconcelos ML, Ruta V, Luo S, Wong A, Demir E, Flores J, Balonze K, Dickson BJ, Axel R (2008) The Drosophila pheromone cVA activates a sexually dimorphic neural circuit. Nature 452(7186):473–477

82. Galizia CG, Munch D, Strauch M, Nissler A, Ma SW (2010) Integrating heterogeneous odor response data into a common response model: a DoOR to the complete olfactome. Chem Senses 35(7):551–563
83. Gordon MD, Scott K (2009) Motor control in a Drosophila taste circuit. Neuron 61 (3):373–384
84. Ha EM, Lee KA, Park SH, Kim SH, Nam HJ, Lee HY, Kang D, Lee WJ (2009) Regulation of DUOX by the G alpha q-phospholipase C beta-Ca2+ pathway in Drosophila gut immunity. Dev Cell 16(3):386–397
85. Hamada FN, Rosenzweig M, Kang K, Pulver SR, Ghezzi A, Jegla TJ, Garrity PA (2008) An internal thermal sensor controlling temperature preference in Drosophila. Nature 454 (7201):217–220
86. Hansson BS, Knaden M, Sachse S, Stensmyr MC, Wicher D (2010) Towards plant-odor-related olfactory neuroethology in Drosophila. Chemoecology 20(2):51–61
87. Ignell R, Root CM, Birse RT, Wang JW, Nassel DR, Winther AME (2009) Presynaptic peptidergic modulation of olfactory receptor neurons in Drosophila. Proc Natl Acad Sci USA 106(31):13070–13075
88. Jiang DW, Zhao LL, Clapham DE (2009) Genome-wide RNAi screen identifies Letm1 as a mitochondrial Ca2+/H + antiporter. Science 326(5949):144–147
89. Joesch M, Plett J, Borst A, Reiff DF (2008) Response properties of motion-sensitive visual interneurons in the lobula plate of *Drosophila melanogaster*. Curr Biol 18(5):368–374
90. Kamikouchi A, Inagaki HK, Effertz T, Hendrich O, Fiala A, Gopfert MC, Ito K (2009) The neural basis of Drosophila gravity-sensing and hearing. Nature 458(7235):165–171
91. Kamikouchi A, Wiek R, Effertz T, Gopfert MC, Fiala A (2010) Transcuticular optical imaging of stimulus-evoked neural activities in the Drosophila peripheral nervous system. Nat Protoc 5(7):1229–1235
92. Lai S-H, Lin YY, Chiang AS, Huang YC (2009) Fluorescence lifetime images of different green fluorescent proteins in fly brain. Proc SPIE 7180:71800D. doi:10. 1117/12. 808974 2009, 7180
93. Liu X, Davis RL (2009) The GABAergic anterior paired lateral neuron suppresses and is suppressed by olfactory learning. Nat Neurosci 12(1):53–59
94. Mamiya A, Beshel J, Xu CS, Zhong Y (2008) Neural representations of airflow in Drosophila mushroom body. PLoS One 3(12):e4063
95. Mao ZM, Davis RL (2009) Eight different types of dopaminergic neurons innervate the Drosophila mushroom body neuropil: anatomical and physiological heterogeneity. Front Neural Circuits 3 (article 5):1–17
96. Masuda-Nakagawa LM, Gendre N, O'Kane CJ, Stocker RF (2009) Localized olfactory representation in mushroom bodies of Drosophila larvae. Proc Natl Acad Sci USA 106 (25):10314–10319
97. Reiff DF, Plett J, Mank M, Griesbeck O, Borst A (2010) Visualizing retinotopic half-wave rectified input to the motion detection circuitry of Drosophila. Nat Neurosci 13 (8):973–978
98. Root CM, Masuyama K, Green DS, Enell LE, Nassel DR, Lee CH, Wang JW (2008) A presynaptic gain control mechanism fine-tunes olfactory behavior. Neuron 59(2):311–321
99. Schnell B, Joesch M, Forstner F, Raghu SV, Otsuna H, Ito K, Borst A, Reiff DF (2010) Processing of horizontal optic flow in three visual interneurons of the drosophila brain. J Neurophysiol 103(3):1646–1657
100. Seelig JD, Chiappe ME, Lott GK, Dutta A, Osborne JE, Reiser MB, Jayaraman V (2010) Two-photon calcium imaging from head-fixed Drosophila during optomotor walking behavior. Nat Methods 7(7):535–540
101. Seki Y, Rybak J, Wicher D, Sachse S, Hansson BS (2010) Physiological and morphological characterization of local interneurons in the Drosophila antennal lobe. J Neurophysiol 104 (2):1007–1019
102. Semmelhack JL, Wang JW (2009) Select Drosophila glomeruli mediate innate olfactory attraction and aversion. Nature 459(7244):218–223

103. Shakiryanova D, Levitan ES (2008) Prolonged presynaptic posttetanic cyclic GMP signaling in Drosophila motoneurons. Proc Natl Acad Sci USA 105(36):13610–13613
104. Silbering AF, Okada R, Ito K, Galizia CG (2008) Olfactory information processing in the Drosophila antennal lobe: anything goes? J Neurosci 28(49):13075–13087
105. Terhzaz S, Finlayson AJ, Stirrat L, Yang JL, Tricoire H, Woods DJ, Dow JAT, Davies SA (2010) Cell-specific inositol 1,4,5 trisphosphate 3-kinase mediates epithelial cell apoptosis in response to oxidative stress in Drosophila. Cell Signal 22(5):737–748
106. Tomchik SM, Davis RL (2009) Dynamics of learning-related cAMP signaling and stimulus integration in the drosophila olfactory pathway. Neuron 64(4):510–521
107. Tsydzik V, Wright NJD (2009) Dopamine modulation of the in vivo acetylcholine response in the Drosophila mushroom body. Dev Neurobiol 69(11):705–714
108. Ueda A, Wu CF (2009) Role of rut Adenylyl cyclase in the ensemble regulation of presynaptic terminal excitability: reduced synaptic strength and precision in a drosophila memory mutant. J Neurogenet 23(1–2):185–199
109. Xu J, Sornborger AT, Lee JK, Shen P (2008) Drosophila TRPA channel modulates sugar-stimulated neural excitation, avoidance and social response. Nat Neurosci 11(6):676–682
110. Xu J, Li M, Shen P (2010) A G-protein-coupled neuropeptide Y-like receptor suppresses behavioral and sensory response to multiple stressful stimuli in Drosophila. J Neurosci 30(7):2504–2512
111. Yao CK, Lin YQ, Ly CV, Ohyama T, Haueter CM, Moiseenkova-Bell VY, Wensel TG, Bellen HJ (2009) A synaptic vesicle-associated Ca2+ channel promotes endocytosis and couples exocytosis to endocytosis. Cell 138(5):947–960
112. Yorozu S, Wong A, Fischer BJ, Dankert H, Kernan MJ, Kamikouchi A, Ito K, Anderson DJ (2009) Distinct sensory representations of wind and near-field sound in the Drosophila brain. Nature 458(7235):201–205
113. Atkin SD, Patel S, Kocharyan A, Holtzclaw LA, Weerth SH, Schram V, Pickel J, Russell JT (2009) Transgenic mice expressing a cameleon fluorescent Ca2+ indicator in astrocytes and Schwann cells allow study of glial cell Ca2+ signals in situ and in vivo. J Neurosci Methods 181(2):212–226
114. Constantin S, Jasoni CL, Wadas B, Herbison AE (2010) Gamma-aminobutyric acid and glutamate differentially regulate intracellular calcium concentrations in mouse gonadotropin-releasing hormone neurons. Endocrinology 151(1):262–270
115. Ducros M, Moreaux L, Bradley J, Tiret P, Griesbeck O, Charpak S (2009) spectral unmixing: analysis of performance in the olfactory bulb in vivo. PLoS One 4(2):e4418
116. Fletcher ML, Masurkar AV, Xing JL, Imamura F, Xiong WH, Nagayama S, Mutoh H, Greer CA, Knopfel T, Chen WR (2009) Optical imaging of postsynaptic odor representation in the glomerular layer of the mouse olfactory bulb. J Neurophysiol 102(2):817–830
117. Furutani N, Hondo M, Tsujino N, Sakurai T (2010) Activation of bombesin receptor subtype-3 influences activity of orexin neurons by both direct and indirect pathways. J Mol Neurosci 42(1):106–111
118. Han SK, Lee K, Bhattarai JP, Herbison AE (2010) Gonadotrophin-releasing hormone (GnRH) exerts stimulatory effects on GnRH neurones in intact adult male and female mice. J Neuroendocrinol 22(3):188–195
119. He J, Ma LM, Kim S, Nakai J, Yu CR (2008) Encoding gender and individual information in the mouse vomeronasal organ. Science 320(5875):535–538
120. Herz J, Siffrin V, Hauser AE, Brandt AU, Leuenberger T, Radbruch H, Zipp F, Niesner RA (2010) Expanding two-photon intravital microscopy to the infrared by means of optical parametric oscillator. Biophys J 98(4):715–723
121. Jasoni CL, Romano N, Constantin S, Lee K, Herbison AE (2010) Calcium dynamics in gonadotropin-releasing hormone neurons. Front Neuroendocrinol 31(3):259–269
122. Ledoux J, Taylor MS, Bonev AD, Hannah RM, Solodushko V, Shui B, Tallini Y, Kotlikoff MI, Nelson MT (2008) Functional architecture of inositol 1,4,5-trisphosphate signaling in restricted spaces of myoendothelial projections. Proc Natl Acad Sci USA 105(28):9627–9632

123. Lee K, Duan W, Sneyd J, Herbison AE (2010) Two slow calcium-activated afterhyperpolarization currents control burst firing dynamics in gonadotropin-releasing hormone neurons. J Neurosci 30(18):6214–6224
124. Qiu DL, Akemann W, Chu CP, Araki R, Knopfel T (2008) Targeted optical probing of neuronal circuit dynamics using fluorescent protein sensors. Neurosignals 16(4): 289–299
125. Qiu DL, Knopfel T (2009) Presynaptically expressed long-term depression at cerebellar parallel fiber synapses. Pflügers Arch Eur J Physiol 457(4):865–875
126. Romano N, Lee K, Abraham IM, Jasoni CL, Herbison AE (2008) Nonclassical estrogen modulation of presynaptic GABA terminals modulates calcium dynamics in gonadotropin-releasing hormone neurons. Endocrinology 149(11):5335–5344
127. Saint Jan DD, Hirnet D, Westbrook GL, Charpak S (2009) External tufted cells drive the output of olfactory bulb glomeruli. J Neurosci 29(7):2043–2052
128. Siffrin V, Radbruch H, Glumm R, Niesner R, Paterka M, Herz J, Leuenberger T, Lehmann SM, Luenstedt S, Rinnenthal JL, Laube G, Luche H, Lehnardt S, Fehling HJ, Griesbeck O, Zipp F (2010) In vivo imaging of partially reversible Th17 cell-induced neuronal dysfunction in the course of encephalomyelitis. Immunity 33(3):424–436
129. Wallace DJ, Borgloh SMZA, Astori S, Yang Y, Bausen M, Kugler S, Palmer AE, Tsien RY, Sprengel R, Kerr JND, Denk W, Hasan MT (2008) Single-spike detection in vitro and in vivo with a genetic Ca2+ sensor. Nat Methods 5(9):797–804
130. Xie XM, Wisor JP, Hara J, Crowder TL, LeWinter R, Khroyan TV, Yamanaka A, Diano S, Horvath TL, Sakurai T, Toll L, Kilduff TS (2008) Hypocretin/orexin and nociceptin/orphanin FQ coordinately regulate analgesia in a mouse model of stress-induced analgesia. J Clin Investig 118(7):2471–2481
131. Andermann ML, Kerlin AM, Reid C (2010) Chronic cellular imaging of mouse visual cortex during operant behavior and passive viewing. Front Cell Neurosci 4 (article 3):1–16
132. Boldogkoi Z, Balint K, Awatramani GB, Balya D, Busskamp V, Viney TJ, Lagali PS, Duebel J, Pasti E, Tombacz D, Toth JS, Takacs IF, Scherf BG, Roska B (2009) Genetically timed, activity-sensor and rainbow transsynaptic viral tools. Nat Methods 6(2):127–130
133. Chesler AT, Le Pichon CE, Brann JH, Araneda RC, Zou DJ, Firestein S (2008) Selective gene expression by postnatal electroporation during olfactory interneuron neurogenesis. PLoS One 3(1):e1517
134. Gourine AV, Kasymov V, Marina N, Tang FG, Figueiredo MF, Lane S, Teschemacher AG, Spyer KM, Deisseroth K, Kasparov S (2010) Astrocytes control breathing through pH-dependent release of ATP. Science 329(5991):571–575
135. Granstedt AE, Szpara ML, Kuhn B, Wang SSH, Enquist LW (2009) Fluorescence-based monitoring of in vivo neural activity using a circuit-tracing pseudorabies virus. PLoS One 4(9):e6923
136. Guo F, Liu BH, Tang FG, Lane S, Souslova EA, Chudakov DM, Paton JFR, Kasparov S (2010) Astroglia are a possible cellular substrate of angiotensin(1–7) effects in the rostral ventrolateral medulla. Cardiovasc Res 87(3):578–584
137. Hong JH, Min CH, Jeong B, Kojiya T, Morioka E, Nagai T, Ikeda M, Lee KJ (2010) Intracellular calcium spikes in rat suprachiasmatic nucleus neurons induced by BAPTA-based calcium dyes. PLoS One 5(3):e9634
138. Hoogland TM, Kuhn B, Gobel W, Huang WY, Nakai J, Helmchen F, Flint J, Wang SSH (2009) Radially expanding transglial calcium waves in the intact cerebellum. Proc Natl Acad Sci USA 106(9):3496–3501
139. Jimenez-Moreno R, Wang ZM, Messi ML, Delbono O (2010) Sarcoplasmic reticulum Ca2+ depletion in adult skeletal muscle fibres measured with the biosensor D1ER. Pflügers Arch Eur J Physiol 459(5):725–735
140. Kuchibhotla KV, Goldman ST, Lattarulo CR, Wu HY, Hyman BT, Bacskai BJ (2008) A beta plaques lead to aberrant regulation of calcium homeostasis in vivo resulting in structural and functional disruption of neuronal networks. Neuron 59(2):214–225

141. Mao T, O'Connor DH, Scheuss V, Nakai J, Svoboda K (2008) Characterization and subcellular targeting of GCaMP-type genetically-encoded calcium indicators. PLoS One 3(3):e1796
142. Mironov SL, Skorova E, Hartelt N, Mironova LA, Hasan MT, Kugler S (2009) Remodelling of the respiratory network in a mouse model of Rett syndrome depends on brain-derived neurotrophic factor regulated slow calcium buffering. J Physiol Lond 587 (11):2473–2485
143. Mues M, Mank M, Griesbeck O, Oboardi F, Flugel A, Kurschus F, Wekerle H (2008) Imaging activation: FRET-based calcium biosensors in T-lymphocytes. J Neuroimmunol 203(2):230
144. Nguyen QT, Schroeder LF, Mank M, Muller A, Taylor P, Griesbeck O, Kleinfeld D (2010) An in vivo biosensor for neurotransmitter release and in situ receptor activity. Nat Neurosci 13(1):127–132
145. Vicencio JM, Ortiz C, Criollo A, Jones AWE, Kepp O, Galluzzi L, Joza N, Vitale I, Morselli E, Tailler M, Castedo M, Maiuri MC, Molgo J, Szabadkai G, Lavandero S, Kroemer G (2009) The inositol 1,4,5-trisphosphate receptor regulates autophagy through its interaction with Beclin 1. Cell Death Differ 16(7):1006–1017
146. Chi NC, Shaw RM, De Val S, Kang G, Jan LY, Black BL, Stainier DYR (2008) Foxn4 directly regulates tbx2b expression and atrioventricular canal formation. Genes Dev 22 (6):734–739
147. Chi NC, Bussen M, Brand-Arzamendi K, Ding C, Olgin JE, Shaw RM, Martin GR, Stainier DYR (2010) Cardiac conduction is required to preserve cardiac chamber morphology. Proc Natl Acad Sci USA 107(33):14662–14667
148. Dorostkar MM, Dreosti E, Odermatt B, Lagnado L (2010) Computational processing of optical measurements of neuronal and synaptic activity in networks. J Neurosci Methods 188(1):141–150
149. Dreosti E, Odermatt B, Dorostkar MM, Lagnado L (2009) A genetically encoded reporter of synaptic activity in vivo. Nat Methods 6(12):883–889
150. Niessing J, Friedrich RW (2010) Olfactory pattern classification by discrete neuronal network states. Nature 465(7294):47–52
151. Orger MB, Kampff AR, Severi KE, Bollmann JH, Engert F (2008) Control of visually guided behavior by distinct populations of spinal projection neurons. Nat Neurosci 11 (3):327–333
152. Richler E, Chaumont S, Shigetomi E, Sagasti A, Khakh BS (2008) Tracking transmitter-gated P2X cation channel activation in vitro and in vivo. Nat Methods 5(1):87–93
153. Smyth JW, Hong TT, Gao DC, Vogan JM, Jensen BC, Fong TS, Simpson PC, Stainier DYR, Chi NC, Shaw RM (2010) Limited forward trafficking of connexin 43 reduces cell–cell coupling in stressed human and mouse myocardium. J Clin Investig 120(1):266–279
154. Sumbre G, Muto A, Baier H, Poo MM (2008) Entrained rhythmic activities of neuronal ensembles as perceptual memory of time interval. Nature 456(7218):102–106
155. Tabor R, Yaksi E, Friedrich RW (2008) Multiple functions of GABA(A) and GABA(B) receptors during pattern processing in the zebrafish olfactory bulb. Eur J Neurosci 28(1):117–127
156. Tabor R, Friedrich RW (2008) Pharmacological analysis of ionotropic glutamate receptor function in neuronal circuits of the zebrafish olfactory bulb. PLoS One 3(1):e1416
157. Wiechert MT, Judkewitz B, Riecke H, Friedrich RW (2010) Mechanisms of pattern decorrelation by recurrent neuronal circuits. Nat Neurosci 13(8):1003–1010
158. Cho D, Kim SA, Murata Y, Lee S, Jae SK, Nam HG, Kwak JM (2009) De-regulated expression of the plant glutamate receptor homolog AtGLR3.1 impairs long-term Ca2+ -programmed stomatal closure. Plant J 58(3):437–449
159. Farkas MH, Mojica ERE, Patel M, Aga DS, Berry JO (2009) Development of a rapid biolistic assay to determine changes in relative levels of intracellular calcium in leaves following tetracycline uptake by pinto bean plants. Analyst 134(8):1594–1600

160. Haruta M, Monshausen G, Gilroy S, Sussman MR (2008) A cytoplasmic Ca2+ functional assay for identifying and purifying endogenous cell signaling peptides in Arabidopsis seedlings: Identification of AtRALF1 peptide. Biochemistry 47(24):6311–6321
161. Iwano M, Entani T, Shiba H, Kakita M, Nagai T, Mizuno H, Miyawaki A, Shoji T, Kubo K, Isogai A, Takayama S (2009) Fine-tuning of the cytoplasmic Ca2+ concentration is essential for pollen tube growth. Plant Physiol 150(3):1322–1334
162. Kosuta S, Hazledine S, Sun J, Miwa H, Morris RJ, Downie JA, Oldroyd GED (2008) Differential and chaotic calcium signatures in the symbiosis signaling pathway of legumes. Proc Natl Acad Sci USA 105(28):9823–9828
163. Michard E, Dias P, Feijo JA (2008) Tobacco pollen tubes as cellular models for ion dynamics: improved spatial and temporal resolution of extracellular flux and free cytosolic concentration of calcium and protons using pHluorin and YC3.1 CaMeleon. Sex Plant Reprod 21(3):169–181
164. Monshausen GB, Messerli MA, Gilroy S (2008) Imaging of the yellow cameleon 3.6 indicator reveals that elevations in cytosolic Ca2+ follow oscillating increases in growth in root hairs of arabidopsis. Plant Physiol 147(4):1690–1698
165. Monshausen GB, Bibikova TN, Weisenseel MH, Gilroy S (2009) Ca2+ regulates reactive oxygen species production and ph during mechanosensing in Arabidopsis roots. Plant Cell 21(8):2341–2356
166. Rincon-Zachary M, Teaster ND, Sparks JA, Valster AH, Motes CM, Blancaflor EB (2010) Fluorescence resonance energy transfer-sensitized emission of yellow cameleon 3.60 reveals root zone-specific calcium signatures in Arabidopsis in response to aluminum and other trivalent cations. Plant Physiol 152(3):1442–1458
167. Sieberer BJ, Chabaud M, Timmers AC, Monin A, Fournier J, Barker DG (2009) A nuclear-targeted cameleon demonstrates intranuclear Ca2+ spiking in medicago truncatula root hairs in response to rhizobial nodulation factors. Plant Physiol 151(3):1197–1206
168. Siegel RS, Xue SW, Murata Y, Yang YZ, Nishimura N, Wang A, Schroeder JI (2009) Calcium elevation-dependent and attenuated resting calcium-dependent abscisic acid induction of stomatal closure and abscisic acid-induced enhancement of calcium sensitivities of S-type anion and inward-rectifying K+ channels in Arabidopsis guard cells. Plant J 59(2):207–220
169. Tanaka K, Swanson SJ, Gilroy S, Stacey G (2010) Extracellular nucleotides elicit cytosolic free calcium oscillations in Arabidopsis. Plant Physiol 154(2):705–719
170. Vahisalu T, Kollist H, Wang YF, Nishimura N, Chan WY, Valerio G, Lamminmaki A, Brosche M, Moldau H, Desikan R, Schroeder JI, Kangasjarvi J (2008) SLAC1 is required for plant guard cell S-type anion channel function in stomatal signalling. Nature 452 (7186):487–491
171. Weinl S, Held K, Schlucking K, Steinhorst L, Kuhlgert S, Hippler M, Kudla J (2008) A plastid protein crucial for Ca2+ -regulated stomatal responses. New Phytol 179(3):675–686
172. Yang YZ, Costa A, Leonhardt N, Siegel RS, Schroeder JI (2008) Isolation of a strong Arabidopsis guard cell promoter and its potential as a research tool. Plant Methods 4 (article 6):1–15
173. Zhu XH, Caplan J, Mamillapalli P, Czymmek K, Dinesh-Kumar SP (2010) Function of endoplasmic reticulum calcium ATPase in innate immunity-mediated programmed cell death. EMBO J 29(5):1007–1018
174. Antigny F, Girardin N, Raveau D, Frieden M, Becq F, Vandebrouck C (2009) Dysfunction of mitochondria Ca2+ uptake in cystic fibrosis airway epithelial cells. Mitochondrion 9(4):232–241
175. Barro-Soria R, Aldehni F, Almaca J, Witzgall R, Schreiber R, Kunzelmann K (2010) ER-localized bestrophin 1 activates Ca2+−dependent ion channels TMEM16A and SK4 possibly by acting as a counterion channel. Pflügers Arch Eur J Physiol 459(3):485–497
176. Baumgartner HK, Gerasimenko JV, Thorne C, Ferdek P, Pozzan T, Tepikin AV, Petersen OH, Sutton R, Watson AJM, Gerasimenko OV (2009) Calcium elevation in mitochondria is the main Ca2+ requirement for mitochondrial permeability transition pore (mPTP) opening. J Biol Chem 284(31):20796–20803

177. Bentley M, Nycz DC, Joglekar A, Fertschai I, Malli R, Graier WF, Hay JC (2010) Vesicular calcium regulates coat retention, fusogenicity, and size of pre-golgi intermediates. Mol Biol Cell 21(6):1033–1046
178. Bursac N, Kirkton RD, McSpadden LC, Liau B (2010) Characterizing functional stem cell-cardiomyocyte interactions. Regen Med 5(1):87–105
179. Campbell RV, Yang YZ, Wang T, Rachamallu A, Li YC, Watowich SJ, Weinman SA (2009) Effects of hepatitis C core protein on mitochondrial electron transport and production of reactive oxygen species, vol 456. Elsevier Academic, San Diego
180. Caporale T, Ciavardelli D, Di Ilio C, Lanuti P, Drago D, Sensi SL (2009) Ratiometric-pericam-mt, a novel tool to evaluate intramitochondrial zinc. Exp Neurol 218 (2):228–234
181. Chigurupati S, Wei ZL, Belal C, Vandermey M, Kyriazis GA, Arumugam TV, Chan SL (2009) The homocysteine-inducible endoplasmic reticulum stress protein counteracts calcium store depletion and induction of CCAAT enhancer-binding protein homologous protein in a neurotoxin model of Parkinson disease. J Biol Chem 284(27):18323–18333
182. Chvanov M, Walsh CM, Haynes LP, Voronina SG, Lur G, Gerasimenko OV, Barraclough R, Rudland PS, Petersen OH, Burgoyne RD, Tepikin AV (2008) ATP depletion induces translocation of STIM1 to puncta and formation of STIM1-ORAI1 clusters: translocation and re-translocation of STIM1 does not require ATP. Pflügers Arch Eur J Physiol 457(2):505–517
183. Claas RF, ter Braak M, Hegen B, Hardel V, Angioni C, Schmidt H, Jakobs KH, Van Veldhoven PP, Heringdorf DMZ (2010) Enhanced Ca2+ storage in sphingosine-1-phosphate lyase-deficient fibroblasts. Cell Signal 22(3):476–483
184. Cohen R, Torres A, Ma HT, Holowka D, Baird B (2009) Ca2+ waves initiate antigen-stimulated Ca2+ responses in mast cells. J Immunol 183(10):6478–6488
185. Csordas G, Varnai P, Golenar T, Roy S, Purkins G, Schneider TG, Balla T, Hajnoczky G (2010) Imaging interorganelle contacts and local calcium dynamics at the ER-mitochondrial interface. Mol Cell 39(1):121–132
186. Da Cruz S, De Marchi U, Frieden M, Parone PA, Martinou JC, Demaurex N (2010) SLP-2 negatively modulates mitochondrial sodium–calcium exchange. Cell Calcium 47(1):11–18
187. Drago I, Giacomello M, Pizzo P, Pozzan T (2008) Calcium dynamics in the peroxisomal lumen of living cells. J Biol Chem 283(21):14384–14390
188. Dror V, Kalynyak TB, Bychkivska Y, Frey MHZ, Tee M, Jeffrey KD, Nguyen V, Luciani DS, Johnson JD (2008) Glucose and endoplasmic reticulum calcium channels regulate HIF-1 beta via presenilin in pancreatic beta-cells. J Biol Chem 283(15):9909–9916
189. Fukano T, Shimozono S, Miyawaki A (2008) Development of microscopic systems for high-speed dual-excitation ratiometric Ca2+ imaging. Brain Cell Biol 36(1–4):43–52
190. Gaspers LD, Thomas AP (2008) Calcium-dependent activation of mitochondrial metabolism in mammalian cells. Methods 46(3):224–232
191. Giacomello M, Drago I, Bortolozzi M, Scorzeto M, Gianelle A, Pizzo P, Pozzan T (2010) Ca2+ hot spots on the mitochondrial surface are generated by Ca2+ mobilization from stores, but not by activation of store-operated Ca2+ channels. Mol Cell 38(2):280–290
192. Girardin NC, Antigny F, Frieden M (2010) Electrophysiological characterization of store-operated and agonist-induced Ca2+ entry pathways in endothelial cells. Pflügers Arch Eur J Physiol 460(1):109–120
193. Grant DM, Zhang W, Mcghee EJ, Bunney TD, Talbot CB, Kumar S, Munro I, Dunsby C, Neil MAA, Katan M, French PMW (2008) Multiplexed FRET to image multiple signaling events in live cells. Biophys J 95(10):L69–L71
194. Gwiazda KS, Yang TLB, Lin YL, Johnson JD (2009) Effects of palmitate on ER and cytosolic Ca2+ homeostasis in beta-cells. Am J Physiol Endocrinol Metab 296(4): E690–E701
195. Harita Y, Kurihara H, Kosako H, Tezuka T, Sekine T, Igarashi T, Ohsawa I, Ohta S, Hattori S (2009) Phosphorylation of nephrin triggers Ca2+ signaling by recruitment and activation of phospholipase C-gamma 1. J Biol Chem 284(13):8951–8962

196. Hasbi A, Fan T, Alijaniaram M, Nguyen T, Perreault ML, O'Dowd BF, George SR (2009) Calcium signaling cascade links dopamine D1-D2 receptor heteromer to striatal BDNF production and neuronal growth. Proc Natl Acad Sci USA 106(50):21377–21382
197. Hewavitharana T, Deng XX, Wang YJ, Ritchie MF, Girish GV, Soboloff J, Gill DL (2008) Location and function of STIM1 in the activation of Ca2+ entry signals. J Biol Chem 283 (38):26252–26262
198. Hou BH, Takanaga H, Griesbeck O, Frommer WB (2009) Osmotic induction of calcium accumulation in human embryonic kidney cells detected with a high sensitivity FRET calcium sensor. Cell Calcium 46(2):130–135
199. Hu Y, Suarez J, Fricovsky E, Wang H, Scott BT, Trauger SA, Han WL, Hu Y, Oyeleye MO, Dillmann WH (2009) Increased enzymatic O-glcnacylation of mitochondrial proteins impairs mitochondrial function in cardiac myocytes exposed to high glucose. J Biol Chem 284(1):547–555
200. Hynes J, O'Riordan TC, Zhdanov AV, Uray G, Will Y, Papkovsky DB (2009) In vitro analysis of cell metabolism using a long-decay pH-sensitive lanthanide probe and extracellular acidification assay. Anal Biochem 390(1):21–28
201. Iguchi M, Akao M, Nakai J, Kato M, Takeda T, Matsumoto-Ida M, Kimura T, Kita T (2009) Direct monitoring of mitochondrial calcium levels in cultured cardiomyocytes using a novel fluorescent indicator protein, GCaMP2-mt. Circulation 120(18):S1059
202. Jiang DW, Chen WQ, Xiao JM, Wang RW, Kong HH, Jones PP, Zhang L, Fruen B, Chen SRW (2008) Reduced threshold for luminal Ca2+ activation of RyR1 underlies a causal mechanism of porcine malignant hyperthermia. J Biol Chem 283(30):20813–20820
203. Jones PP, Jiang DW, Bolstad J, Hunt DJ, Zhang L, Demaurex N, Chen SRW (2008) Endoplasmic reticulum Ca2+ measurements reveal that the cardiac ryanodine receptor mutations linked to cardiac arrhythmia and sudden death alter the threshold for store-overload-induced Ca2+ release. Biochem J 412:171–178
204. Jousset H, Malli R, Girardin N, Graier WF, Demaurex N, Frieden M (2008) Evidence for a receptor-activated Ca2+ entry pathway independent from Ca2+ store depletion in endothelial cells. Cell Calcium 43(1):83–94
205. Kazuno AA, Munakata K, Tanaka M, Kato N, Kato T (2008) Relationships between mitochondrial DNA subhaplogroups and intracellular calcium dynamics. Mitochondrion 8(2):164–169
206. Kazuno AA, Munakata K, Kato N, Kato T (2008) Mitochondrial DNA-dependent effects of valproate on mitochondrial calcium levels in transmitochondrial cybrids. Int J Neuropsychopharmacol 11(1):71–78
207. Kettlewell S, Cabrero P, Nicklin SA, Dow JAT, Davies S, Smith GL (2009) Changes of intra-mitochondrial Ca2+ in adult ventricular cardiomyocytes examined using a novel fluorescent Ca2+ indicator targeted to mitochondria. J Mol Cell Cardiol 46(6):891–901
208. Khuon S, Liang L, Dettman RW, Sporn PHS, Wysolmerski RB, Chew TL (2010) Myosin light chain kinase mediates transcellular intravasation of breast cancer cells through the underlying endothelial cells: a three-dimensional FRET study. J Cell Sci 123 (3):431–440
209. Kim MS, Usachev YM (2009) Mitochondrial Ca2+ cycling facilitates activation of the transcription factor NFAT in sensory neurons. J Neurosci 29(39):12101–12114
210. Kim TJ, Seong JH, Ouyang MX, Sun J, Lu SY, Hong JP, Wang N, Wang YX (2009) Substrate rigidity regulates Ca2+ oscillation via RhoA pathway in stem cells. J Cell Physiol 218(2):285–293
211. Klee M, Pallauf K, Alcala S, Fleischer A, Pimentel-Muinos FX (2009) Mitochondrial apoptosis induced by BH3-only molecules in the exclusive presence of endoplasmic reticular Bak. EMBO J 28(12):1757–1768
212. Kong H, Jones PP, Koop A, Zhang L, Duff HJ, Chen SRW (2008) Caffeine induces Ca2+ release by reducing the threshold for luminal Ca2+ activation of the ryanodine receptor. Biochem J 414:441–452

213. Korzeniowski MK, Szanda G, Balla T, Spat A (2009) Store-operated Ca2+ influx and subplasmalemmal mitochondria. Cell Calcium 46(1):49–55
214. Lissandron V, Podini P, Pizzo P, Pozzan T (2010) Unique characteristics of Ca2+ homeostasis of the trans-Golgi compartment. Proc Natl Acad Sci USA 107(20):9198–9203
215. Liu XL, Gong H, Li XN, Zhou W (2008) Monitoring calcium concentration in neurons with cameleon. J Biosci Bioeng 105(2):106–109
216. Luciarti DS, Gwiazda KS, Yang TLB, Kalynyak TB, Bychkivska Y, Frey MHZ, Jeffrey KD, Sampaio AV, Underhill TM, Johnson JD (2009) Roles of IP3R and RyR Ca2+ channels in endoplasmic reticulum stress and beta-cell death. Diabetes 58(2):422–432
217. Luik RM, Wang B, Prakriya M, Wu MM, Lewis RS (2008) Oligomerization of STIM1 couples ER calcium depletion to CRAC channel activation. Nature 454(7203):538–541
218. Malli R, Naghdi S, Romanin C, Graier WF (2008) Cytosolic Ca2+ prevents the subplasmalemmal clustering of STIM1: an intrinsic mechanism to avoid Ca2+ overload. J Cell Sci 121(19):3133–3139
219. McCombs JE, Gibson EA, Palmer AE (2010) Using a genetically targeted sensor to investigate the role of presenilin-1 in ER Ca2+ levels and dynamics. Mol Biosyst 6(9):1640–1649
220. McGinty J, Soloviev VY, Tahir KB, Laine R, Stuckey DW, Hajnal JV, Sardini A, French PMW, Arridge SR (2009) Three-dimensional imaging of Forster resonance energy transfer in heterogeneous turbid media by tomographic fluorescent lifetime imaging. Opt Lett 34(18):2772–2774
221. Medvedeva YV, Kim MS, Usachev YM (2008) Mechanisms of prolonged presynaptic Ca2+ signaling and glutamate release induced by TRPV1 activation in rat sensory neurons. J Neurosci 28(20):5295–5311
222. Meyer JS, Tullis G, Pierret C, Spears KM, Morrison JA, Kirk MD (2009) Detection of calcium transients in embryonic stem cells and their differentiated progeny. Cell Mol Neurobiol 29(8):1191–1203
223. Micaroni M, Perinetti G, Di Giandomenico D, Bianchi K, Spaar A, Mironov AA (2010) Synchronous intra-Golgi transport induces the release of Ca2+ from the Golgi apparatus. Exp Cell Res 316(13):2071–2086
224. Naghdi S, Waideck-Weiermair M, Fertschai I, Poteser M, Graier WF, Malli R (2010) Mitochondrial Ca2+ uptake and not mitochondrial motility is required for STIM1-Orai1-dependent store-operated Ca2+ entry. J Cell Sci 123(15):2553–2564
225. Nakamura N, Yamazawa T, Okubo Y, Iino M (2009) Temporal switching and cell-to-cell variability in Ca2+ release activity in mammalian cells. Mol Syst Biol 5 (article 247):1–10
226. Norman JP, Perry SW, Reynolds HM, Kiebala M, Bentley KLD, Trejo M, Volsky DJ, Maggirwar SB, Dewhurst S, Masliah E, Gelbard HA (2008) HIV-1 tat activates neuronal ryanodine receptors with rapid induction of the unfolded protein response and mitochondrial hyperpolarization. PLoS One 3(11):e3731
227. Ohta T, Imagawa T, Ito S (2008) Novel gating and sensitizing mechanism of capsaicin receptor (TRPV1) – tonic inhibitory regulation of extracellular sodium through the external protonation sites on TRPV1. J Biol Chem 283(14):9377–9387
228. Ohta T, Imagawa T, Ito S (2009) Involvement of transient receptor potential vanilloid subtype 1 in analgesic action of methylsalicylate. Mol Pharmacol 75(2):307–317
229. Ong DST, Mu TW, Palmer AE, Kelly JW (2010) Endoplasmic reticulum Ca2+ increases enhance mutant glucocerebrosidase proteostasis. Nat Chem Biol 6(6):424–432
230. Palty R, Silverman WF, Hershfinkel M, Caporale T, Sensi SL, Parnis J, Nolte C, Fishman D, Shoshan-Barmatz V, Herrmann S, Khananshvili D, Sekler I (2010) NCLX is an essential component of mitochondrial Na+/Ca2+ exchange. Proc Natl Acad Sci USA 107(1):436–441
231. Park KS, Wiederkehr A, Kirkpatrick C, Mattenberger Y, Martinou JC, Marchetti P, Demaurex N, Wollheim CB (2008) Selective actions of mitochondrial fission/fusion genes on metabolism-secretion coupling in insulin-releasing cells. J Biol Chem 283(48):33347–33356
232. Park KS, Poburko D, Wollheim CB, Demaurex N (2009) Amiloride derivatives induce apoptosis by depleting ER Ca2+ stores in vascular endothelial cells. Br J Pharmacol 156 (8):1296–1304

233. Perocchi F, Gohil VM, Girgis HS, Bao XR, McCombs JE, Palmer AE, Mootha VK (2010) MICU1 encodes a mitochondrial EF hand protein required for Ca2+ uptake. Nature 467 (7313):291–296
234. Poburko D, Liao CH, van Breemen C, Demaurex N (2009) Mitochondrial regulation of sarcoplasmic reticulum Ca2+ content in vascular smooth muscle cells. Circ Res 104(1): 104–112
235. Prorok J, Kovacs PP, Kristof AA, Nagy N, Tombacz D, Toth JS, Ordog B, Jost N, Virag L, Papp JG, Varro A, Toth A, Boldogkoi Z (2009) Herpesvirus-mediated delivery of a genetically encoded fluorescent Ca2+ sensor to canine cardiomyocytes. J Biomed Biotechnol Article 361795:1–12
236. Rafei M, Berchiche YA, Birman E, Boivin MN, Young YK, Wu JH, Heveker N, Galipeau J (2009) An engineered GM-CSF-CCL2 fusokine is a potent inhibitor of CCR2-driven inflammation as demonstrated in a murine model of inflammatory arthritis. J Immunol 183(3):1759–1766
237. Ravier MA, Cheng-Xue R, Palmer AE, Henquin JC, Gilon P (2010) Subplasmalemmal Ca2+ measurements in mouse pancreatic beta cells support the existence of an amplifying effect of glucose on insulin secretion. Diabetologia 53(9):1947–1957
238. Reeves MA, Bellinger FP, Berry MJ (2010) The neuroprotective functions of selenoprotein M and its role in cytosolic calcium regulation. Antioxid Redox Signal 12(7):809–818
239. Rothermel M, Brunert D, Klupp BG, Luebbert M, Meilenleiter TC, Hatt H (2009) Advanced tracing tools: functional neuronal expression of virally encoded fluorescent calcium indicator proteins. J Neurovirol 15(5–6):458–465
240. Salgado A, Ordaz B, Sampieri A, Zepeda A, Glazebrook P, Kunze D, Vaca L (2008) Regulation of the cellular localization and function of human transient receptor potential channel 1 by other members of the TRPC family. Cell Calcium 43(4):375–387
241. Sano R, Annunziata I, Patterson A, Moshiach S, Gomero E, Opferman J, Forte M, d'Azzo A (2009) GM1-ganglioside accumulation at the mitochondria-associated ER membranes links ER stress to Ca2 + -dependent mitochondrial apoptosis. Mol Cell 36(3):500–511
242. Shigetomi E, Kracun S, Sofroniew MV, Khakh BS (2010) A genetically targeted optical sensor to monitor calcium signals in astrocyte processes. Nat Neurosci 13(6):759–766
243. Suarez J, Hu Y, Makino A, Fricovsky E, Wang H, Dillmann WH (2008) Alterations in mitochondrial function and cytosolic calcium induced by hyperglycemia are restored by mitochondrial transcription factor A in cardiomyocytes. Am J Physiol Cell Physiol 295(6):C1561–C1568
244. Suzuki T, Orba Y, Okada Y, Sunden Y, Kimura T, Tanaka S, Nagashima K, Hall WW, Sawa H (2010) The human polyoma JC virus agnoprotein acts as a viroporin. PLoS Pathog 6(3): e1000801
245. Tomosugi W, Matsuda T, Tani T, Nemoto T, Kotera I, Saito K, Horikawa K, Nagai T (2009) An ultramarine fluorescent protein with increased photostability and pH insensitivity. Nat Methods 6(5):351–353
246. Viero C, Kraushaar U, Ruppenthal S, Kaestner L, Lipp P (2008) A primary culture system for sustained expression of a calcium sensor in preserved adult rat ventricular myocytes. Cell Calcium 43(1):59–71
247. Voccoli V, Colombaioni L (2009) Mitochondrial remodeling in differentiating neuroblasts. Brain Res 1252:15–29
248. Waldeck-Weiermair M, Zoratti C, Osibow K, Balenga N, Goessnitzer E, Waldhoer M, Malli R, Graier WF (2008) Integrin clustering enables anandamide-induced Ca2+ signaling in endothelial cells via GPR55 by protection against CB1-receptor-triggered repression. J Cell Sci 121 (10):1704–1717
249. Waldeck-Weiermair M, Malli R, Naghdi S, Trenker M, Kahn MJ, Graier WF (2010) The contribution of UCP2 and UCP3 to mitochondrial Ca2+ uptake is differentially determined by the source of supplied Ca2+. Cell Calcium 47(5):433–440
250. Wegierski T, Steffl D, Kopp C, Tauber R, Buchholz B, Nitschke R, Kuehn EW, Walz G, Kottgen M (2009) TRPP2 channels regulate apoptosis through the Ca2+ concentration in the endoplasmic reticulum. EMBO J 28(5):490–499

251. Willoughby D, Wachten S, Masada N, Cooper DMF (2010) Direct demonstration of discrete Ca2+ microdomains associated with different isoforms of adenylyl cyclase. J Cell Sci 123(1):107–117
252. Wu GS, Xie X, Lu ZH, Ledeen RW (2009) Sodium–calcium exchanger complexed with GM1 ganglioside in nuclear membrane transfers calcium from nucleoplasm to endoplasmic reticulum. Proc Natl Acad Sci USA 106(26):10829–10834
253. Wu PC, Fann MJ, Kao LS (2010) Characterization of Ca2 + signaling pathways in mouse adrenal medullary chromaffin cells. J Neurochem 112(5):1210–1222
254. Wu ZF, Zhang J, Zhao BL (2009) Superoxide anion regulates the mitochondrial free Ca2+ through uncoupling proteins. Antioxid Redox Signal 11(8):1805–1818
255. Xi Q, Adebiyi A, Zhao GL, Chapman KE, Waters CM, Hassid A, Jaggar JH (2008) IP3 constricts cerebral arteries via IP3 receptor-mediated TRPC3 channel activation and independently of sarcoplasmic reticulum Ca2+ release. Circ Res 102(9):1118–1126
256. Xia S, Li XG, Johnson T, Seidel C, Wallace DP, Li R (2010) Polycystin-dependent fluid flow sensing targets histone deacetylase 5 to prevent the development of renal cysts. Development 137(7):1075–1084
257. Xu CY, Xu WJ, Palmer AE, Reed JC (2008) BI-1 regulates endoplasmic reticulum Ca2+ homeostasis downstream of bcl-2 family proteins. J Biol Chem 283(17):11477–11484
258. Yang DM, Teng HC, Chen KH, Tsai ML, Lee TK, Chou YC, Chi CW, Chiou SH, Lee CH (2009) Clodronate-induced cell apoptosis in human thyroid carcinoma is mediated via the P2 receptor signaling pathway. J Pharmacol Exp Ther 330(2):613–623
259. Young KW, Bampton ETW, Pinon L, Bano D, Nicotera P (2008) Mitochondrial Ca2+ signalling in hippocampal neurons. Cell Calcium 43(3):296–306
260. Zhang H, Das S, Li QZ, Dragatsis I, Repa J, Zeitlin S, Hajnoczky G, Bezprozvanny I (2008) Elucidating a normal function of huntingtin by functional and microarray analysis of huntingtin-null mouse embryonic fibroblasts. BMC Neurosci 9
261. Zhang H, Sun SY, Herreman A, De Strooper B, Bezprozvanny I (2010) Role of presenilins in neuronal calcium homeostasis. J Neurosci 30(25):8566–8580
262. Zhdanov AV, Ward MW, Taylor CT, Souslova EA, Chudakov DM, Prehn JHM, Papkovsky DB (2010) Extracellular calcium depletion transiently elevates oxygen consumption in neurosecretory PC12 cells through activation of mitochondrial Na+ Ca2+ exchange. Biochim Biophys Acta Bioenerg 1797(8):1627–1636
263. Shindo A, Hara Y, Yamamoto TS, Ohkura M, Nakai J, Ueno N (2010) Tissue–tissue interaction-triggered calcium elevation is required for cell polarization during xenopus gastrulation. PLoS One 5(2):e8897
264. Cai L, Dalal CK, Elowitz MB (2008) Frequency-modulated nuclear localization bursts coordinate gene regulation. Nature 455(7212):485–490
265. Zhou YB, Meraner P, Kwon HT, Machnes D, Oh-hora M, Zimmer J, Huang Y, Stura A, Rao A, Hogan PG (2010) STIM1 gates the store-operated calcium channel ORAI1 in vitro. Nat Struct Mol Biol 17(1):112–116
266. Mank M, Griesbeck O (2008) Genetically encoded calcium indicators. Chem Rev 108 (5):1550–1564
267. McCombs JE, Palmer AE (2008) Measuring calcium dynamics in living cells with genetically encodable calcium indicators. Methods 46(3):152–159
268. Garaschuk O, Griesbeck O (2010) Monitoring calcium levels with genetically encoded indicators. In: Verkhratsky A, Petersen OH (Eds) Calcium measurement methods. Humana Press, a part of Springer Science + Business Media, pp 101–117
269. Demaurex N, Poburko D, Frieden M (2009) Regulation of plasma membrane calcium fluxes by mitochondria. Biochim Biophys Acta Bioenerg 1787(11):1383–1394
270. Rizzuto R, Marchi S, Bonora M, Aguiari P, Bononi A, De Stefani D, Giorgi C, Leo S, Rimessi A, Siviero R, Zecchini E, Pinton P (2009) Ca2+ transfer from the ER to mitochondria: When, how and why. Biochim Biophys Acta Bioenerg 1787(11): 1342–1351

271. el Dein OS, Gallerne C, Deniaud A, Brenner C, Lemaire C (2009) Role of the permeability transition pore complex in lethal inter-organelle crosstalk. Front Biosci Landmark 14:3465–3482
272. Pinton P, Giorgi C, Siviero R, Zecchini E, Rizzuto R (2008) Calcium and apoptosis: ER-mitochondria Ca2+ transfer in the control of apoptosis. Oncogene 27(50):6407–6418
273. Griffiths EJ (2009) Mitochondrial calcium transport in the heart: physiological and pathological roles. J Mol Cell Cardiol 46(6):789–803
274. Laude AJ, Simpson AWM (2009) Compartmentalized signalling: Ca2+ compartments, microdomains and the many facets of Ca2+ signalling. FEBS J 276(7):1800–1816
275. Contreras L, Drago I, Zampese E, Pozzan T (2010) Mitochondria: the calcium connection. Biochim Biophys Acta Bioenerg 1797(6–7):607–618
276. Tallini YN, Brekke JF, Shui B, Doran R, Hwang SM, Nakai J, Salama G, Segal SS, Kotlikoff MI (2007) Propagated endothelial Ca2+ waves and arteriolar dilation in vivo – measurements in Cx40(BAC)-GCaMP2 transgenic mice. Circ Res 101(12):1300–1309
277. Saito T, Nakatsuji N (2001) Efficient gene transfer into the embryonic mouse brain using in vivo electroporation. Dev Biol 240(1):237–246
278. Tabata H, Nakajima K (2001) Efficient in utero gene transfer system to the developing mouse brain using electroporation: visualization of neuronal migration in the developing cortex. Neuroscience 103(4):865–872
279. Heim N, Garaschuk O, Friedrich MW, Mank M, Milos RI, Kovalchuk Y, Konnerth A, Griesbeck O (2007) Improved calcium imaging in transgenic mice expressing a troponin C-based biosensor. Nat Methods 4(2):127–129
280. Davidson BL, Breakefield XO (2003) Viral vectors for gene delivery to the nervous system. Nat Rev Neurosci 4(5):353–364
281. Matz MV, Fradkov AF, Labas YA, Savitsky AP, Zaraisky AG, Markelov ML, Lukyanov SA (1999) Fluorescent proteins from nonbioluminescent Anthozoa species. Nat Biotechnol 17(10):969–973
282. Mizuno H, Sawano A, Eli P, Hama H, Miyawaki A (2001) Red fluorescent protein from Discosoma as a fusion tag and a partner for fluorescence resonance energy transfer. Biochemistry 40(8):2502–2510
283. Chudakov DM, Matz MV, Lukyanov S, Lukyanov KA (2010) Fluorescent proteins and their applications in imaging living cells and tissues. Physiol Rev 90(3):1103–1163
284. Shcherbo D, Shemiakina II, Ryabova AV, Luker KE, Schmidt BT, Souslova EA, Gorodnicheva TV, Strukova L, Shidlovskiy KM, Britanova OV, Zaraisky AG, Lukyanov KA, Loschenov VB, Luker GD, Chudakov DM (2010) Near-infrared fluorescent proteins. Nat Methods 7(10):827–829
285. Heim N (2005) Genetically encoded calcium indicators based on troponin C and fluorescent proteins. PhD thesis, Ludwig-Maximillians-University Munich, Munich
286. Carlson HJ, Cotton DW, Campbell RE (2010) Circularly permuted monomeric red fluorescent proteins with new termini in the beta-sheet. Protein Sci 19(8):1490–1499
287. Li YK, Sierra AM, Ai HW, Campbell RE (2008) Identification of sites within a monomeric red fluorescent protein that tolerate peptide insertion and testing of corresponding circular permutations. Photochem Photobiol 84(1):111–119

Action Potentials in Heart Cells

Lars Kaestner, Qinghai Tian, and Peter Lipp

Abstract Action potentials are a basic and fast communication mode within or in between neuronal and muscular cells in the human body. The rhythmic initiation and structured propagation of action potentials in the heart are most essential for a vital organism. The use of genetically encoded biosensors based on fluorescent proteins allows a non-invasive biocompatible way to read-out action potentials in cardiac myocytes. This comprises the physiological situation as well as pathophysiological models of the diseased heart. Although the approaches to design such biosensors date back to the time when the first fluorescent protein-based FRET sensors were constructed, it took 15 years until first reliable sensors became available. In this chapter, it is shown in cardiac myocytes how fluorescent protein-based action potential measurements can be used in pharmacological screening applications as well as in basic biomedical research. Potentials and limitations will be discussed and perspectives of possible future developments will be provided.

Keywords Cardiac action potential · Membrane potential sensor · Mermaid · QT-screen

Contents

L. Kaestner (✉), Q. Tian, and P. Lipp
Institute for Molecular Cell Biology, Medical Faculty, Saarland University, Building 61, 66421 Homburg, Saarland, Germany
e-mail: lars_kaestner@me.com

G. Jung (ed.), *Fluorescent Proteins II*, Springer Ser Fluoresc (2012) 12: 163–182,
DOI 10.1007/4243_2011_28
Published online: 18 June 2011

1 Introduction to Action Potentials in Cardiac Myocytes

Every heart beat is initiated by an electrical excitation. The formation and the propagation of this signal inside the heart can be measured by an electrocardiogram (ECG). Such an ECG recording detects the concert of the electrical signals of all excitable cells in the heart. This electrical transient is called action potential and is initiated either externally (i.e. from neighbouring cells) or occurs spontaneously.

Once the cells are isolated for scientific or pharmacological investigations, this excitation can be substituted by a field stimulation using short electrical pulses (≤ 5 ms) [1]. This electrical stimulation activates fast sodium channels followed by voltage-activated calcium channels that allow calcium to enter the cytoplasm. This calcium translates the electrical signal into a mechanical response of each cell and eventually into the entire heart beat. The calcium channels are counterbalanced by the activity of the electrogenic sodium/calcium exchanger. A plethora of potassium channels majorly shape the repolarisation phase of each action potential. Various physiological and pathophysiological conditions can modulate further molecular entities such as the sodium/potassium pump, chloride channels and additional potassium channels [2]. This rather complex concert of players give rise to a typical shape of an action potential that may vary from species to species. It is classified into four phases followed by a resting phase, all of which are visualised in Fig. 1. Depending on the heart rate, the action potential duration is in the range of hundreds of milliseconds.

Investigating and monitoring the alterations of cardiac action potentials is insofar important as most of the mayor cardiac diseases are also manifested in the action potential. This includes all forms of arrhythmia, cardiac infarction and indirectly coronary heart diseases.

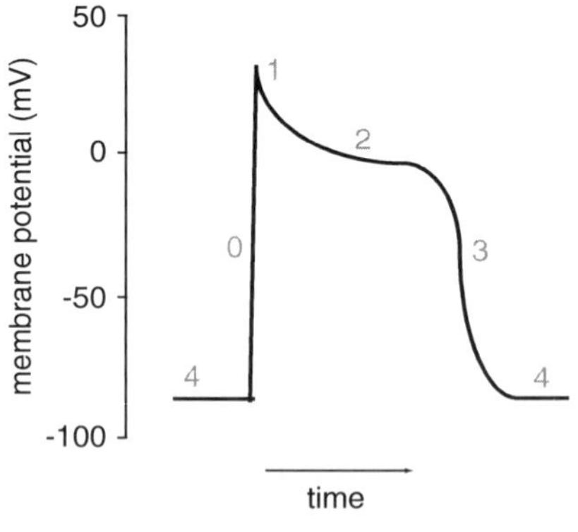

Fig. 1 Shape of a ventricular action potential. *Phase 0*: rapid depolarisation, *Phase 1*: initial repolarisation, *Phase 2*: plateau phase due to delay in repolarisation, *Phase 3*: repolarisation, *Phase 4*: resting membrane potential (associated with the diastole in the heart)

2 Measuring and Screening Cellular Action Potentials: Principles and Properties

Action potentials of muscles and nerves have been measured since the middle of the nineteenth century, e.g. [3–5], using metal electrodes. The transition to the cellular level was achieved by Cole and Curtis in the 1930s [6] by the development of the voltage-clamp technique. A quantum leap was the introduction of the patch-clamp technique by Neher and Sakmann [7, 8] which still is the Gold-standard for measurements of cellular action potentials. The big advantage of this approach is that the entire cell can be controlled, i.e. clamped to a given potential or alternatively to a given current (voltage-clamp and current-clamp mode, respectively). By controlling the intra- and extracellular solutions in combination with pharmacological manipulations, ionic currents underlying the action potential could be studied in great detail. However, there are some considerable disadvantages: (1) cells need to be mechanically disturbed by the glass pipette; (2) no spatial information concerning the subcellular potential distribution becomes available and (3) moving cells like cardiomyocytes within a beating heart cannot be approached. All these limitations can be overcome using contact-free optical read-outs. All optical sensors for membrane potential investigations developed so far, independent of whether they rely on small molecule dyes or genetically encoded chromophores, are based on membrane potential-induced changes of fluorescent properties. These sensors are described and classified in the following subchapters. Fluorescence read-out has general advantages over detection based on absorption [9], but for a high spatial (diffraction limited optical resolution) and temporal resolution ($\leq$1 ms) over a duration of minutes, the number of emitting photons becomes the limiting factor. Therefore, dark field read-outs based on abnormal dispersion could be an alternative concept meeting those challenges.

During the last few years, there has been a growing demand for screening of action potentials especially in cardiac myocytes. This was caused by the need to probe the proarrhythmic potential following drug administration regardless whether cardiac related or not [10]. The proarrhythmic potential appears as a prolongation of the QT-interval in the ECG. This is the temporal distance between the start of the QRS-complex (dominant peak in the ECG) and the end of the T-wave. Recognition of potential QT alterations at the earliest possible state is now an essential requirement in the drug discovery and development workflow. The cellular equivalent of the QT-interval is the action potential duration [11]. As a result of such regulations, several successful methodologies to automate action potential measurements (or electrophysiological properties in general) based on the patch-clamp technique were developed. A number of these approaches were turned into commercial products, e.g. Ionworks Quattro or PatchXpress 7000A from Molecular Devices (Sunnyvale, USA), NPC-16 Patchliner or SyncroPatch 96 from Nanion (Munich, Germany), Flyscreen from Flyion (Tübingen, Germany) or QPatch form Sophion (Copenhagen, Denmark). However, all these systems are designed to investigate immortalised cell lines rather than adult cardiac myocytes, which allow a well-defined handling routine. When these cells are not adherent, i.e. in suspension, they adopt a perfectly spherical shape that is ideal for such robotic approaches. These cell lines are useful tools if general cellular properties or properties of isolated ionic currents are to be investigated. If it comes to more specific physiological or pathophysiological questions, primary isolated cells are a much more appropriate tool [12]. This holds true especially when excitable cells such as cardiomyocytes or neurones are to be investigated. Due to easier isolation and handling, neonatal cardiac cells are often preferred in studies investigating specialised cell types. Recently, we were able to show that with respect to action potential properties, neonatal and adult cardiac myocytes depict a totally different pharmacological profiling. Therefore, it can be concluded that neonatal cells are not a valid model for the investigation of action potentials in the adult situation. On the other hand, a highly reproducible isolation procedure for adult cardiomyocytes is available [13], and cells can be cultured for 1 week without major de-differentiation, thereby enabling genetical manipulation of the cells such as expressing genetically encoded sensors and fusion proteins [1, 14]. However, the shape and fragility of isolated cardiomyocytes do not allow a reliable handling in suspension, as it is necessary for the commercially available automated cellular electrophysiology set-ups. Therefore, contact-free optical read-out would be advantageous. So far, optical action potential measurements were approached either by photometry, e.g. [15], detecting the signal of an entire cell or by random-access point scans [16] or for spatially resolved data by confocal line scans [17]. This restriction to line scans was necessary for a very high acquisition speed of approximately 1 kHz. For confocal and semi-confocal approaches, this can be performed by three principle scanning technologies: (1) resonant scanners as commercialised by Nikon (Tokyo, Japan) and Leica (Mannheim, Germany); (2) acousto-optical deflector (AOD) scanning as implemented in the VT_{eye} (VisiTech Int., Sunderland, UK); (3) real slit scanning as implemented in the LSM5Life (Zeiss, Jena, Germany).

With the latest developments of the sCMOS technology [18], camera acquisition rates in combination with high quantum efficiencies ($\sim$0.6 for front illuminated sensors) have reached a level that may allow sufficiently fast data acquisition while simultaneously maintaining spatial resolution to distinguish individual cells in a cell-based assay. This key technology in combination with new generations of voltage sensors (see below) may outperform the classical patch-clamp approaches in screening action potentials.

3 Optical Measurements of Action Potentials: An Overview

Depending on the particular mechanism of detecting voltage changes, optical sensors can be classified into four groups:

1. Dyes displaying a potential-dependent change in their transmembrane distribution, which leads to a large fluorescence change (typically a 100% change

per 100 mV). However, the redistribution process takes too long, usually seconds or even minutes, to be used for action potential recording [19]. Examples are rhodamine-based dyes to measure mitochondrial membrane potentials [20].

2. Sensors in which potential-dependent changes in the conjugated double-bond structures take place. This leads to sensing of the electromagnetic field changes in their vicinity that translates into changes of their fluorescence properties. This type of intramolecular electron redistribution can take place within microseconds; so the dyes are fast enough to follow action potentials. The electron rearrangement takes place inside a small molecule – therefore, its magnitude leads to a small signal, usually in the range of 10% per 100 mV. Typical dyes, such as RH-237, di-4-ANEPPS and di-8-ANEPPs, are widely used in cellular cardiology [15, 21–24]. Especially di-8-ANEPPS is a popular dye because it shows a comparably low rate of internalisation into cells [25]. Additionally, it can be used in a ratiometric read-out [26, 27], which is fundamental when accounting for contraction artefacts [28, 29].
3. Probes relying exclusively on the Stark effect, such as ANNINE dyes [30, 31], which have also been used to measure spatially resolved action potentials in cardiac myocytes [17].
4. Genetically encoded voltage sensors, which are introduced in great detail in Sect. 4.

Figure 2 depicts a time scale with milestones of developments in optical membrane potential sensors. It covers the entire time range for the genetically encoded sensors from the identification of GFP as a protein to *in vivo* application of such sensors.

In summary, small molecule dyes have a number of superb properties, but suffer from the unspecific staining of all cell types. A possible alternative is a genetic targeting using tissue-specific promotors. In addition to the fluorescent protein-based voltage sensors (see Sect. 4), it is worthwhile to mention the combination of genetic targeting and conventional organic chromophores [32]. A proof of principle was reported by combining a membrane-targeted phosphatase which was able to cleave the hydrophilic phosphate group of a precursor dye, leading to a membrane-bound voltage-sensitive dye [33].

4 Fluorescence Protein-Based Approaches to Sense Membrane Potentials

A further hybrid approach uses the expression of a membrane-bound GFP as FRET donor in combination with dipicrylamine (DPA), a synthetic voltage-sensing molecule, as FRET acceptor [34]. This hybrid approach has the advantage of genetic targeting but suffers from the common disadvantages of small molecule dyes such as the limitations for long-term observation. There are controversial

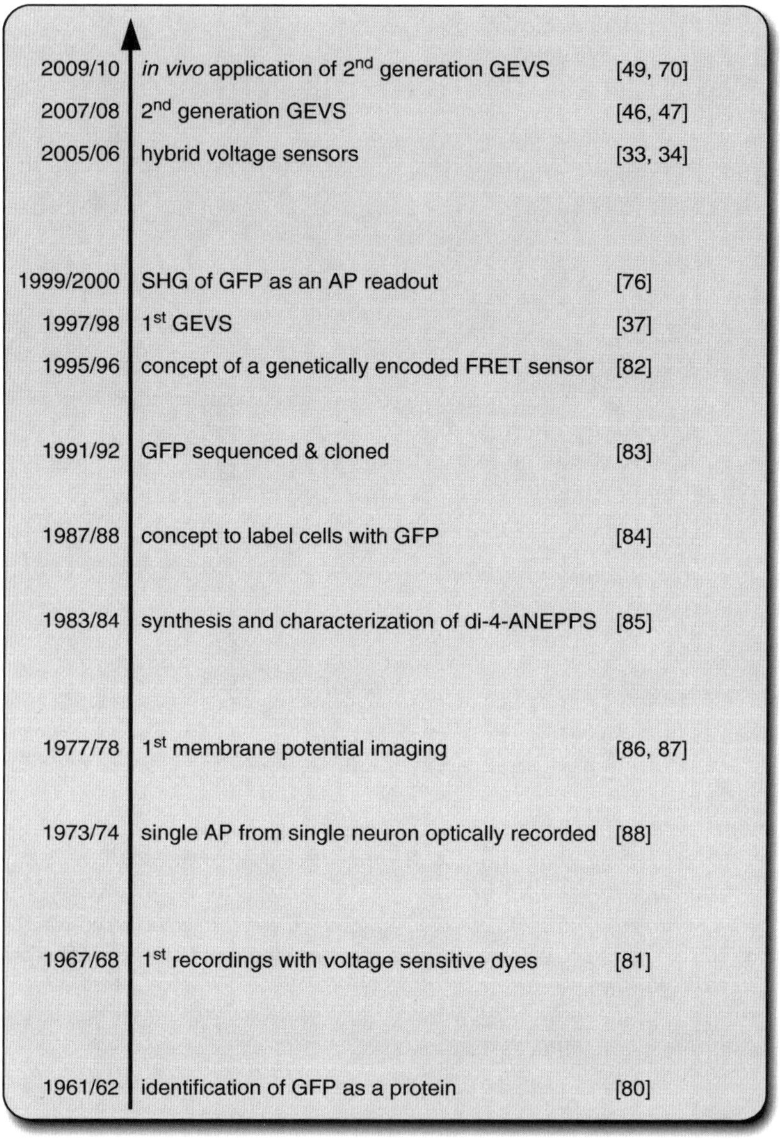

Fig. 2 Half a century of scientific innovations to bring genetically encoded voltage sensors to the present level. Abbreviations used in this figure: *GEVS* genetically encoded voltage sensor, *SHG* second harmonic generation, *GFP* green fluorescent protein, *FRET* Förster resonance energy transfer, *AP* action potential

reports of the usability of this construct. While DiFranco and colleagues applied it successfully in the transverse tubules in mouse skeletal muscle fibres [35], Sjulson and Miesenböck showed that – due to DPA-induced increase in membrane capacitance – it was not possible to detect action potentials in the *Drosophila* antennal lobe [36].

To revisit voltage sensors solely comprising genetically encoded proteins, we need to go back in time. The first of such sensors (called FlaSh) was a wtGFP fused to the C-terminus of the *Drosophila* Shaker K^+-channel [37]. This and its follow-up sensors were composed of a voltage-sensing domain in combination with one or several fluorescent proteins. Later, FlaSh was improved in terms of dynamics and voltage range [38]. The second voltage sensor generated independently was based on the fourth transmembrane segment (S4) of the voltage-gated K^+ channel $K_v2.1$ coupled to a CFP/YFP FRET pair in sequence and was named VSFP1 [39]. This sensor was followed by a circular permutated version of the fluorescent protein [40]. The third sensor type was called SPARC and comprised of a fused GFP between domains I and II of the rat skeletal muscle Na^+ channel [41]. These three sensors are subsumised under the first generation of voltage sensors [42]. They all display a rather bad membrane localisation [43], and the fluorescence intensity change per 100 mV potential change is limited to only 0.5–5% [37, 39, 41]. While the first generation of voltage sensors are based on voltage-activated ion channels or domains thereof, the second generation of voltage sensors comprised self-contained voltage sensor domains, such as the voltage-sensing domain of the *Ciona intestinalis* voltage sensor-containing phosphatase (Ci-VSP) [44] or voltage sensor domain only proteins (VSOPs) [45].

The Ci-VSP domain was chosen by two groups of the same institution (Brain Science Institute, RIKEN, Japan) that performed independent approaches to develop what is now termed second generation voltage sensors: VSFP2.1 [46] and Mermaid [47]. These initial papers in mammalian cells reported ratio changes of approximately 9% and 20% per 100 mV at around 34°C for VSFP2.1 and Mermaid, respectively. Since Mermaid is up to now the only genetically encoded potential sensor that was successfully applied in cardiac cells and tissue, it will be described in greater detail in a separate Sect. 6.

Based on the VSFP2.1 design, further developments were performed. Linker optimisation led to VSFP2.3, which was characterised biophysically [48] and in an *in vivo* proof of principle in the cortex of mice [49]. In a recent study using spectrally resolved data, VSFP2.3 and Mermaid displayed similar ratio changes of around 13% per 100 mV under putatively similar conditions [50].

A further line of developments explored alterations in the fluorescent entities: circular permutated fluorescent proteins [51] and red-shifted variants [52] with positive proof of principles but moderate intensity changes were generated.

The second generation of genetically encoded voltage sensors share the property of a fast and a slow kinetic component. The fast component corresponds to the sensing currents in the voltage-sensing domain, while the slow component reflects the voltage-dependent conformational change [53]. The Knöpfel group performed seminal work in linker and fluorescent read-out optimisation and introduced a novel probe named VSFP3.1 [54] in which the kinetic properties of the slow sensing component was dramatically shifted towards faster sensing. Accordingly, they designed the fastest genetically encoded voltage sensor so far. This approach is based on a reaction of the voltage-sensing domain without major conformational changes of a single fluorescence probe without taking advantage of FRET. This results in intensity changes of approximately 0.5% per 100 mV potential change [54]. When considering the overall properties of VSFP3.1, it appears to be of limited use, and we would not yet regard it as a third generation voltage sensor as claimed by the Knöpfel group [55].

5 Concepts to Transduce Sensors in End-Differentiated Cells

Genetically encoded voltage sensors similar to other genetically coded sensors require that the genetic information, coding for the sensor, will be incorporated into the cell of interest. Popular chemical transfection methods such as catalysing DNA cross-membrane transport using calcium phosphate, polycations or dendrimers as well as physical transfection approaches like electroporation, microinjection or particle guns provide only very low transfection rates in end-differentiated cardiac myocytes. Therefore, they are ousted by biological transfection methods. Nevertheless, even the biological transfection methods such as the popular lipofection, which leads to a 5–100-fold increased transfection rate compared to chemical methods, fail in cardiac myocytes. Viral gene transfer was identified as the transduction method of choice for end-differentiated cell such as cardiac myocytes. To perform viral transduction, we evaluated various viruses: Semliki-Forest virus [56], Lentivirus [57] and Adenovirus [58].

In our hands in cardiac myocytes, Semliki-Forest virus transduction leads to a transduction rate lower than 5%. In neurons, a fast expression of the transducted gene was achieved within 6 h, but it killed all cells within 24 h [59]. In contrast, Lentiviral gene transfer required about a week for protein expression. This is far too long for established culture systems of cardiac myocytes [1, 14]. Therefore, Lentiviruses may well serve for gene transfer in animals, but they are not suitable for gene transfer in primary cultured cells such as cardiac myocytes. Using an adenoviral gene transfer approach, protein expression reached sufficient levels within 24 h for most genetically encoded biosensors tested, e.g. Mermaid – see below or the Ca^{2+} sensors inverse pericam, YC3.6 and TN-XL [1, 59]. The adenoviral-mediated expression is fairly stable over at least 1 week, and we could not find adverse effects of virus transduction and induced gene expression. Therefore, the adenoviral approach appears suitable for high content screening applications in primary isolated cardiomyocytes and enables monitoring of chronic applications of test substances [59].

For various scientific questions, it is advantageous to perform the sensor transduction and expression not just *in vitro* but already *in vivo*. A very elaborate approach is the breeding of transgenic animals with a cardiac-specific expression of the sensor as it was performed with Mermaid in zebrafish (see below) or for the Ca^{2+} sensor GCaMP2 in mice [60]. Another approach is to transduce the sensor into individual animals, again with the help of a virus. Although there are reports to perform adenoviral-mediated gene silencing in the heart [61], in our hands the expression of the sensor could not be achieved in the heart neither after injection in neonatal hearts nor in adult hearts of rat and mice. This is most probably due to the immune responses of the animals. Therefore, we regard adeno-associated viruses (AAVs) as an alternative. Since the AAV infection starts with a receptor-mediated endocytosis, the different serotypes in combination with distinct intracellular processing give rise to specific tissue tropism [62]. Additionally, tissue-specific expression can be fostered by cloning of the appropriate tissue-specific promotor into the AAV genome. This is a popular methodology in concepts of gene therapy [63], but was also shown for cardiac-specific expression of fluorescent proteins [64, 65].

6 "Mermaid": Characterisation and Application in Cardiac Myocytes

"Mermaid" is a novel sensor introduced in 2008 [47] in a row of developments of continuously improving voltage sensors (see Sect. 4). The FRET-based approach was realised with the voltage sensor containing phosphatase of sea squirt *C. intestinalis* as the base for the conformational change similar to what was originally introduced by Murata et al. [44]. In the Miyawaki laboratory, this sensing unit was fused to two mutated coral-derived fluorescent proteins [47, 66]. These proteins are the monomeric Umi Kinoko (mUKG) and the monomeric Kusabira Orange (mKOκ) derived from *Sarcophyton* sp. and *Fulgoraria concinna*, respectively. A big drawback of previous genetically encoded membrane potential sensors was their limited plasma membrane localisation [43]. Here we show a comparison between the small molecule dye di-8-ANEPPS that provides almost exclusive membrane staining [14] and Mermaid (Fig. 3). Panels (a) and (b) depict the subcellular localisation of the Mermaid and di-8-ANEPPS, respectively. Panel (c) depicts the overlay of images (a) and (b), and panel (d) a pixel-based colocalisation analysis of the two sensors.

We found that approximately 83% of the mermaid protein was located at the plasma membrane. The entities of Mermaid non-colocalising can be divided into two pools: (1) sensor proteins that are not localised at the plasma membrane, but instead in the perinuclear region, representing Mermaid in the Golgi apparatus or other organelles such as the sarcoplasmic reticulum; (2) sensor located in the plasma membrane that cannot be accessed by di-8-ANEPPS. We could recently show that transversal membrane tubules in cardiomyocytes in culture subsequently pinch off [14]. This was achieved with a membrane-localising fluorescent fusion protein (GPI-GFP) under very similar conditions as applied here (rat ventricular myocytes, same viral expression system). If parts of the transversal membrane tubules are disconnected to the outer plasma membrane, they cannot be accessed by di-8-ANEPPS staining, but a (fluorescent) protein can still be expressed.

The second important property is the signal alteration upon potential change. Mermaid displayed relative ratio changes between 7% and 30% per 100 mV

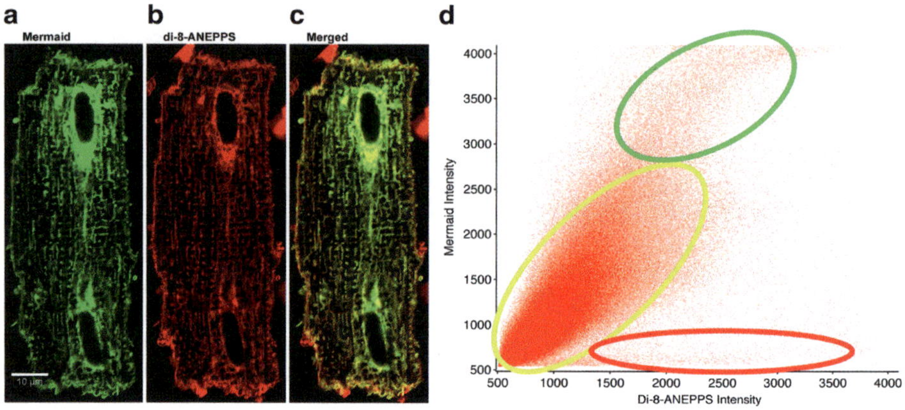

Fig. 3 Membrane localisation of Mermaid. Images (**a**) and (**b**) visualises a cardiac myocyte with a preferential detection of Mermaid and di-8-ANEPPS, respectively. Image (**c**) is merged out of the previous images, while panel (**d**) providing a pixel-based colocalisation analysis. The *yellow* surrounded area depicts the colocalising pixels. The *red ellipsis* points out pixels from membrane fragments outside the Mermaid expressing cell on the edge of the image, while the *green* surrounded area refers to pixels of exclusive Mermaid localisation (mainly in the perinuclear region). The "red component" in these pixels is caused by the spectral cross talk. For spectral information and further global unmixing, refer to [29]

membrane change (in the −80 mV to +20 mV interval depending on the cell type and/or recording conditions) [29, 47] and therefore compares well with small molecular sensors such as RH-237 or di-8-ANEPPS [20]. We even noticed a 25% higher change of the relative fluorescence ratio compared to the ratiometric read-out mode of di-8-ANEPPS [29]. We have to note that these values refer to a simple ratio of the two spectral channels allocated to the FRET donor and acceptor. Calculation of the real FRET efficiency or the apparent FRET efficiency [67] was achieved so far neither by us nor by others.

As stated by the designers of Mermaid, "it is not easy to bring out the best in Mermaid", (Miyawaki A, Tsutsui H, 2008, personal communication). This refers to the practical aspect of using the combination of the novel coral-derived pair of fluorescent proteins that do not fit to spectral recording conditions of the popular and widespread fluorescent protein pair CFP and YFP (including their more inert derivatives). Especially the excitation in the range between 400 and 440 nm results in substantial bleaching of mKOκ.

The relatively slow (in terms of frame rate) FLIM technologies [68] in combination with the entirety of the photophysical properties of Mermaid did not allow to record action potentials based on fluorescence lifetime changes. However, we performed fluorescence lifetime measurements of cells at a resting membrane potential. Since we do not know the fluorescence lifetime of mUKG in the absence of the acceptor, the fluorescence lifetime in the donor channel does not provide information whether FRET already occurs in the resting state or not.

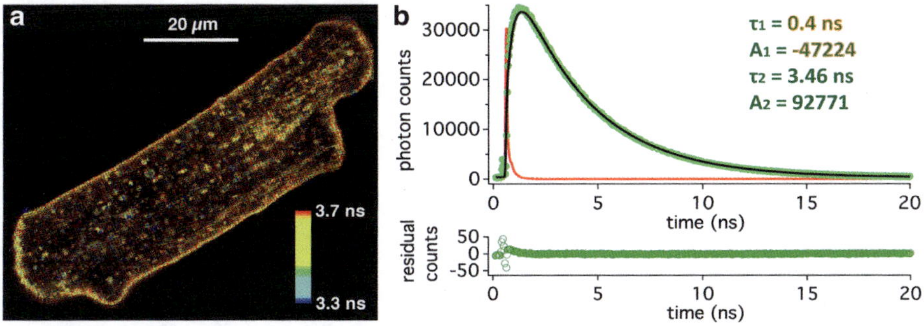

Fig. 4 Fluorescence lifetime imaging (FLIM) of a Mermaid expressing cardiac myocyte. (**a**) A FLIM image of ventricular cardiomyocyte. The fluorescence was recorded in the spectral region of 570–640 nm, where one detects exclusively the emission of the FRET acceptor. When fitting the measured time resolved photons [*light green circles* in (**b**)] we considered the instrument response function [*red curve* in (**b**)]. Nonetheless, when performing a double exponential fit [*black line* in (**b**)], we received two time constants of 0.4 and 3.46 ns, with surprisingly a negative amplitude for τ_1 (−47,224) and a positive amplitude for τ_2 (92,771). Since τ_1 is not caused by the instrument response function, it must account for the "switching on" behaviour of the FRET between the Mermaid proteins. We could find such a negative amplitude in approximately 50% of the cells investigated. When the negative amplitude was absent, the fluorescence lifetime could be nicely fitted with a monoexponential decay. Therefore, it can be concluded that under certain conditions, there is a FRET in cardiac myocytes under resting conditions of the membrane potential. It is still open what parameters determine this distribution. However, to our knowledge this is not related to Mermaids expression level

Therefore, we analysed at the Mermaid fluorescence lifetime in the spectral region of the acceptor mKOκ (570–640 nm). As depicted in Fig. 4, we could fit the time course only well when enabling negative values for one of the two amplitudes in a double exponential approach. The decay time constant was in the range of 0.3–0.4 ns. Since we already considered the instrument response function, this negative amplitude must refer to a photophysical induced "switch on" process. According to FRET theory, the time constant corresponding to the "switching on" process of the acceptor should be equal to the decay time of the donor. The concentration of the excited acceptor $[A^*]$ can be written as:

$$[A^*] = \frac{[D^*]_0 k_{\mathrm{T}}}{\frac{1}{\tau_{\mathrm{D}}} - \frac{1}{\tau_{\mathrm{A}}^0}} \left[e^{-\frac{t}{\tau_{\mathrm{A}}^0}} - e^{-\frac{t}{\tau_{\mathrm{D}}}} \right],$$

where $[D^*]_0$ is the concentration of the excited donor at the time $t = 0$, k_{T} is the transfer rate, τ_{D} is the donor lifetime and τ_{A}^0 the acceptor lifetime at the time $t = 0$ [69].

However, this consideration applies only when (1) the donor has a monoexponential decay time, (2) there is no direct excitation of the acceptor and (3) an isotropic dynamic average applies. For Mermaid, none of the three conditions are

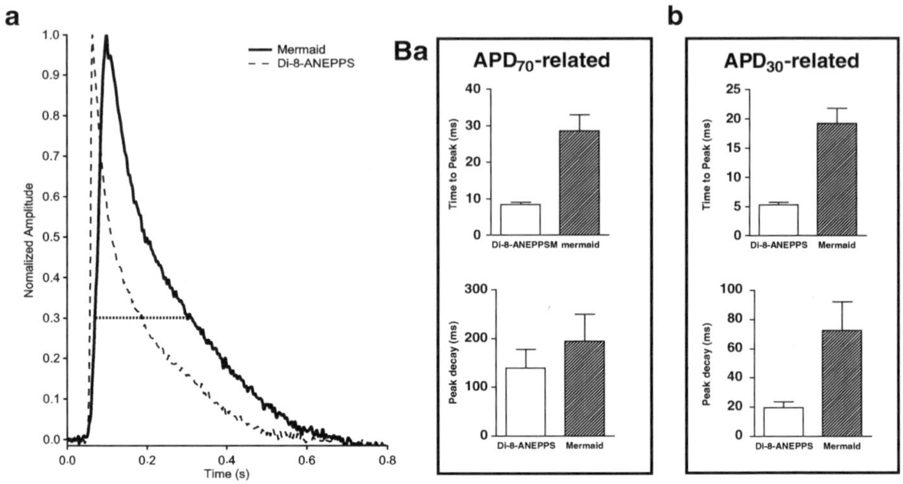

Fig. 5 Comparing kinetic properties of di-8-ANEPPS and Mermaid. Panel (**a**) compares normalised action potentials between di-8-ANEPPS-stained cardiomyocytes (*dashed line*) and Mermaid expressing cardiomyocytes (*solid line*). The cells were measured at 35°C and the potentials are averages of five cells. The *horizontal dotted line* indicates the APD_{70} line. Panel (**b**) depicts the statistical analysis [time to peak (*top*) and peak decay (*bottom*)] for APD_{70} (Ba) and APD_{30} (Bb)

fulfilled because (1) the decay time of mUKG is double exponential (1.25 and 3.08 ns) [29], (2) at the applied excitation band (450–490 nm) mKOκ is directly excited [47] and (3) the fluorescent proteins are connected intramolecularly and therefore the dipole orientational factor is more less fixed. Nevertheless, to our understanding the negative amplitude we observed is compatible with the "switch on" behaviour of a FRET acceptor mKOκ.

Furthermore, Mermaid displays a rather slow kinetic, as the time constant for the switch on is in the range of 10–20 ms at about 35°C, and at 19°C a slow component around 0.5 s appears [29, 47] (compare also Fig. 5). Nevertheless, in neurones it was possible to detect a 100 Hz stimulation frequency by Fourier analysis [47].

Moreover, pharmaceutical prolongation of action potentials in adult cardiomyocytes was obvious to detect [29]. This paves the way for a potential application: pharmacological safety screens (compare Sect. 2).

A second application area arose with the transgenic zebrafish (*Danio rerio*) heart expressing Mermaid under a cardiac-specific promoter [70]. In this work, the first *in vivo* imaging of voltage dynamics in a whole heart was reported, using a genetically encoded voltage sensor. Figure 6 presents the key results. Although this study is eminent to have the first optical *in vivo* potential recordings of the entire heart, transfer from zebra fish cardiac physiology to relevant tasks of mammalian and human cardiology is rather equivocal. Nonetheless, this animal model provides a huge potential to address questions in developmental biology as well as the high regenerative capability of cardiac tissue in zebra fish. Despite the fact that we regard Mermaid as the best genetically encoded potential sensor so far, there is

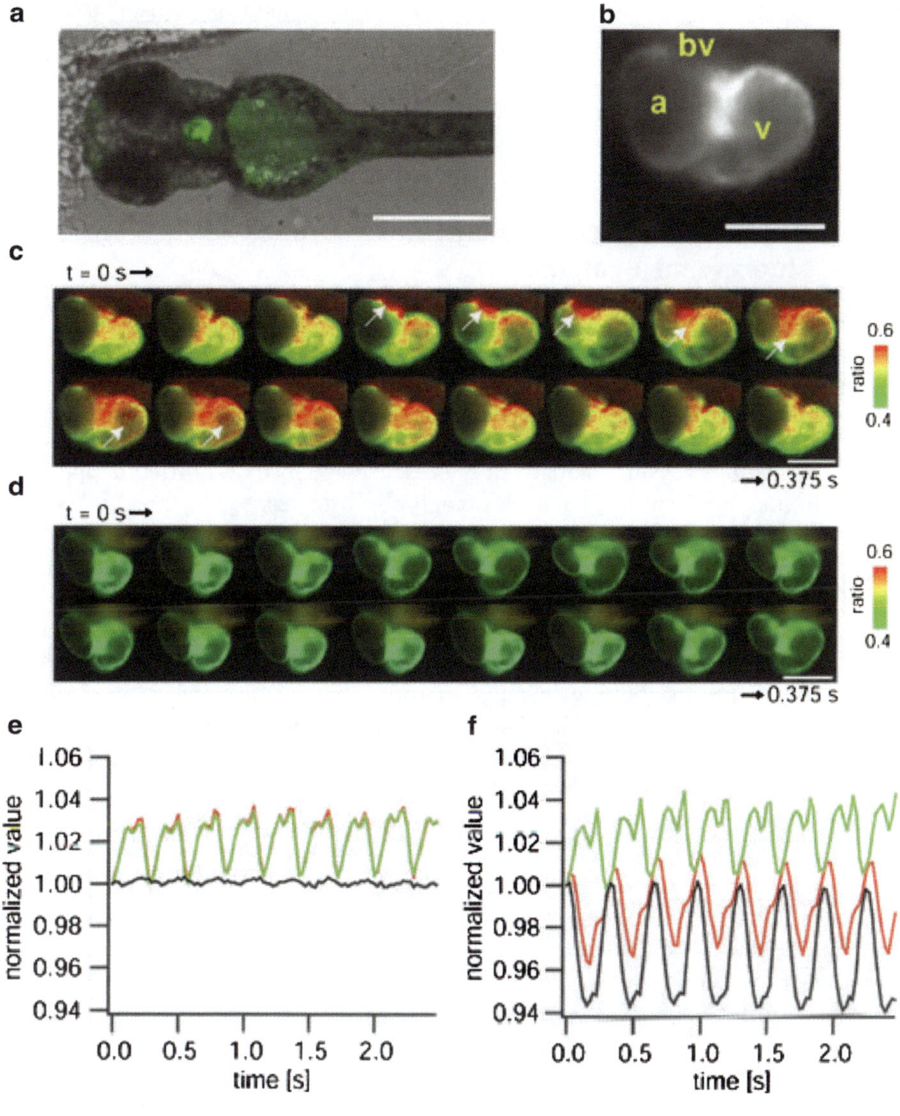

Fig. 6 Voltage dynamics of a beating heart in zebrafish. (**a**) Transgenic zebra fish (cmlc2::mermaid). An unanaesthetised, unrestrained fish was imaged ventrally. A donor fluorescence image (mUKG channel; in *green*) was imposed with transmission image (in *grey*). A 4× objective was used here. (**b**) A representative donor channel fluorescence image. *a* atrium, *v* ventricle, *bv* blood vessel. (**c**, **d**) pseudocoloured ratio (mKOκ/mUKG) images representing a single cycle of heart contraction in the cmlc::mermaid (**c**) and mermaid-null (**d**) transgenic fish. Propagation of excitation from the junction of vessel and atrium to the ventricle was highlighted with *arrows* in (**c**). (**e**, **f**) Plot of total fluorescence intensity from a whole beating heart versus time in the mermaid-null (**e**) and mermaid (**f**) transgenic fish. *Green* = mUKG (donor) channel. *Red* = mKOκ (acceptor) channel. *Black* = ratio (mKOκ/mUKG). *Bar* = 500 μm (**a**) and 100 μm (**b–d**). This figure is reprinted from Tsutsui et al. [70], Copyright 2010, with permission from John Wiley and Sons

still room for improvements, especially in terms of its kinetic behaviour (see above). This could be approached by the design of smaller sensors and sensing units inside beta-barrels instead of moving beta-barrels against each other (see Sect. 7).

7 A Perspective Based on Non-Linear Properties of Fluorescent Proteins

As described in Sect. 3, all previous attempts of genetically encoded voltage sensors are based on native voltage-sensing proteins that rely in their function on conformational changes, such as voltage-gated ion channels or voltage-dependent phosphatases. Their mechanical action towards fluorescent proteins is used to induce and subsequently measure changes in fluorescence intensity. A different approach would be the exploration of the chromophores with respect to direct effects of the membrane potential on the chromophore itself. The Stark effect caused by electric field changes is used in voltage sensor concepts for small molecular dyes, e.g. [30]. However, for chromophores of fluorescent proteins, this effect is too small to be detected by fluorescence microscopy. This leads us to other properties of the chromophores that have been vastly ignored in the development of biosensors in general. These are the non-linear properties allowing the formation of second harmonic generation (SHG) in response to femtosecond pulsed infrared light. The fact that fluorescent proteins induce measurable SHG and are able to modulate SHG upon external stimulation was shown recently [71, 72]. The process of SHG in protein chromophores can be modulated by external electrical fields. The change of the static dipole moment upon excitation, which influences the second-order polarisability, has already been determined for GFP [73]. Also the direction of the transition dipole moment is known [74, 75]. These changes are fast enough to follow changes of the transmembrane potential. The structural changes of the anionic chromophore in GFP can be described by two mesomeric structures in Fig. 7a.

One may think that all that considerations are rather theoretical, but indeed it has been shown that based on SHG alterations even native GFP could be used to measure membrane potential alterations in neurons of *Caenorhabditis elegans* [76].

Since it is possible to measure membrane potential-induced SHG changes of GFP *in vivo*, a development and optimisation of fluorescent proteins towards favourable properties for SHG changes could lead to a most favourable genetically encoded membrane potential sensor based on a completely novel concept that has conceptional advantages over FRET-based sensors (see above). To make such an approach successful, efforts are required in (1) improving the sensor; (2) innovative recording modes in combination with novel excitation and detection technologies.

1. As for the sensor, point mutations are one possibility to improve the chromophore's electronic properties to optimise the SHG signal. Orientational alterations with respect to the external electric field, i.e. the membrane potential, can

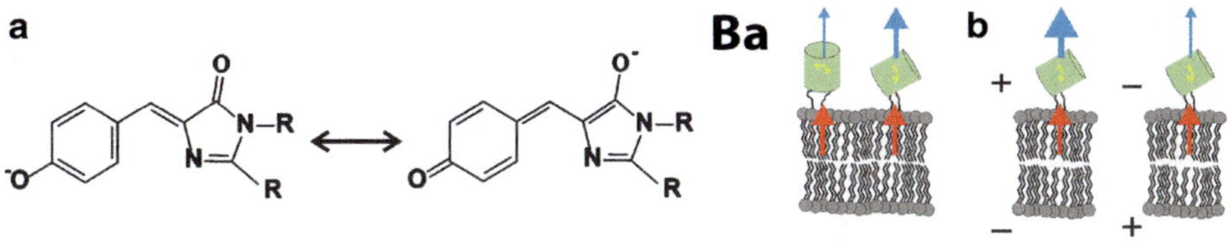

Fig. 7 (**a**) The mesomeric phenolic (*left*) and quinoid (*right*) structure describe the electronic chromophore properties. Alternating electric field displaces the electronic cloud asymmetrically, which results in a SHG signal. (**b**) Favourable orientations of the chromophore can be realised using circularly permuted proteins (Ba). If it turns out that an orientation of the chromophore's long axis parallel to the surface normal leads to reliable SHG signals, then a strong dependence of the SHG-signal on the transmembrane potential can be expected (Bb)

be used to further improve the sensor by modifications of the primary structure of the fluorescent protein: In a geometry where the chromophore's long axis is parallel to the external electric field, i.e. in circularly permuted proteins [77, 78], the intensity of the SHG signal should be sensitive to changes of the electric field. Such a concept is visualised in Fig. 7b.

2. Such a sensor – if used to image cardiac action potentials – requires much higher acquisition speeds than the recordings of *C. elegans* [76]. Since focused laser light is required, photometry and video-imaging as performed in Figs. 5 and 6 are no option. However, multibeam two-photon scanners [79] commercialised by LaVision BioTec (Bielefeld, Germany) enable voxel dwell times prolonged two orders of magnitude compared to single point scanners. This would still allow imaging rates of several hundred frames per second. Since the intensity of the SHG scales with the square of laser pulse energy, laser sources with short pulses below 50 fs would be advantageous and perspectively might be provided by emerging fibre laser technologies. In addition to that, the microscope needs to fulfil two requirements: (1) high sensitive detection of SHG changes and (2) detection of two-photon-induced chromophore fluorescence for calibration/normalisation.

The sum of the improvements may lead to the next generation of genetically encoded membrane potential sensors. Since they rely on effects of the electrical field on the chromophore, the kinetics should be much faster as in previous sensors. They have the intrinsic property for normalisation, because SHG and fluorescence are generated by the same molecular structure. Finally, if properly attached to the membrane (see above), the field changes should induce sufficiently high modulation (contrast) in the SHG.

8 Conclusions

Presently, the development and characterisation of genetically encoded voltage sensors are boosted (1997–2007: approximately one paper a year; 2008–2010: more than five papers a year). The major improvements of the sensors are accompanied by an increasing number of scientists recognising the tremendous potential of such genetically encoded probes. However, all paper on genetically encoded voltage sensors focus on the engineering, characterisation and proof of principle, but do not report their application in physiology or pathophysiology-driven studies. The direct comparison between small molecule dyes and genetically encoded sensors was so far avoided. However, when all of these probes are applied in a ratiometric manner, their performance is in a comparable range [29].

Although the development of the sensors was driven by neuroscience applications, the examples presented here on cardiac cells and cardiac tissues pave the way for an application beyond basic science, into pharmaceutical industry, especially for safety screens. Since genetically encoded voltage sensors can still be regarded to be in the early stages of their development, a further improvement can be expected. With these improvements, the applicability of genetically encoded voltage sensors will rise and make it a standard method in electrophysiology.

References

1. Viero C, Kraushaar U, Ruppenthal S, Kaestner L, Lipp P (2008) A primary culture system for sustained expression of a calcium sensor in preserved adult rat ventricular myocytes. Cell Calcium 43:59–71
2. (1991) The Sicilian gambit. A new approach to the classification of antiarrhythmic drugs based on their actions on arrhythmogenic mechanisms. Task Force of the working group on Arrhythmias of the European Society of Cardiology. Circulation 84:1831–1851
3. Matteucci C (1842) Sur un phenomene physiologique produit par les muscles en contraction. Ann Chim Phys 6:339–341
4. Bois-Reymond (ed) (1848) Untersuchungen über Thierische Elektricität. Verlag von G. Reimer, Berlin
5. Hv H (1867) Handbuch der physiologischen Optik. Leopold Voss, Leipzig
6. Cole KS (1979) Mostly membranes (Kenneth S. Cole). Annu Rev Physiol 41:1–24
7. Hamill OP, Marty A, Neher E, Sakmann B, Sigworth FJ (1981) Improved patch-clamp techniques for high-resolution current recording from cells and cell-free membrane patches. Pflugers Arch 391:85–100
8. Sigworth FJ (1986) The patch clamp is more useful than anyone had expected. Fed Proc 45: 2673–2677
9. Kaestner L, Lipp P (2007) Towards imaging the dynamics of protein signalling. In: Shorte SL, Frischknecht F (eds) Imaging cellular and molecular biological functions. Springer, Berlin, pp 289–312
10. ICH. (2005) S7B nonclinical evaluation of the potential for delayed ventricular repolarization (QT interval prolongation) by human pharmaceuticals. FDA & EMEA, pp 1–10
11. Arrigoni C, Crivori P (2007) Assessment of QT liabilities in drug development. Cell Biol Toxicol 23:1–13
12. Lipp P, Kaestner L (2006) Image based high content screening – a view from basic science. In: Hüser J (ed) High-throughput screening in drug discovery. Wiley VCH, Weinheim, pp 129–149
13. Kaestner L, Scholz A, Hammer K, Vecerdea A, Ruppenthal S, Lipp P (2009) Isolation and genetic manipulation of adult cardiac myocytes for confocal imaging. J Vis Exp 31

14. Hammer K, Ruppenthal S, Viero C, Scholz A, Edelmann L, Kaestner L, Lipp P (2010) Remodelling of Ca(2+) handling organelles in adult rat ventricular myocytes during long term culture. J Mol Cell Cardiol 49:427–437
15. Hardy ME, Pollard CE, Small BG, Bridgland-Taylor M, Woods AJ, Valentin JP, Abi-Gerges N (2009) Validation of a voltage-sensitive dye (di-4-ANEPPS)-based method for assessing drug-induced delayed repolarisation in beagle dog left ventricular midmyocardial myocytes. J Pharmacol Toxicol Meth 60:94–106
16. Bullen A, Saggau P (1999) High-speed, random-access fluorescence microscopy: II fast quantitative measurements with voltage-sensitive dyes. Biophys J 76:2272–2287
17. Bu G, Adams H, Berbari EJ, Rubart M (2009) Uniform action potential repolarization within the sarcolemma of in situ ventricular cardiomyocytes. Biophys J 96:2532–2546
18. Coates C, Fowler B, Holst G (2009) Scientific CMOS technology - a high-performance imaging breakthrough. In: FI Andor Technology PLC, PCO AG (ed) www.scmos.com. Andor Technology PLC, Fairchild imaging, PCO AG, Belfast, Milpitas, Kehlheim, pp 1–14
19. Sims PJ, Waggoner AS, Wang CH, Hoffman JF (1974) Studies on the mechanism by which cyanine dyes measure membrane potential in red blood cells and phosphatidylcholine vesicles. Biochemistry 13:3315–3330
20. Haugland RP (2002) Handbook of fluorescent probes and research products. Molecular Probes, Eugene
21. Huser J, Lipp P, Niggli E (1996) Confocal microscopic detection of potential-sensitive dyes used to reveal loss of voltage control during patch-clamp experiments. Pflugers Arch 433: 194–199
22. Lipp P, Huser J, Pott L, Niggli E (1996) Spatially non-uniform Ca^{2+} signals induced by the reduction of transverse tubules in citrate-loaded guinea-pig ventricular myocytes in culture. J Physiol 497(Pt 3):589–597
23. Dibb KM, Clarke JD, Horn MA, Richards MA, Graham HK, Eisner DA, Trafford AW (2009) Characterization of an extensive transverse tubular network in sheep atrial myocytes and its depletion in heart failure. Circ Heart Fail 2:482–489
24. Sill B, Hammer PE, Cowan DB (2009) Optical mapping of Langendorff-perfused rat hearts. J Vis Exp. doi:10.3791/1138
25. Loew LM (1996) Potentiometric dyes: imaging electrical activity of cell membranes. Pure Appl Chem 68:1405 1409
26. Bedlack RS Jr, Wei M, Loew LM (1992) Localized membrane depolarizations and localized calcium influx during electric field-guided neurite growth. Neuron 9:393–403
27. Gross E, Bedlack RS Jr, Loew LM (1994) Dual-wavelength ratiometric fluorescence measurement of the membrane dipole potential. Biophys J 67:208–216
28. Hardy ME, Lawrence CL, Standen NB, Rodrigo GC (2006) Can optical recordings of membrane potential be used to screen for drug-induced action potential prolongation in single cardiac myocytes? J Pharmacol Toxicol Meth 54:173–182
29. Kaestner L, Tian Q, Lipp P (2011) Cardiac Safety Screens: Molecular, cellular and optical advancements. In: Lin C P, Ntziachistos V (eds) SPIE biomedical optics III, vol. 8089, SPIE, Munich, in press
30. Kuhn B, Fromherz P, Denk W (2004) High sensitivity of Stark-shift voltage-sensing dyes by one- or two-photon excitation near the red spectral edge. Biophys J 87:631–639
31. Fromherz P, Hubener G, Kuhn B, Hinner MJ (2008) ANNINE-6plus, a voltage-sensitive dye with good solubility, strong membrane binding and high sensitivity. Eur Biophys J 37: 509–514
32. Hinner MJ, Hbener G, Fromherz P (2004) Enzyme-induced staining of biomembranes with voltage-sensitive fluorescent dyes. J Phys Chem B 108:2445–2453
33. Hinner MJ, Hubener G, Fromherz P (2006) Genetic targeting of individual cells with a voltage-sensitive dye through enzymatic activation of membrane binding. Chembiochem 7: 495–505

34. Chanda B, Blunck R, Faria LC, Schweizer FE, Mody I, Bezanilla F (2005) A hybrid approach to measuring electrical activity in genetically specified neurons. Nat Neurosci 8: 1619–1626
35. DiFranco M, Capote J, Quinonez M, Vergara JL (2007) Voltage-dependent dynamic FRET signals from the transverse tubules in mammalian skeletal muscle fibers. J Gen Physiol 130: 581–600
36. Sjulson L, Miesenbock G (2008) Rational optimization and imaging *in vivo* of a genetically encoded optical voltage reporter. J Neurosci 28:5582–5593
37. Siegel MS, Isacoff EY (1997) A genetically encoded optical probe of membrane voltage. Neuron 19:735–741
38. Guerrero G, Siegel MS, Roska B, Loots E, Isacoff EY (2002) Tuning FlaSh: redesign of the dynamics, voltage range, and color of the genetically encoded optical sensor of membrane potential. Biophys J 83:3607–3618
39. Sakai R, Repunte-Canonigo V, Raj CD, Knopfel T (2001) Design and characterization of a DNA-encoded, voltage-sensitive fluorescent protein. Eur J Neurosci 13:2314–2318
40. Knopfel T, Tomita K, Shimazaki R, Sakai R (2003) Optical recordings of membrane potential using genetically targeted voltage-sensitive fluorescent proteins. Methods 30:42–48
41. Ataka K, Pieribone VA (2002) A genetically targetable fluorescent probe of channel gating with rapid kinetics. Biophys J 82:509–516
42. Baker BJ, Mutoh H, Dimitrov D, Akemann W, Perron A, Iwamoto Y, Jin L, Cohen LB, Isacoff EY, Pieribone VA, Hughes T, Knopfel T (2008) Genetically encoded fluorescent sensors of membrane potential. Brain Cell Biol 36:53–67
43. Baker BJ, Lee H, Pieribone VA, Cohen LB, Isacoff EY, Knopfel T, Kosmidis EK (2007) Three fluorescent protein voltage sensors exhibit low plasma membrane expression in mammalian cells. J Neurosci Meth 161:32–38
44. Murata Y, Iwasaki H, Sasaki M, Inaba K, Okamura Y (2005) Phosphoinositide phosphatase activity coupled to an intrinsic voltage sensor. Nature 435:1239–1243
45. Ramsey IS, Moran MM, Chong JA, Clapham DE (2006) A voltage-gated proton-selective channel lacking the pore domain. Nature 440:1213–1216
46. Dimitrov D, He Y, Mutoh H, Baker BJ, Cohen L, Akemann W, Knopfel T (2007) Engineering and characterization of an enhanced fluorescent protein voltage sensor. PLoS ONE 2:e440
47. Tsutsui H, Karasawa S, Okamura Y, Miyawaki A (2008) Improving membrane voltage measurements using FRET with new fluorescent proteins. Nat Methods 5:683–685
48. Lundby A, Akemann W, Knopfel T (2010) Biophysical characterization of the fluorescent protein voltage probe VSFP2.3 based on the voltage-sensing domain of Ci-VSP. Eur Biophys J 39:1625–1635
49. Akemann W, Mutoh H, Perron A, Rossier J, Knopfel T (2010) Imaging brain electric signals with genetically targeted voltage-sensitive fluorescent proteins. Nat Methods 7:643–649
50. Mutoh H, Perron A, Dimitrov D, Iwamoto Y, Akemann W, Chudakov DM, Knopfel T (2009) Spectrally-resolved response properties of the three most advanced FRET based fluorescent protein voltage probes. PLoS ONE 4:e4555
51. Gautam SG, Perron A, Mutoh H, Knopfel T (2009) Exploration of fluorescent protein voltage probes based on circularly permuted fluorescent proteins. Front Neuroeng 2:14
52. Perron A, Mutoh H, Launey T, Knopfel T (2009) Red-shifted voltage-sensitive fluorescent proteins. Chem Biol 16:1268–1277
53. Villalba-Galea CA, Sandtner W, Dimitrov D, Mutoh H, Knopfel T, Bezanilla F (2009) Charge movement of a voltage-sensitive fluorescent protein. Biophys J 96:L19–L21
54. Lundby A, Mutoh H, Dimitrov D, Akemann W, Knopfel T (2008) Engineering of a genetically encodable fluorescent voltage sensor exploiting fast Ci-VSP voltage-sensing movements. PLoS ONE 3:e2514
55. Perron A, Mutoh H, Akemann W, Gautam SG, Dimitrov D, Iwamoto Y, Knopfel T (2009) Second and third generation voltage-sensitive fluorescent proteins for monitoring membrane potential. Front Mol Neurosci 2:5

56. Lundstrom K (2003) Semliki forest virus vectors for gene therapy. Expert Opin Biol Ther 3: 771–777
57. Delenda C (2004) Lentiviral vectors: optimization of packaging, transduction and gene expression. J Gene Med 6(Suppl 1):S125–S138
58. Russell WC (2000) Update on adenovirus and its vectors. J Gen Virol 81:2573–2604
59. Kaestner L, Ruppenthal S, Schwarz S, Scholz A, Lipp P (2009) Concepts for optical high content screens of excitable primary isolated cells for molecular imaging. In: Licha K, Lin CP (eds) SPIE biomedical optics, vol 7370. SPIE, Munich, pp 737–745
60. Tallini YN, Ohkura M, Choi BR, Ji G, Imoto K, Doran R, Lee J, Plan P, Wilson J, Xin HB, Sanbe A, Gulick J, Mathai J, Robbins J, Salama G, Nakai J, Kotlikoff MI (2006) Imaging cellular signals in the heart *in vivo*: cardiac expression of the high-signal Ca^{2+} indicator GCaMP2. Proc Natl Acad Sci USA 103:4753–4758
61. Christensen G, Minamisawa S, Gruber PJ, Wang Y, Chien KR (2000) High-efficiency, long-term cardiac expression of foreign genes in living mouse embryos and neonates. Circulation 101:178–184
62. Buning H, Perabo L, Coutelle O, Quadt-Humme S, Hallek M (2008) Recent developments in adeno-associated virus vector technology. J Gene Med 10:717–733
63. Schultz BR, Chamberlain JS (2008) Recombinant adeno-associated virus transduction and integration. Mol Ther 16:1189–1199
64. Du L, Kido M, Lee DV, Rabinowitz JE, Samulski RJ, Jamieson SW, Weitzman MD, Thistlethwaite PA (2004) Differential myocardial gene delivery by recombinant serotype-specific adeno-associated viral vectors. Mol Ther 10:604–608
65. Schirmer JM, Miyagi N, Rao VP, Ricci D, Federspiel MJ, Kotin RM, Russell SJ, McGregor CG (2007) Recombinant adeno-associated virus vector for gene transfer to the transplanted rat heart. Transpl Int 20:550–557
66. Karasawa S, Araki T, Nagai T, Mizuno H, Miyawaki A (2004) Cyan-emitting and orange-emitting fluorescent proteins as a donor/acceptor pair for fluorescence resonance energy transfer. Biochem J 381:307–312
67. Wlodarczyk J, Woehler A, Kobe F, Ponimaskin E, Zeug A, Neher E (2008) Analysis of FRET signals in the presence of free donors and acceptors. Biophys J 94:986–1000
68. Kaestner L, Lipp P (2009) Fluoreszenzproteine "fühlen" wo sie sind. BIOforum 32:17–18
69. Valeur B (2002) Molecular fluorescence. Wiley VCH, Winheim
70. Tsutsui H, Higashijima S, Miyawaki A, Okamura Y (2010) Visualizing voltage dynamics in zebrafish heart. J Physiol 588:2017–2021
71. Asselberghs I, Flors C, Ferrighi L, Botek E, Champagne B, Mizuno H, Ando R, Miyawaki A, Hofkens J, Van der Auweraer M, Clays K (2008) Second-harmonic generation in GFP-like proteins. J Am Chem Soc 130:15713–15719
72. De Meulenaere E, Asselberghs I, de Wergifosse M, Botek E, Spaepen S, Champagne B, Vanderleyden J, Clays K (2009) Second-order nonlinear optical properties of fluorescent proteins for second-harmonic imaging. Journal of Materials Chemistry, 19:7514–7519
73. Bublitz G, King B, Boxer S (1998) Electronic structure of the chromophore in green fluorescent protein. J Am Chem Soc 120:9370
74. Rosell FI, Boxer SG (2003) Polarized absorption spectra of green fluorescent protein single crystals: transition dipole moment directions. Biochemistry 42:177–183
75. Shi X, Basran J, Seward HE, Childs W, Bagshaw CR, Boxer SG (2007) Anomalous negative fluorescence anisotropy in yellow fluorescent protein (YFP 10C): quantitative analysis of FRET in YFP dimers. Biochemistry 46:14403–14417
76. Khatchatouriants A, Lewis A, Rothman Z, Loew L, Treinin M (2000) GFP is a selective non-linear optical sensor of electrophysiological processes in Caenorhabditis elegans. Biophys J 79:2345–2352
77. Topell S, Hennecke J, Glockshuber R (1999) Circularly permuted variants of the green fluorescent protein. FEBS Lett 457:283–289

78. Baird GS, Zacharias DA, Tsien RY (1999) Circular permutation and receptor insertion within green fluorescent proteins. Proc Natl Acad Sci USA 96:11241–11246
79. Kaestner L, Lipp P (2007) Non-linear and ultra high-speed imaging for explorations of the murine and human heart. In: Popp J, von Bally G (eds) Optics in life science, vol 6633. SPIE, Munich, pp 66330K-1–66330K-10
80. Shimomura O, Johnson FH, Saiga Y (1962) Extraction, purification and properties of aequorin, a bioluminescent protein from the luminous hydromedusan, Aequorea. J Cell Comp Physiol 59:223–239
81. Tasaki I, Watanabe A, Sandlin R, Carnay L (1968) Changes in fluorescence, turbidity, and birefringence associated with nerve excitation. Proc Natl Acad Sci USA 61:883–888
82. Heim R, Tsien RY (1996) Engineering green fluorescent protein for improved brightness, longer wavelengths and fluorescence resonance energy transfer. Curr Biol 6:178–182
83. Prasher DC, Eckenrode VK, Ward WW, Prendergast FG, Cormier MJ (1992) Primary structure of the *Aequorea victoria* green-fluorescent protein. Gene 111:229–233
84. Zimmer M (2005) Glowing genes: a revolution in biotechnology. Prometheus Books, Amherst
85. Fluhler E, Burnham VG, Loew LM (1985) Spectra, membrane binding, and potentiometric responses of new charge shift probes. Biochemistry 24:5749–5755
86. Salzberg BM, Grinvald A, Cohen LB, Davila HV, Ross WN (1977) Optical recording of neuronal activity in an invertebrate central nervous system: simultaneous monitoring of several neurons. J Neurophysiol 40:1281–1291
87. Grinvald A, Salzberg BM, Cohen LB (1977) Simultaneous recording from several neurones in an invertebrate central nervous system. Nature 268:140–142
88. Salzberg BM, Davila HV, Cohen LB (1973) Optical recording of impulses in individual neurones of an invertebrate central nervous system. Nature 246:508–509

Part III
Advanced Bioanalytical Applications

Probing Structure and Dynamics of the Cell Membrane with Single Fluorescent Proteins

Anna Pezzarossa, Susanne Fenz, and Thomas Schmidt

Abstract In recent years, our picture of the cell membrane has changed from a homogenous, fluid matrix to a highly dynamic and compartmentalized structure. This structuring influences the dynamical behavior of proteins embedded in the cell membrane. Since cell signaling is largely mediated through membrane-bound multi-protein complexes, there might be a direct link between membrane structure and signaling. In this chapter, we first review recent single molecule studies proving the existence of microdomains in the membrane of different cultured cell types and in a living organism by fluorescently labeling individual Ras proteins and studying their mobility. In the second part, we report on a study of the mobility of G protein-coupled receptors and G proteins in the model system *Dictyostelium discoideum*. The insights allowed building up a mechanistic model of the early steps of chemotaxis exemplifying the implication of membrane domains in cell signaling.

Keywords Chemotaxis · Diffusion · G protein-coupled receptor · Membrane domains · Ras protein · Single molecule microscopy

Contents

A. Pezzarossa, S. Fenz, and T. Schmidt (✉)
Physics of Life Processes, Institute of Physics, Leiden University, Leiden, The Netherlands
e-mail: schmidt@physics.leidenuniv.nl

G. Jung (ed.), *Fluorescent Proteins II*, Springer Ser Fluoresc (2012) 12: 185–212,
DOI 10.1007/4243_2011_24
Published online: 7 April 2011

1 Introduction

1.1 Membrane

Biological membranes form the outer layer of cells. They consist of an asymmetric lipid bilayer with phospholipids, cholesterol, and proteins as the main components. Glycosilated lipids and proteins built up a dense mesh protecting the cell against mechanical and chemical damage from outside. From the inside, the membrane is stabilized by the actin-rich cell cortex. The plasma membrane separates the contents (organelles, cytosol, cytoskeleton) and processes (e.g., protein synthesis, transport, and recycling) in the cell interior from the surrounding environment. At the same time, it plays a role in cell–cell communication, cell–cell recognition, and transport processes. The membrane represents the major regulatory platform for the initiation of early signaling events.

The role of lipids in membrane assembly and dynamics was the base for the Singer–Nicolson fluid mosaic model in the 1970s [1]. In this model, the bilayer is represented as a passive, neutral two-dimensional solvent in which active membrane proteins are homogeneously distributed and free to move. Some striking observations in epithelial cells proved this picture to be incorrect. Lipids were found to form functional microdomains insoluble in some detergents [2, 3]. Basic research on multicomponent lipid bilayers and vesicles containing two kinds of lipids with different phase transition temperatures and cholesterol revealed that even lipid mixtures exhibited complex phase separation behavior [4–7]. Based on these results, it was predicted that the spatial and temporal structure of the membrane might play an important regulatory role also in living cells. Although the observations on cells and biomimetic systems are still to be integrated into a global picture, it is accepted that lateral inhomogeneities are necessary for the function of biological membranes [8]. At present, the consensus is to describe the plasma membrane as a highly complex, organized structure at length scale in the order of tens of nanometers with a broad range of dynamic processes [8–10].

Genetic engineering of proteins made it possible to label membrane proteins with fluorescent proteins in living cells. In combination with single molecule fluorescence microscopy, it represents a powerful tool to study structure and dynamics of the cell membrane under physiological conditions.

1.2 Diffusion

The term diffusion refers to the Brownian motion of particles in a fluid medium, driven by thermal energy. We can picture the cell membrane as bilayer of lipids enriched in protein either embedded (integral proteins) or surface-adsorbed (peripheral proteins). Both proteins and lipids diffuse along the plane of the membrane (translational diffusion). A complete description of the motion in a membrane should also consider rotational diffusion, which accounts for the orientation of the molecules. In what follows, we will concentrate on translational diffusion only.

The lateral diffusion of a particle in a medium, characterized by a diffusion constant D, is described by Fick's second law:

$$\frac{\mathrm{d}}{\mathrm{d}t}p(\vec{x},t) = D \cdot \nabla^2 p(\vec{x},t), \tag{1}$$

where $\nabla^2 p(\vec{x},t)\mathrm{d}\vec{x}$ describes the probability of finding a particle, which started at $\vec{x}_0$ at time $t = 0$, within the volume $[\vec{x}, \vec{x} + \mathrm{d}\vec{x}]$ at time t. Solving in $\vec{x}$ yields:

$$p(\vec{x},t)\mathrm{d}\vec{x} = \left(\frac{1}{\sqrt{(4\pi Dt)^d}}\right)\exp\left(-\frac{(\vec{x}-\vec{x}_0)^2}{4Dt}\right)\mathrm{d}\vec{x}, \tag{2}$$

where d is the number of dimensions and $\vec{x}_0$ the starting position.

For analysis of actual data, it is advantageous to analyze cumulative distribution functions. Integration of $p(\vec{x},t)$ in (2) gives the cumulative probability distribution function $P_{\mathrm{cum}}(x^2, t)$, for particles that move up to the squared distance x^2 in time t [11, 12]. $P_{\mathrm{cum}}(x^2, t)$ reads for one, two, and three dimensions:

$$P_{\mathrm{cum}}^{d=1}(x^2,t) = \mathrm{erf}\left(\sqrt{\frac{x^2}{4Dt}}\right), \tag{3}$$

$$P_{\mathrm{cum}}^{d=2}(x^2,t) = 1 - \exp\left(-\frac{x^2}{4Dt}\right), \tag{4}$$

$$P_{\mathrm{cum}}^{d=3}(x^2,t) = \mathrm{erf}\left(\sqrt{\frac{x^2}{4Dt}}\right) - \sqrt{\frac{x^2}{\pi Dt}}\exp\left(-\frac{x^2}{4Dt}\right). \tag{5}$$

Those cumulative distribution functions are characterized by the typical area the molecule will cover during its diffusional path, the mean squared displacement (MSD):

$$\mathrm{MSD}\,(t) = \langle x^2(t)\rangle = \int (\vec{x}-\vec{x}_0)^2 p(\vec{x},t)\mathrm{d}\vec{x} = 2dDt. \tag{6}$$

The MSD increases linearly with time between observation for regular diffusion, the Brownian walk.

A biomolecule diffusing in a two-dimensional membrane will experience multiple forces and boundaries affecting its motion, causing a deviation from a free diffusion Brownian trajectory. Physical barriers such as other immobile membrane proteins or cytoskeletal fences can transiently or permanently restrict the lateral diffusion of proteins in the plasma membrane. In this scenario, the diffusional behavior of a molecule becomes more complex and it cannot be described by (6). It has been observed that deviations from the linear regime occur often in cell membranes. This motion is modeled according to anomalous or sub-diffusion:

$$\mathrm{MSD}\,(t) = \Gamma t^{1-\epsilon}; \epsilon \geq 0, \tag{7}$$

in which ε is the anomalous diffusion exponent [13]. In this model, the diffusion coefficient Γ appears to decrease with time. In case the molecules are restricted to diffuse within a square area with reflecting boundaries of side-length L, we speak of confined diffusion [14], described by:

$$\mathrm{MSD}\,(t) = \frac{L^2}{3}\left[1 - \exp\left(\frac{-12Dt}{L^2}\right)\right]. \tag{8}$$

In this scenario, we would observe free diffusion behavior on short time scales, when the particle cannot yet "feel" the barrier, but on longer time scales the mean square displacement levels off to an asymptotic value as illustrated in Fig. 1.

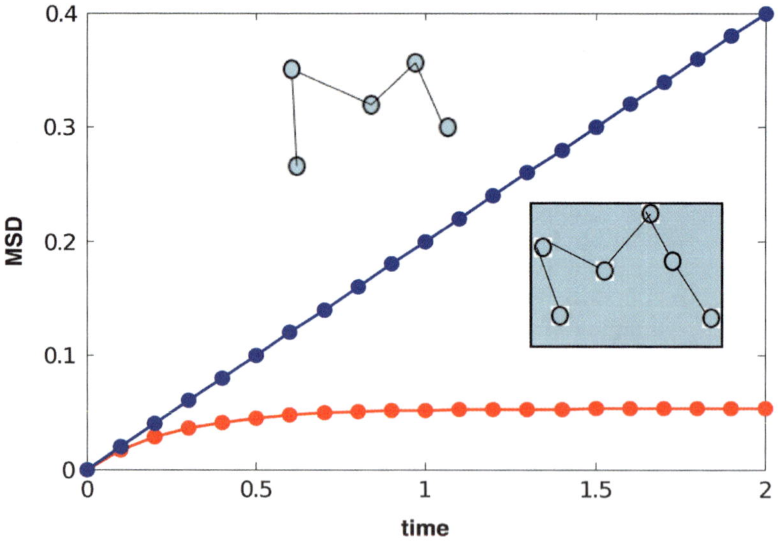

Fig. 1 Comparison between free (*blue curve*) and confined (*red curve*) diffusion. Graphs are created assuming identical diffusion coefficients

Many models have been proposed, which take into account all the possible situation a molecule can experience. It is worth to mention the hop diffusion model in which the molecules are temporarily trapped by a permeable barrier that restricts the available diffusion space to a few nanometers. There is a non-zero probability of barrier crossing and hopping to the adjacent domain [15]. This latter model, however, is not easily accessible experimentally, due to the time resolution limitation of current microscopy setups.

2 Single-Molecule Fluorescence Microscopy

Since its development, single-molecule fluorescence microscopy has been recognized as an ideal tool to observe dynamics in cells. Conventional microscopy is suitable to observe large objects, but to observe small biomolecules such as proteins, its resolution is not sufficient. At the beginning of the 1990s it became possible to detect single fluorophores and subsequently, thanks also to genetical engineering, to track objects in a living cells labeled with a fluorescent probe [16, 17, 93].

To understand how it is possible to visualize subwavelength structures, the main concepts of fluorophore localization will be given in what follows. The image of a fluorescent emitter is described by its diffraction pattern, known as the point spread function (PSF) given by an Airy disk of width $w = 0.61\ \lambda/\mathrm{NA}$, where λ is the imaging wavelength and NA the numerical aperture. In this scenario, the resolution limit is given by the Abbe limit or Rayleigh criterion, according to which two objects closer than the width of the Airy disk cannot be resolved. Using high NA oil objectives (NA = 1.4–1.5), in the visible region of light, this distance is typically 200 nm. Two emitters separated less than this distance would therefore be undistinguishable. An isolated fluorophore, on the other hand, can be localized with high precision. Its position is determined from the maximum of the PSF. Typically, this precision is in the order of tens of nanometers, depending on the signal-to-noise ratio. There are two ways to isolate the objects of interest: spatially [17] or spectrally [18]. Spatial isolation is achieved by dilution until there is only one fluorophore left within the Airy disk. The other option is to label the objects with spectrally well-separated fluorophores. In the latter case, multicolor imaging allows to distinguish between two objects closer than the width of their Airy disk.

The image of an arbitrary object is described as the sum of point light sources, localized at $\vec{s}_i$. For practical purposes, the PSF is approximated by a Gaussian, yielding the image $I(\vec{s})$:

$$I(\vec{s}) = \sum_i N_i \frac{1}{2\pi w^2} \exp\left(- \frac{(\vec{s} - \vec{s}_i)^2}{2w^2} \right), \tag{9}$$

where N_i is the number of emitted photons and w the width of the Gaussian. The Gaussian PSF predicts that the localization accuracy σ_i for an individual emitter is given by:

$$\sigma_i = w/\sqrt{N_i}. \tag{10}$$

Taking into consideration background noise b and pixelation (with pixel size a), σ_i is given by [19]:

$$\sigma_{si} = \sqrt{\left(\frac{w_i^2}{N_i} + \frac{a^2/12}{N_i} + \frac{8\pi w_i^4 b^2}{a^2 N_i^2}\right)}. \tag{11}$$

A more rigorous expression for σ_i that also takes into account the real PSF of a microscope and other noise sources has been derived by Ober et al. [20].

Prior to further data analysis, accurate image processing is needed to obtain the position of each single-molecule peak. The first step is to remove any static or slowly varying background. A reliable background image can be obtained by applying several different algorithms: high-order polynomial fitting, low-spatial-frequency filtering, mean-image calculation. The background-subtracted images are subsequently filtered by cross-correlation with the PSF of the microscope, approximated by a Gaussian. To reliably detect signals above the background, a threshold is set and only signals that are above the noise by a factor η, determined according to the experimental conditions, are further processed. Finally, each single-molecule peak is fitted to the PSF, and information about position, signal intensity, spatial width of the signal, and background signal are retrieved. In the following sections, we illustrate two widely used methods to obtain information on single-molecule dynamics from such position data.

2.1 Tracking

The single-molecule positions obtained from PSF fitting are the starting point for single-particle tracking. To generate trajectories, we need to correlate M mobile particles in an image to N mobile molecules in the subsequent image. A probabilistic algorithm is used to connect the positions of molecules in two frames, i and j, of a movie. The probability that a single particle k in image I_i with diffusion coefficient D is identical with particle l in image I_j is given by:

$$p(k,l) = \exp\left(-\frac{\left(\vec{s}_{k,i} - \vec{s}_{l,j}\right)^2}{4D\Delta t}\right). \tag{12}$$

Δt is the time lag between the recording of the two images I_i and I_j.

A transitional matrix is built up out of (12) which includes the probabilities of all possible connections between all M molecules in I_i and N molecules in I_j. Trajectories are constructed by optimizing for the combination of all connections with the highest total probability, i.e., where $\sum_{k,j} \log(p(k,l))$ is maximized. The algorithm implemented in our laboratory further includes corrections to deal with photobleaching and molecules that move into the field of view from outside. Starting from the trajectories, we construct the experimental cumulative probability to find the MSD.

Generating trajectories for all particles constitutes a NP-complete, or "traveling salesman" problem. Exact solutions can be calculated only for problems with a limited amount of particles (<20). However, there are approximative solutions that are close to the optimal. To calculate the approximate solution, we use Vogel's algorithm from operations research. The order in which the connectivity between subsequent images is drawn is determined by the maximal probability difference in connectivity for each molecule k in image i, $\{p(k,l)\}_l$. This simple sort-draw algorithm significantly reduces the computational costs of the analysis.

Single-particle tracking allows to directly visualize the diffusion behavior of single molecules drawn from a potentially inhomogeneous ensemble such as proteins diffusing in an inhomogeneous cell membrane. However, it suffers from fluorophore photobleaching that results in short trajectories [21]. This limitation can be overcome using quantum dots or gold nanoparticles, which permit longer observation times. Another drawback of this technique is that some prior knowledge on the dynamic of the system, i.e., its diffusion constant D, is needed to cope with the probabilistic nature of the tracking problem described in (12) [16]. This is particularly significant at high concentration of molecules when trajectories can be accidentally mixed.

2.2 *Particle Image Correlation Spectroscopy*

An alternative method to determine the P_{cum} without prior trajectory analysis was recently developed in our laboratory. This method does not require to assign individual traces and is consequently able to deal with high numbers of molecule as long as the Rayleigh criterium is fulfilled and individual molecules can be identified [22]. This algorithm is based on a correlation function, similarly as in spatiotemporal image correlation spectroscopy (STICS) [23].

In an isotropic medium, the cumulative correlation function of pairwise distances, C_{cum}, depends on two parameters: distance x and a time lag Δt. Thus, considering M molecules in image I_i and N molecules in image I_j, C_{cum} is given by:

$$C_{\mathrm{cum}}(x^2, \Delta t) = \frac{\sum_{k=1}^{M} \langle n(\vec{s}_{k,i}, x^2) \rangle_{\Delta t}}{M}, \tag{13}$$

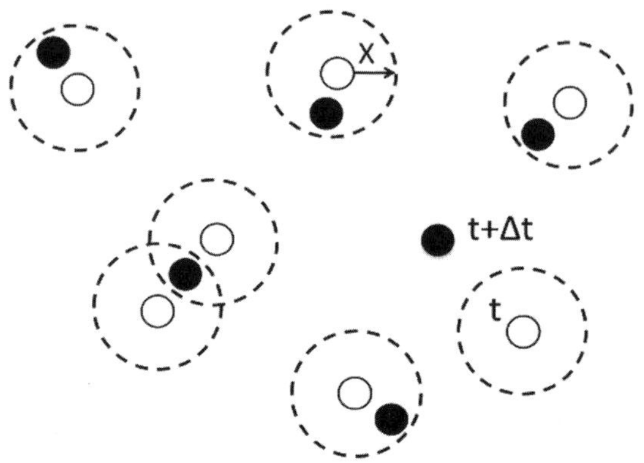

Fig. 2 PICS algorithm. For each molecule in image I_i (*white circles*), the number of molecules in image I_j (*black circles*) closer than x^2 is counted, 5 in this example. The peak on the left which lies within the overlap of two circles will be counted twice. Hence, the contribution due to diffusion is 4, whereas 1 count is due to random spatial proximity of the molecules

where $\vec{s}_{k,i}$ is the position of molecule k in image I_i and $n(\vec{s}_{k,i}, x^2)$ the number of particles in image I_j that lie in a circle of radius x around $\vec{s}_{k,i}$. The algorithm is illustrated in Fig. 2.

For each molecule in image I_i, the number of molecules in image I_j closer than x^2 is counted. Subsequently, the contributions from all molecules in I_i are summed and averaged over all image pairs. C_{cum} contains both temporal (e.g., diffusion) and spatial correlations (e.g., random proximity of the molecules).

The first contribution contains the information on the diffusion dynamics of the molecule and is equal to P_{cum} to find a diffusion step with size smaller than x^2 in the time lag Δt. In a two-dimensional case, for Brownian diffusion, with diffusion coefficient D, P_{cum} $(x^2,\Delta t)$ is given by (4). For the second contribution, a correction term that accounts for random proximity needs to be calculated to correct for spatial correlation. Assuming that the particles are distributed uniformly and independently with a density c, the probability to find N molecules in a circle with radius x is given by a Poisson distribution with mean and variance equal to $c\pi x^2$. Combining the two contribution, P_{cum} is calculated from the experimental C_{cum} distribution:

$$P_{\mathrm{cum}}(x^2, \Delta t) = C_{\mathrm{cum}}(x^2, \Delta t) - c \cdot \pi x^2. \tag{14}$$

2.3 *Fitting the Probability Distribution Function*

In both the methods outlined above, the dynamic information is contained in the cumulative probability distribution function. In the scenario of a single diffusing species, P_{cum} is given by (4). However, this model is often not adequate to describe the dynamics of biomolecules in living cells. In many biologically relevant processes, it is often necessary to include more than one diffusing population. In case of two mobile components, characterized by diffusion coefficients D_1 and D_2 respectively, P_{cum} assumes a double exponential form

$$P_{\mathrm{cum}}(x^2, \Delta t) = 1 - \left[\alpha \exp\left(-\frac{x^2}{4D_1\Delta t}\right) + (1-\alpha)\exp\left(-\frac{x^2}{4D_2\Delta t}\right)\right], \tag{15}$$

where α denotes the population size. In single-molecule experiments, the probability distributions are constructed either from the trajectories or by particle image correlation spectroscopy (PICS) as described above. This model has been applied successfully to describe protein dynamics in living cells [24].

2.4 Fluorescent Proteins for Life Cell Imaging

Single-molecule fluorescence microscopy is a powerful tool to address many biologically challenging questions both in vivo and in vitro. For its application though, the object under investigation (protein, DNA) must be specifically labeled by an appropriate fluorescent probe. Several fluorophore classes have emerged recently to label protein and intracellular structures, including (but not limited to) genetically encoded fluorescent proteins (FP), quantum dots (QD), and synthetic dyes.

An optimal probe for single-molecule tracking should meet several requirements: it should be expressed without toxicity in the system under investigation, it should be brighter than the cell's autofluorescence or have a different wavelength to be unequivocally detected, and should have sufficient photostability to be imaged for long time. The method that offers the least interference with a cell biological functions and allows the observation of dynamics in living cells is still fusion with a FP. In order to choose the best FP for a designed experiment, several photophysical parameters have to be taken into account: the saturation intensity (I_s), photobleaching time (τ_{bl}), and maximal photon emission rate (k_∞).

These parameters will be briefly described in this section and experimental results for the most common FP [eYFP (yellow), eGFP (green), eCFP (cyan), and DsRed (red)] will be given (see Table 1).

Following a standard two-level energy level diagram of a fluorophore, the dependence of τ_{bl} on the excitation intensity I is given by $\tau_{\mathrm{bl}} = \tau_{\mathrm{bl}}^{\infty}(1 + I_s/I)$,

Table 1 Photophysical properties of autofluorescent proteins, measured in a page gel compared to flavin-di-nucleotide, which was measured at high concentrations (>100 nM) [25]

Fluorophore	λ_{exc} (nm)	I_s (kW/cm^2)	k^∞ (photons/ms)	$\tau_{\mathrm{bl}}^{\infty}$ (ms)	ϕ_{bl} ($\times 10^{-5}$)
eCFP	458	48 ± 12	6,000 ± 3000	<1	>19
eGFP	488	13 ± 6	2,900 ± 200	2.8 ± 0.2	6.9 ± 0.5
eYFP	514	6 ± 1	3,100 ± 100	3.5 ± 0.5	5.5 ± 0.5
DsRed	532	50 ± 10	18,000 ± 2,000	0.4 ± 0.1	15 ± 3
flavin	514	35 ± 10	500 ± 100		

where τ_{bl}^{∞} is the photobleaching time limit. The detected signal (S_{det}) one can expect for a given experimental arrangement depends on the detector efficiency (η_{det}), on the integration time (t), the excitation intensity and wavelength, the chemical environment, and last on the photobleaching yield. Taking these parameters into account, we obtain the following expression for S_{det}:

$$S_{det} = \eta_{det} k_{\infty} \tau_{bl}^{\infty} \left[1 - \exp\left(\frac{-t}{\tau_{bl}^{\infty}(1 + I_s/I)} \right) \right]. \tag{16}$$

When photobleaching is negligible ($\tau_{bl} \gg t$), the equation converts into the simpler form:

$$S_{det} = \frac{\eta_{det} k_{\infty} t}{1 + I_s/I}. \tag{17}$$

Typical values for individual FPs when excited for 5 ms at an intensity of 1 kW/cm^2 are 100–200 counts detected on a back-illuminated CCD camera.

In a living cell, the main source of autofluorescence is given by flavinoids, which is located in the yellow-green spectral region. A spectral comparison between flavinoids and FPs shows an almost complete overlap of the flavine emission with eCFP, eGFP, and eYFP, where the excitation overlap with eCFP and eGFP is high, that with eYFP is low (see Fig. 3).

Taking into account the high concentration of flavinoids in a cell (10^7 molecules/cell), even excitation in the tail of their spectrum will cause a large autofluorescence background; therefore, FP with a red-shifted spectra are to be preferred.

From the above considerations, it appears that the suitability for single molecule microscopy is given by: eYFP > eGFP » eCFP (see Table 1). DsRed is not considered here due to its photobleaching rate, which is tenfold higher than the other FPs, which overwhelms the advantages given by the spectral separation from flavins. Many new variants of GFP have been created using genetic engineering, which cover almost the whole light spectrum [26, 27]. Most of those, however, are dimeric or even tetrameric, which is potentially toxic. Functional monomers can be engineered, but this results in a loss of brightness or photostability. Dimmer proteins require either higher laser power or longer exposure time to be imaged, which is limiting for fast in vivo experiments. For single molecule experiments though eYFP is still the preferred choice [26].

To overcome the limitations of autofluorescent protein, several alternative labeling techniques have been proposed [28]. Most of these rely on the possibility of specifically coupling a synthetic fluorophore to the biomolecule of interest through a chemical reaction. One widely used strategy is to create genetically modified proteins carrying a target domain (peptide tag) which is selectively posttranslationally labeled with a small complementary functionalized fluorophore. An important class of these peptide tags is represented by peptidyl carrier protein (PCP) and acyl carrier protein (ACP) domains [29, 30]. Those domains are modified by

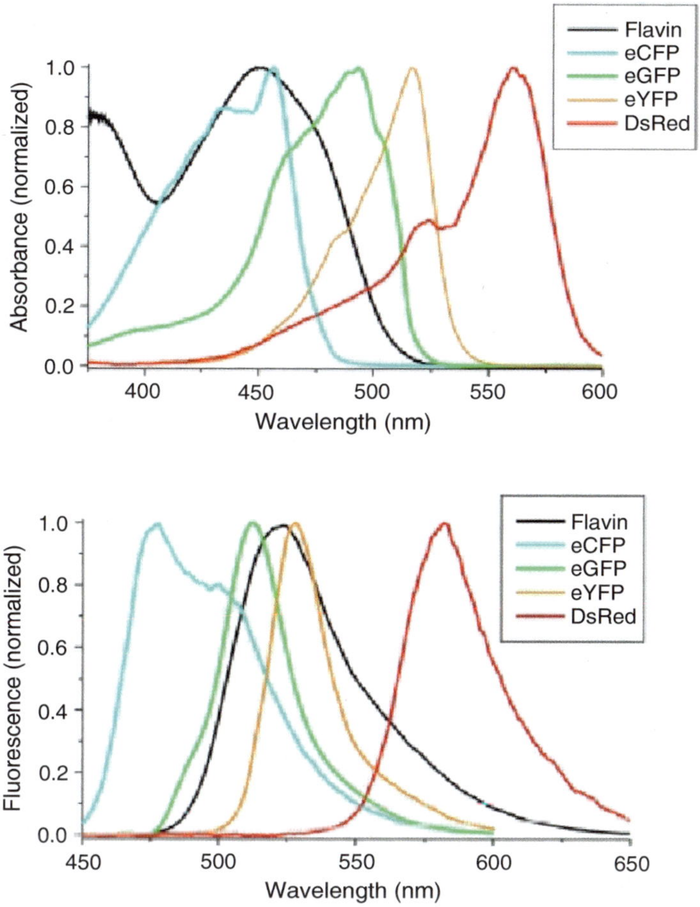

Fig. 3 Spectral comparison of flavin-di-nucleotide to fluorescent proteins. *Top*: Normalized absorption spectra. *Bottom*: Normalized emission spectra [25]

phosphopantetheinyl transferases (PPTases) of bacterial origin, such as the 4′-phosphopantetheinyl transferase (SFP) and the ACP synthase. The labeling reaction requires a fluorophore functionalized with a phosphopantetheinyl (Ppant) prosthetic group, derived from coenzyme A. In the reaction, the prosthetic group is covalently attached to a specific residue in the peptide tag by the enzyme (see Fig. 4).

Another commonly used peptide tag, HaLo, is based on an bacterial enzyme haloalkane dehalogenas. The enzymatic reaction relies on the formation of an ester bond between the fusion protein and a HaLo linker bound to a fluorescent reporter [31].

A different approach is based on the ability of particular sequences, rich in histidine residues, to bind transition metal complexes. One of those, the hexahistidine tag (His6), has been largely studied; however, this technique suffers from high

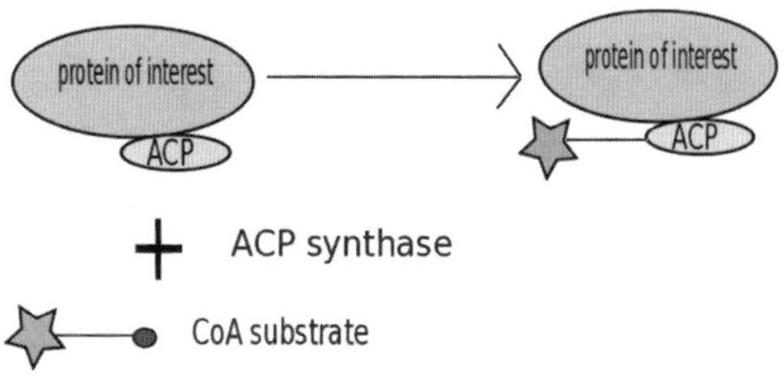

Fig. 4 ACP labeling reaction

dissociation constant (>300 nM), which makes it difficult to use for fluorescence experiments [32].

New opportunities arise from with the introduction of unnatural amino acid in the protein of interest, which allows direct or indirect introduction of fluorophores [33, 34]. The advantage of these techniques is that they allow for a wider choice of fluorophores, making it possible to reach a better spectral separation from cellular autofluorescence. Moreover, they are usually more photostable than FPs. On the other hand, most of these techniques suffer from nonspecific labeling and blinking (as in the case of QDots or nanocrystals), thus limiting their applicability. Moreover, they require specifically functionalized fluorophores, whose synthesis is quite laborious and which are potentially toxic for cells.

3 H-Ras Mobility in Membranes: From Cultured Cells to the Living Vertebrate

3.1 The Ras Family

Ras GTPases are small (~20 kDa), lipid-anchored membrane proteins involved in signal transduction. They play a major role in regulating cell growth, proliferation, and differentiation. Mutations in the Ras proteins are related to oncogenesis and cancer. Three different isoforms are expressed in all mammalian cells: H-Ras, K-Ras, and N-Ras. They are localized mainly in the inner leaflet of the plasma membrane, but they are found as well in the endoplasmic reticulum and in the Golgi apparatus [35–38]. Ras proteins effectively work as molecular switches, cycling between two possible states: a GTP-bound "on" state and a GDP bound "off" state which promotes association and activation of effector proteins. For signaling, they all interact with the same set of effectors, yet generating different output.

Ras proteins share 90% sequence homology, characterized by a highly conserved region in the N-terminus and showing a significant divergence in the C-terminus, which is referred to as hypervariable region, encoding for the membrane-anchoring domain [39]. Based on these structural observations, it was proposed that functional differences may result, at least partially, from a differential membrane compartmentalization of the isoforms. Several biochemical evidences were obtained which support this model: first, cholesterol depletion in the plasma membrane hyperactivates downstream Ras signaling in an isoform-specific manner, inhibiting H-Ras but not K-Ras [40]. Second, cell fractioning experiments showed that the isoforms target different membrane domains [41, 42]. Biochemical and electron microscopy (EM) studies suggested that H-Ras, but not K-Ras, was associated with cholesterol and phospholipids-enriched domains, forming nanoclusters of proteins with diameter in the range of 6–11 nm [41, 43, 44]. Another hypothesis was that membrane anchor domain orientation plays a role in isoform diversity. But the orientation alone cannot explain the functional differences observed [45].

It had also been shown that H-Ras partitioning is regulated by its GTP-bound state, suggesting a complex model in which H-Ras segregation is activation dependent. However, biochemical studies were not able to provide an insight on domain distribution and dynamics, and EM studies are limited to thin sample sheets and cannot be performed on living cells. Ideally, direct visualization of those objects *in vivo* is needed to fully understand inner leaflet plasma membrane organization.

3.2 Mobility Studies of Membrane-Anchored Proteins Reveal Membrane Domains in Cultured Cells

The plasma membrane of mammalian cells is heterogenous in structure and contains different types of domains, varying in size from a few to several hundreds of nanometers. Most of these domains have been extensively studied in the exoplasmatic membrane of cells, because of its easier accessibility (for reviews, see [46, 47]). Only more recent work has been done on the cytoplasmic leaflet. The proteins of the Ras family and their membrane anchor motif (CAAX) are ideal to study the partitioning and dynamics property of the inner leaflet.

To investigate the presence of membrane domains *in vivo*, and to obtain dynamic information, single-molecule fluorescence microscopy was used in our group to track the membrane targeting domain of two Ras isoforms (H-Ras and K-Ras) fused with eYFP. These results were compared to those obtained for the membrane anchor of a member of the family of Src-kinase, Lck [24, 48].

The H-Ras targeting sequence consists of ten amino acids and contains three cysteines to which one S-prenyl and two S-acyl groups are attached posttranslationally [35, 38]. Biochemical studies showed its presence in the detergent-resistant membrane (DRM) fraction. Instead, the K-Ras membrane anchor, consisting of a farnesyl moiety in conjunction with a polybasic domain, does not co-precipitate with DRM [43]. On the other hand, Lck is anchored via two cysteine-linked S-acyl groups and one glycine-linked myristoyl group at the N-terminus of the protein, which confers a high affinity for cholesterol-rich membrane, analogous to H-Ras. The idea was to compare the mobility of these different anchors and retrieve from it information on inner leaflet membrane structure and the possible implication of domains in Ras signaling. Surprisingly, all those proteins showed similarity in their diffusional behavior.

The eYFP fusion constructs were tracked as described in the previous section, and from the traces analysis information on their dynamics in the membrane and confinement in domains was obtained.

All the anchors exhibited a biphasic behavior, represented by two distinct fractions, a fast moving one which included the majority of molecules [between 60% (H-Ras) and 84% (K-Ras)], and a slow diffusing one. Mean square displacement vs. time curves showed a linear increase with time, indicating free diffusion of the fast fraction for all the different anchors with slightly different diffusion coefficients: $D_{\mathrm{HRas}} = 1.13\ \mu\mathrm{m}^2/\mathrm{s}$, $D_{\mathrm{KRas}} = 1.00\ \mu\mathrm{m}^2/\mathrm{s}$, and $D_{\mathrm{Lck}} = 1.30\ \mu\mathrm{m}^2/\mathrm{s}$. The remaining fraction of molecules showed instead in all different experiments confined behavior to domains of ~200 nm in diameter with an initial diffusion coefficient of $D_{\mathrm{slow}} \sim 0.2\ \mu\mathrm{m}^2/\mathrm{s}$.

These results proved that if nanoclusters are present, anchors with high affinity (H-Ras and Lck) were not significantly slowed down when compared to K-Ras. These results did not exclude the presence of cholesterol-dependent Ras nanoclusters in the inner leaflets. Due to their supposedly very small size (<70 nm), it was not possible to visualize them. However, if they are present, their effects would be negligible in terms of protein mobility. Other studies using fluorescence recovery after photobleaching (FRAP) [41] and electron microscopy (EM) [44] confirmed a two state model in which 30% of H-Ras is present in 20 nm cholesterol-dependent domains. Although these domains are much smaller than what can be detected with SPT, the two observations are closely related. One possible explanation could be that these 20 nm domains are temporally trapped to the actin cytoskeleton and to scaffolding proteins, leading to observed 200 nm domains. To conclude, trapping in 200 nm domains was observable with SPT, but no isoform-specific effect on mobility was observable on the 10–60 ms time scale of the measurement.

A further study was conducted on full-length H-Ras protein to explore the relationship between Ras mobility and activation. Evidences on activation-dependent mobility were suggested by the results obtained by FRAP experiments [41]. A constitutive active mutant GFP-H-Ras(V12) showed an increase in its lateral mobility with the expression level in a saturable manner, suggesting association with saturable domains. To further investigate this relationship, we performed single-molecule tracking experiments on H-Ras and on two mutants, one constitutive active, eYFP-H-Ras(V12), and one inactive, eYFP-H-Ras(N17) *in vivo* [49]. We observed two mobile fractions for both mutants, a fast one showing free diffusion and a slow population that showed free diffusion for the inactive mutant and confinement to ~200 nm domains for the active one. Similarly, the eYFP-H-Ras (wt) slow fraction showed a mobility change upon insulin stimulation. Before activation, the H-Ras (wt) slow fraction appeared to diffuse freely, but after 5 min of insulin treatment a confinement of ~200 nm was observed (see Fig. 5).

Most likely the actin cytoskeleton plays a role in H-Ras trapping, as it is known that insulin induces rearrangements of the cytoskeleton [50–53]. The SPT results largely agreed with results from single molecule fluorescence resonance energy transfer (FRET) [54], which showed the involvement of actin in immobilization of active H-Ras.

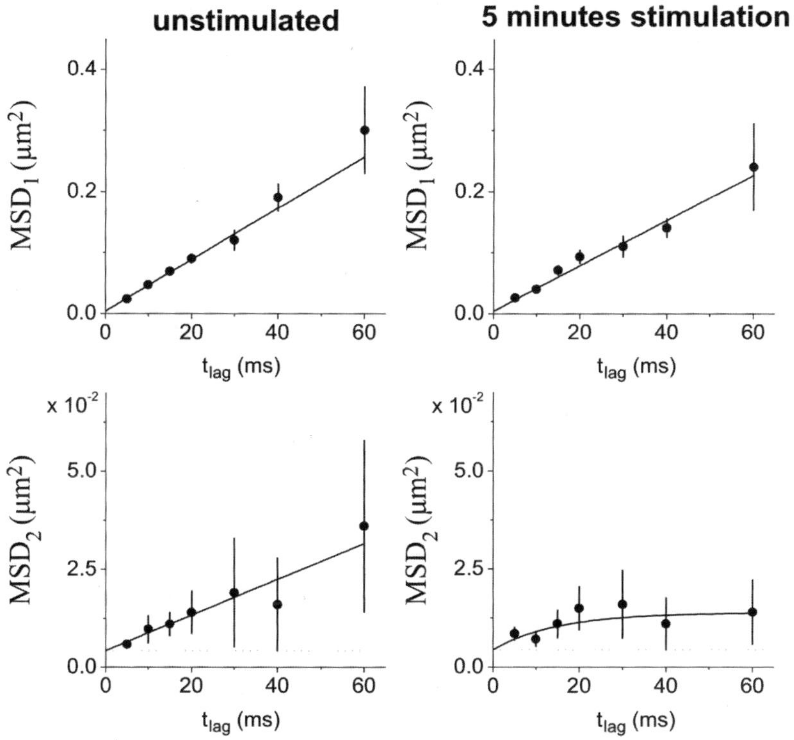

Fig. 5 Diffusion characteristics of eYFP-H-Ras(wt) before stimulation (*left*) and 5 min after stimulation with insulin (*right*). *Top*: Mean square displacement of the fast-diffusing fraction, MSD vs. timelag. *Bottom*: Mean square displacement of the slow diffusing fraction, MSD vs. timelag [49]

Those results proved a major role for the plasma membrane as a platform for Ras signaling. A mechanistic model for Ras microlocalization and activation had been suggested by Rotblat et al. [55], in which the interactions that regulate H-Ras membrane affinity also regulates its segregation in nanodomains. Recently, an increasing amount of evidence has proved a role for the growth factor receptor Gal-1 in H-Ras nanocluster formation [56]. These latter studies confirm the importance of microlocalization in signal transduction and identify new fundamental players that drive cluster formation.

3.3 H-Ras Studies in Zebrafish Embryos Prove Membrane Structuring in Living Organisms

A significant amount of experimental work has been done on cultured cells to study membrane structuring and its effects on cellular processes. Only recently, this approach has been extended to the living vertebrate organism, zebrafish [57–60].

Zebrafish was chosen because their embryos are small, transparent, and easy to manipulate, making it the ideal model system to study [61–63]. In our group, the membrane anchor of H-Ras was genetically fused to a YFP, YFP-C10H-Ras, and was used to demonstrate the possibility of measuring protein mobility in living organism [64].

In this study, results from *in vitro* measurements, performed on a cultured zebrafish cell line (ZF4), were compared to those obtained *ex vivo* on primary

Table 2 Comparison of the obtained diffusion characteristic for the different experimental conditions. α fast fraction size, D_{fast}, D_{slow} diffusion coefficient fast and slow fraction, respectively, L_1 and L_2 domain size fast and slow fraction, respectively

	α (%)	D_{fast} (μm²/s)	L_1 (μm)	D_{slow} (μm²/s)	L_2 (μm)
In vitro	70	0.67	0.79	0.06	0.16
Ex vivo	67	0.72	0.55	0.05	0.21
In vivo	75	0.51	Free	0.04	0.12

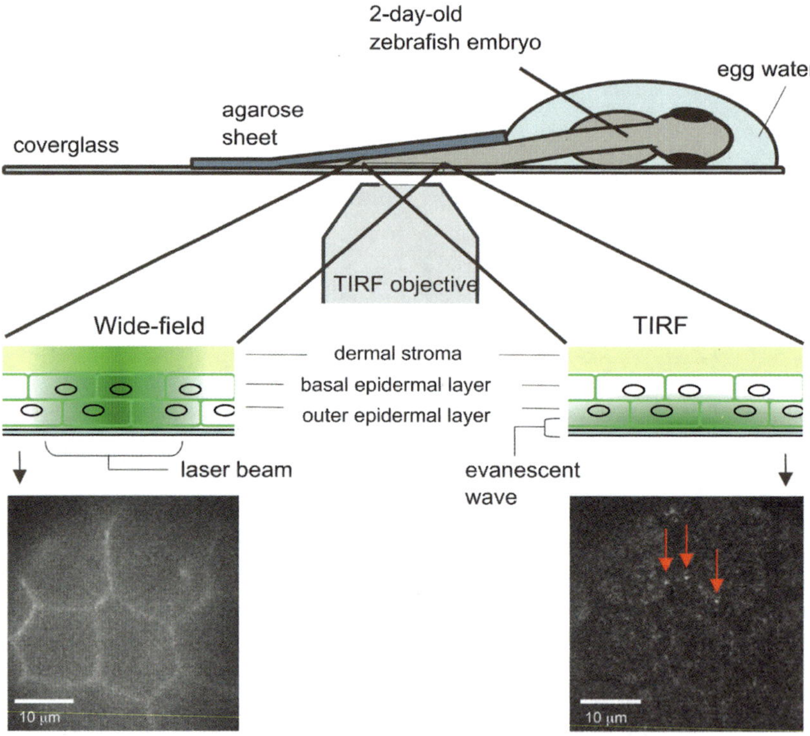

Fig. 6 Single molecule microscopy *in vivo*. The zebrafish embryo is placed on a coverslip and the tail region is covered with a sheet of agarose 0.75 mm thick. The sample was mounted on a microscopy setup suitable for wide field and TIRF microscopy. *Left*: Wide field picture, displaying membrane localization of the fluorescent signal. *Right*: TIRF image of the same region. *Arrows* indicate three examples of the YFP-H-Ras fluorescent peaks which can be attributed to single YFP molecules

embryonic stem cells and *in vivo* in 2 days old zebrafish embryo. In all the experiments, two populations in protein mobility were observed. Both *in vitro* and *ex vivo* the fast population appeared confined, but remarkably, *in vivo* data could be fitted according to a free diffusion model. The slow fraction instead showed a more stable behavior through all the different experiments, being always confined in small domains with comparable diffusion coefficients (see Table 2).

In vitro and *ex vivo* experiments were performed using epifluorescence microscopy, while *in vivo* experiments needed a different experimental setup because of the high level of out-of-focus fluorescence. A total internal reflection fluorescence (TIRF) microscopy approach was used, in which an evanescent light field excites the molecules within 100 nm from the coverglass (see Fig. 6).

Using this approach, it was possible, for the first time, to record individual molecules in a living organism. To summarize, the existence of two kinds of domains was observed, a smaller one with a diameter between 120 and 210 nm, present in all the experiments, and a bigger one, 550–790 nm found only in *in vitro* and *ex vivo* experiments [64].

This study showed that the two populations model of protein diffusion and their confinement in domains applies to living organisms as well as to cultured cells, and could play a physiological role in cell signaling. However, it also showed that large differences exist between cultured cells and living organism. Thus, to make physiologically relevant considerations on a system, studies in living organism are required.

4 Receptor and G Protein Mobility in *Dictyostelium discoideum*

The soil-living amoeba *Dictyostelium discoideum* is a widespread model organism for chemotaxis. Its completely sequenced 34 Mb-genome [65] contains many genes that are homologous to those in higher vertebrate species, thus representing an ideal model organism. Gene engineering and GFP technology provide a versatile toolbox to study vital processes such as cell motility, chemotaxis, or signal transduction in *D. discoideum*. This section highlights recent studies that yielded insight into the role of G protein-coupled receptor (GPCR) and G protein mobility in chemotaxis by combining fluorescent protein technology and high resolution microscopy.

4.1 Chemotaxis in *D. discoideum*

Nature developed fascinating strategies to ensure survival of a species. One example is the amoeba *D. discoideum*. Under optimal conditions, *D. discoideum* lives as an individual cell, but it has the ability to aggregate into a multicellular structure containing ~10^5 amoeba when the environmental conditions deteriorate. The resulting aggregate undergoes cell differentiation and morphogenesis leading to a so-called pseudoplasmodium or slug that has the ability to sense temperature and light.

These capabilities allow the pseudoplasmodium to find its way to the soil surface where it undergoes a final transformation into a fruiting body emitting robust spores. Taken away by the wind or animals, these spores bear the potential to germinate into a new protozoa under propitious environmental conditions. For an illustration of the complete cycle, see Fig. 7.

Chemotaxis, the directed cell movement in a chemical gradient, is at the basis of this complex process. An amoeba that experiences unfavorable conditions, e.g., starvation, secretes the chemoattractant cAMP. Neighboring amoeba detect cAMP, change their shape building up a leading and a trailing edge, migrate with the help of pseudopodia toward the source, and in turn also secret cAMP. Thereby, streamlines of amoeba are built up as a first level of organization.

On a molecular level, chemotaxis is triggered by the interaction of the chemoattractant with a GPCR embedded in the membrane of the chemotaxing cell. A crucial GPCR in *D. discoideum* chemotaxis is the cAMP receptor1 (cAR1). It is not only the first receptor expressed as a consequence of disadvantageous conditions but also the most sensitive to cAMP. For this reason, the function of cAR1 was studied extensively [66–69]. As implied by the name GPCR, these transmembrane receptors are associated with G proteins on their cytosolic side. G proteins are heterotrimers consisting of a G_{α}, and a heterodimeric $G_{\beta\gamma}$ subunit. The common view of G protein signaling implies that upon ligand binding to the receptor, the associated G proteins exchange guanine di-phosphate for guanine tri-phosphate in the G_{α} subunit. This leads to destabilization of the complex and the dissociation of the G protein in its G_{α} and $G_{\beta\gamma}$ subunit. In *D. discoideum* chemotaxis, both G_{α} and $G_{\beta\gamma}$ subsequently

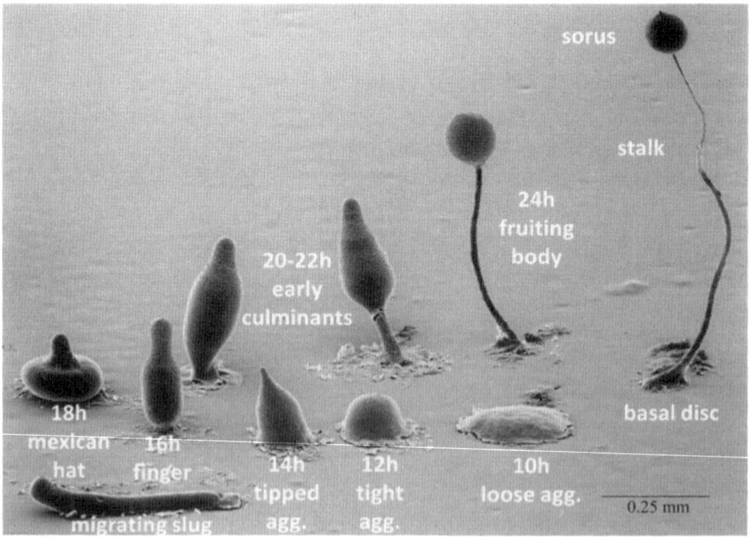

Fig. 7 Scanning electron micrograph of the developmental stages of the amoeba *D. discoideum*. Image copyright M. J. Grimson and R. L. Blanton, Biological Sciences Electron Microscopy Laboratory, Texas Tech University

activate different intracellular pathways that finally lead to cell polarization and directed migration toward the cAMP gradient.

Chemotaxis is triggered over a wide range of chemoattractant gradients and mean concentrations down to remarkably shallow gradients of 2% across the cell body. Thus, it is highly interesting to identify the ubiquitous molecular mechanisms that translate various extracellular gradients steadily in a highly polarized cellular phenotype exhibiting biased migration toward the source of the gradient. Several models were proposed for the early chemotaxis stages of gradient sensing followed by cell polarization: the pilot pseudopodia model [70, 71], the first hit model [72], several positive feedback loop models [73–75], the local excitation, global inhibition model [76, 77], and the balanced inactivation model [78]. Recently, increasing experimental evidence was found that questions the concept that signaling guides the generation of well-placed pseudopods [79–81]. *D. discoideum* cells move by default and constantly generate pseudopods in a probabilistic manner. Moreover, new pseudopods protrude preferentially from existing ones, but can be retracted after review. These new insights demand a reevaluation of the molecules important to chemotaxis with respect to their influence on pseudopod generation frequency, persistency, or the probabilistic distribution of their placement. To conclude, more quantitative information about the localization, mobility, and reaction rates of major players like cAR1 and the G protein subunits are required to elucidate their interaction with, e.g., the actin cytoskeleton that pushes the pseudopodia forward. These findings will allow to achieve a more realistic picture of chemotaxis.

4.2 GPCR and G Protein Exist as a Precoupled Complex

G proteins are important signal transducing molecules in cells. They not only are involved in chemotaxis, but also regulate many vital processes such as transcription and secretion in mammalian cells. Extensive research has been pursued since their discovery in the 1960s yielding the identification of more than 1,000 mammalian GPCRs and the atomic structure of three of them [82–84]. Nevertheless, it is not yet known how ligand-induced conformational changes of the GPCR activate the G protein. Another open question is whether GPCR association with the G protein before receptor stimulation is required or even stable. In 2005, Nobles et al. could prove the existence of a precoupled pentamer consisting of a GPCR dimer of either α2A adrenergic or muscarinic M4 receptor and the trimeric G protein in living HEK cells applying FRET [85]. More recently, we used single-molecule widefield microscopy (SMM) to deduce information about the GPCR/G protein complex from the diffusion behavior of cAR1 and the G protein subunits in *D. discoideum* [67]. Despite its low axial resolution of ~1 µm, SMM is ideally suited to study processes that are confined to two dimensions, like diffusion of transmembrane or membrane-bound proteins, with high spatial (~40 nm) and temporal (~50 ms) resolution. For this purpose, *D. discoideum* cells were stably transformed with

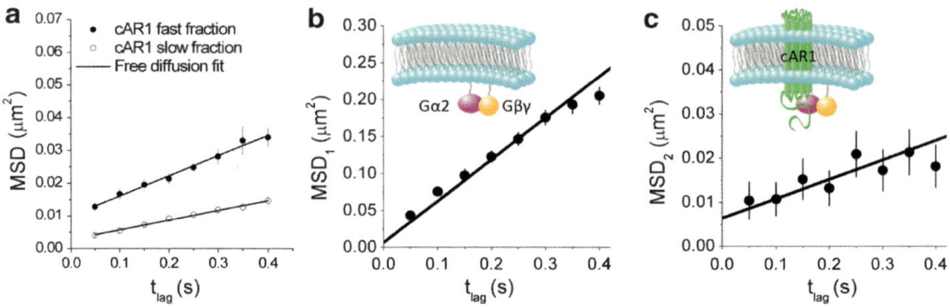

Fig. 8 Mobility of cAR1-YFP and G_β-YFP. (**a**) cAR1 exists in two populations exhibiting free diffusion. (**b**) Mean square displacement vs. time plot for the fast fraction of G_β-YFP. (**c**) Mean square displacement vs. time plot for the slow fraction of G_β-YFP [67]

Table 3 Results of cAR1 and G protein diffusion as measured in unstimulated *D. discoideum* cells. D_{FF} and D_{SF}, diffusion constants of the fast and slow fraction respectively. α, fast fraction

	D_{FF} (μm²/s)	D_{SF} (μm²/s)	α (%)	Diffusion type
cAR1	0.015 ± 0.002	0.007 ± 0.001	45 ± 6	Free
$G_{\alpha 2}$	0.14 ± 0.01	0.015 ± 0.001	68 ± 4	Free
$G_{\beta\gamma}$	0.15 ± 0.01	0.011 ± 0.001	68 ± 3	Free

cAR1-YFP, $G_{\alpha 2}$-YFP, or $G_{\beta\gamma}$-YFP constructs. Particle image correlation analysis [22] of the single molecule data allowed for the calculation of the cumulative probability of diffusion steps on different time lags. The results are summarized in Fig. 8 and Table 3. Neither cAR1 nor the G protein subunits were found to diffuse homogenously. Their P_{cum}s could be best fitted by a diffusion model taking two diffusing species into account (see Sect. 2.3). A global fit of all P_{cum}s at different time lags provided the MSDs of the fast and the slow fraction, as well as the fast fraction size α. Both cAR1 and the G protein subunits showed free diffusion on all observed timescales (50–400 ms; see Fig. 8a). The high similarity of the diffusion behavior of $G_{\alpha 2}$ and $G_{\beta\gamma}$ suggested that all membrane-bound G proteins were $G_{\alpha 2\beta\gamma}$ heterotrimers. Moreover, comparison of the diffusion constants of the slow moving fractions of $G_{\alpha 2}$ and $G_{\beta\gamma}$ with the fast moving fraction of cAR1 provided a strong argument that ~30% of the G proteins exist in a precoupled complex with cAR1 and therefore move at the same speed. This hypothesis could be further supported by alternative biochemical assays [67].

4.3 *Polarized Mobility of GPCR and G Protein upon Agonist Stimulation*

The same study [67] monitored the effect of global and gradient stimulation with cAMP on *D. discoideum* cells expressing either $G_{\alpha 2}$-YFP or $G_{\beta\gamma}$-YFP to identify the

Table 4 Results of G protein diffusion as measured in globally stimulated *D. discoideum* cells. D_{FF} and D_{SF}, diffusion constants of the fast and slow fraction respectively. α, fast fraction [67]

	D_{FF} (μm^2/s)	Diffusion type	D_{SF} (μm^2/s)	α (%)	Diffusion type
$G_{\alpha 2}$	0.19 ± 0.02	Confined	0.015 ± 0.001	68 ± 4	Free
$G_{\beta\gamma}$	0.16 ± 0.02	Confined	<0.001	59 ± 3	Immob.

molecular mechanism leading to cell polarization. It had been shown previously that the cAR1 receptors and the G proteins are homogenously distributed in the membrane of *D. discoideum*. Moreover, this homogeneous configuration was not affected by stimulation with cAMP [66, 69, 86, 87]. Thus, a simple picture, in which cell polarization is caused by asymmetric receptor or G protein distribution in the cell membrane, is not applicable.

The scenario described in Sect. 4.2 changed dramatically upon global stimulation with 10 μM cAMP (see Table 4). The diffusion behavior of the fast fraction of both the $G_{\alpha 2}$ and the $G_{\beta\gamma}$ subunit changed from free diffusion to confined diffusion within domains of ~600 nm diameter. Moreover, the slow fraction of the $G_{\beta\gamma}$ subunit increased from 32% to 41% and was immobilized ($D < 0.001$ μm^2/s). Both effects were shown to be F actin dependent by disrupting the actin cytoskeleton by adding latrunculin A. These observations led to the conclusion that the observed domains are imposed by the actin cytoskeleton, and that $G_{\beta\gamma}$ interacts either directly or indirectly with the F actin meshwork.

When exposed to a gradient of cAMP (~0.4 nM/μm), the diffusion behavior of the G protein subunits became polarized. The proteins at the trailing edge diffused like in the unstimulated situation (see Sect. 4.2). Both the fast and the slow fraction showed free diffusion. The proteins at the leading edge, on the other hand, exhibited the same diffusion characteristics as in the case of global stimulation: confined fast fractions of $G_{\alpha 2}$ and $G_{\beta\gamma}$ and immobilized as well as increased slow fraction of $G_{\beta\gamma}$. These results support the hypothesis that F actin plays an important regulatory role in maintaining cell polarity during chemotaxis.

Earlier single molecule studies investigated the cAMP/cAR1 dissociation rate in chemotaxing cells and found polarized behavior in the receptor off-rate [69]. cAMP/receptor complexes at the leading edge dissociated faster that those at the trailing edge. Since the steady-state amount of cAMP binding was measured to be almost equal at both locations, the association rates at the leading edge had to be increased, too. As a result, cAMP binding and unbinding and thus receptor and G protein activation are cycled faster at the leading edge of the cell. A possible explanation that why G protein reactivation is facilitated at the leading edge was provided by de Keijzer et al., who showed that the mobility of cAR1 is increased at the leading edge in comparison to the trailing edge [66]. Thus, activated cAR1 receptors that dissociated from the receptor/G protein complex are able to activate many more G proteins than at the trailing edge. Such a local amplification step leads to a final fivefold linear amplification of the external cAMP gradient to an intracellular gradient in active G proteins constituting the first step in diffusion-controlled asymmetric signaling.

4.4 *Mechanistic Model of Early Chemotactic Signaling in D. discoideum*

Combining existing knowledge about *D. discoideum* chemotaxis with new insights from single molecule fluorescence microscopy, we suggested a mechanistic model of early chemotaxis events [67]. In the resting state of the cell, the membrane is populated by cAR1 and complexed cAR1-$G_{\alpha 2\beta\gamma}$. The complexed G protein diffuses one order of magnitude slower than the free G protein [67, 68]. The cytosol provides a pool of $G_{\beta\gamma}$ [68, 78] and heterotrimeric G proteins. Upon cAMP binding to cAR1, the equilibrium between the $G_{\alpha 2\beta\gamma}$ heterotrimer and the complexed cAR1-$G_{\alpha 2\beta\gamma}$ gets disrupted by the formation of an activated cAR1-$G_{\alpha 2\beta\gamma}$ complex. This complex

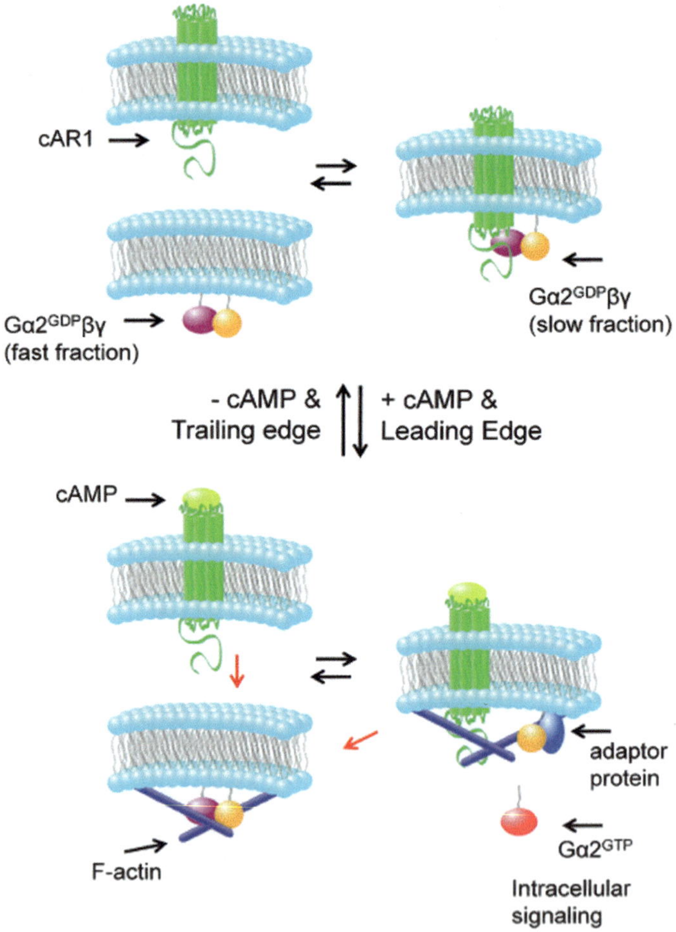

Fig. 9 Model describing the dynamic cAR1/G protein interaction at the leading and trailing edge [67]. For details, see text

subsequently dissociates into the activated receptor, and the free $G_{\alpha 2}$ and $G_{\beta\gamma}$ subunits. The stimulation-dependent dissociation of the G protein in its subunits was observed by Janetopoulos et al. with the help of FRET [88]. It takes approximately 0.4–1 s^{-1} until cAMP dissociates from cAR1 [89]. During that time, the activated receptor can activate more G proteins. Since the receptors at the leading edge diffuse considerably faster than at the trailing edge, an intracellular gradient in active G proteins is generated that is significantly larger than the external chemoattractant gradient [66]. G protein activation initiates multiple intracellular signaling cascades leading to actin cytoskeleton reorganization [90]. The tightening of the membrane-associated F actin is reflected in the diffusion properties of the G protein subunits. The fast fraction of both subunits becomes confined in actin-dependent 600 nm domains. This process is conceptually similar to the diffusion barriers at the leading edge of moving fish epidermal keratocytes [91]. Simultaneously, the slow fraction of $G_{\beta\gamma}$ gets immobilized by binding to F actin at the leading edge; at the trailing edge, $G_{\beta\gamma}$ diffuses into the cytosol. This allows for two scenarios: F actin either functions as a scaffold of $G_{\beta\gamma}$ signaling or impairs the discussed inhibitory function of $G_{\beta\gamma}$ [78].

$G_{\beta\gamma}$ immobilization might be another example for the widely used clustering of signaling components in a larger protein complex, a so-called signalosome. The cytoskeleton together with anchoring and scaffolding proteins was found to play a crucial role in the maintenance of these signaling complexes [92]. Figure 9 illustrates the dynamic cAR1/G protein interactions at the leading and the trailing edge.

5 Conclusion

In this chapter, we first reviewed the basic concepts of diffusion and single molecule microscopy. It was shown how combining fluorescent protein technology and high resolution microscopy lead to new insights into relevant biological processes. Two applications were introduced in detail. First, extensive work on the model proteins of the Ras family elucidated the role of membrane partitioning in protein dynamics and its influence on signaling. Moreover, we showed how this technique was extended to study protein dynamics in living organism. Second, single molecule fluorescence microscopy allowed to characterize the mobility of the GPCR cAR1 and its associated G protein in *D. discoideum*. On the basis of these data, a mechanistic model was developed to shed light on the molecular basis of cAR1-cAMP mediated chemotaxis in *D. discoideum*.

References

1. Singer SJ, Nicolson GL (1972) The fluid mosaic model of the structure of cell membranes. Science 175:720–731
2. Simons K, Ikonen E (1997) Functional rafts in cell membranes. Nature 387:569–572
3. Simons K, van Meer G (1998) Lipid sorting in epithelial cells. Biochemistry 27:6197–6202

4. Brown D, London E (2000) Structure and function of sphingolipid- and cholesterol-rich membrane rafts. J Biol Chem 275:17221–17224
5. Dietrich C, Volovyk Z, Levi M, Thompson N, Jacobson K (2001) Partitioning of thy-1, gm1, and cross-linked phospholipid analogs into lipid rafts reconstituted in supported model membrane monolayers. Proc Natl Acad Sci USA 98:10642–10647
6. Dietrich C, Bagatolli L, Volovyk Z, Thompson N, Levi M, Jacobson K, Gratton E (2001) Lipid rafts reconstituted in model membranes. Biophys J 80:1417–1428
7. Korlach J, Schwille P, Webb W, Feigenson G (1999) Characterization of lipid bilayer phases by confocal microscopy and fluorescence correlation spectroscopy. Proc Natl Acad Sci USA 96:8461–8466
8. Jacobson K, Mouritsen OG, Anderson RGW (2007) Lipid rafts: at a crossroad between cell biology and physics. Nat Cell Biol 9:7–14
9. Engelman DM (2005) Membranes are more mosaic than fluid. Nature 438:578–580
10. Kusumi A, Suzuki K (2005) Toward understanding the dynamics of membraneraft- based molecular interactions. Biochim Biophys Acta 1746:234–251
11. Almeida PFF, Vaz WLC (1995) Lateral diffusion in membranes. In: Lipowsky R, Sackmann E (eds) Handbook of biological physics, vol 1. Elsevier/North Holland, Amsterdam, pp 305–357
12. Anderson CM, Georgiou GN, Morrison IEG, Stevenson GVW, Cherry RJ (1992) Tracking of cell surface receptors by fluorescence digital imaging microscopy using a charge-coupled device camera. J Cell Sci 101:415–425
13. Feder TJ, Brust-Mascher I, Slattery JP, Baird B, Webb WWW (1996) Constrained diffusion or immobile fraction on cell surfaces: a new interpretation. Biophys J 70:2767–2773
14. Kusumi A, Sako J, Yamamoto M (1993) Confined lateral diffusion of membrane receptors as studied by single particle tracking (nanovid microscopy). Effects of calcium-induced differentiation in cultured epithelial cells. Biophys J 65:2021–2040
15. Wieser S, Moertelmaier M, Fuertbauer E, Stockinger H, Schutz GJ (2007) (Un)confined diffusion of CD59 in the plasma membrane determined by high-resolution single-molecule microscopy. Biophys J 92:3719–3728
16. Schmidt T, Schutz GJ, Gruber HJ, Schindler H (1995) Characterization of photophysics and mobility of single molecules in a fluid lipid membrane. J Phys Chem 99:17662–17668
17. Schmidt T, Schutz GJ, Baumgartner W, Gruber HJ, Schindler H (1996) Imaging of single molecule diffusion. Proc Natl Acad Sci USA 93:2926–2929
18. Funatsu T, Harada Y, Tokunaga M, Saito K, Yanagida T (1995) Imaging of single fluorescent molecules and individual ATP turnovers by single myosin molecules in aquous solution. Nature 374:555–559
19. Thompson RE, Larson DR, Webb WW (2002) Precise nanometer localization analysis for individual fluorescent probes. Biophys J 82:2775–2783
20. Abraham AV, Ram S, Chao J, Ward ES, Ober RJ (2009) Quantitative study of single molecule location estimation techniques. Opt Exp 17:23352–23373
21. Qian H, Sheetz MP, Elson EL (1991) Single particle tracking. Analysis of diffusion and flow in two-dimensional system. Biophys J 60:910–921
22. Semrau S, Schmidt T (2007) Particle image correlation spectroscopy (PICS): retrieving nanometer-scale correlations from high-density single molecule position data. Biophys J 92:613–621
23. Hebert B, Costantino S, Wiseman PW (2005) Spatiotemporal image correlation spectroscopy (STICS) theory, verification, and application to protein velocity mapping in living CHO cells. Biophys J 88:3601–3614
24. Lommerse PHM, Blab GA, Cognet L, Harms GS, Snaar-Jagalska EB, Spaink HP, Schmidt T (2004) Single-molecule imaging of lipid-anchored proteins reveals domains in the cytoplasmic leaflet of the cell membrane. Biophys J 86:609–616
25. Harms GS, Cognet L, Lommerse PHM, Blab GA, Schmidt T (2001) Autofluorescent proteins in single-molecule research: applications to live cell imaging microscopy. Biophys J 80:2396–2408

26. Shaner NC, Steinbach PA, Tsien RY (2005) A guide to choosing fluorescent proteins. Nat Methods 2:905–909
27. Zhang J, Campbell R, Ting AY, Tsien RY (2002) Creating new fluorescent probes for cell biology. Nat Rev Cell Biol 3:906–918
28. Kapanidis AN, Weiss S (2002) Fluorescent probes and bioconjugation chemistries for single molecule fluorescence analysis of biomolecule. J Chem Phys 117:10953–10964
29. Yin J, Lin AJ, Golan DE, Walsh CT (2006) Site-specific protein labeling by Sfp phosphopantetheinyl transferase. Nat Protoc 1:280–285
30. Zhou Z, Cironi P, Lin AJ, Xu Y, Hrvatin S, Golan DE, Silver PA, Walsh CT, Yin J (2007) Genetically encoded short peptide tags for orthogonal protein labeling by Sfp and AcpS phosphopantetheinyl transferases. ACS Chem Biol 2:337–346
31. Los G, Anson B, Zimprich C, Karassina N, January C, Buleit, RF (2004) The HaloTag: a novel technology for protein labeling in living cells: effect of the HaloTagTM system on hERG channel activity. FENS Forum 2004, Federation of European Neuroscience Societies (FENS), Berlin, 196
32. Kapanidis AN, Ebright YW, Ebright RH (2001) Site-specific incorporation of fluorescent probes into protein: hexahistidine-tag-mediated fluorescent labeling using (Ni2+:Nitrilotriacetic acid)n-fluorochrome conjugates. J Am Chem Soc 123:12123–12125
33. Cornish VW, Hahn KM, Schultz PG (1996) Site-specific protein modification using a ketone handle. J Am Chem Soc 118:8150–8151
34. Wang L, Magliery TJ, Liu DR, Schultz PG (2000) A new functional suppressor tRNA/ Aminoacyl-tRNA synthetase pair for the in vivo incorporation of unnatural amino acids into proteins. J Am Chem Soc 122:5010–5011
35. Hancock JF, Paterson H, Marshall CJ (1990) A polybasic domain or palmitoylation is required in addition to the CAAX motif to localize p21ras to the plasma membrane. Cell 63:133–139
36. Hancock JF, Cadwallader K, Paterson H, Marshall CJ (1991) A CAAX or a CAAL motif and a second signal are sufficient for plasma membrane targeting of ras proteins. EMBO J 10:4033–4039
37. Choy E, Chiu VK, Silletti J, Feoktistov M, Morimoto T, Michaelson D, Ivanov IE, Philips MR (1999) Endomembrane trafficking of ras: the CAAX motif targets proteins to the ER and Golgi. Cell 98:69–80
38. Apolloni A, Prior IA, Lindsay M, Parton RG, Hancock JF (2000) H-ras but not K-ras traffics to the plasma membrane through the exocytic pathway. Mol Cell Biol 20:2475–2487
39. Alberts B, Johnson A, Lewis J, Raff M, Roberts K, Walter P (2007) Molecular biology of the cell. Garland Science, New York
40. Roy S, Luetterforst R, Harding A, Apolloni A, Etheridge M, Stang E, Rolls B, Hancock JF, Parton RG (1999) Dominant-negative caveolin inhibits H-ras function by disrupting cholesterol-rich plasma membrane domains. Nat Cell Biol 1:98–105
41. Niv H, Gutman O, Kloog Y, Henis YI (2002) Activated K-Ras and H-Ras display different interactions with saturable nonraft sites at the surface of live cells. J Cell Biol 157:865–872
42. Parmryd I, Adler J, Patel R, Magee AI (2003) Imaging metabolism of phosphatidylinositol 4,5-bisphosphate in T-cell GM1- enriched domains containing Ras proteins. Exp Cell Res 285:27–38
43. Prior IA, Harding A, Yan J, Sluimer J, Parton RG, Hancock JF (2001) GTP-dependent segregation of H-ras from lipid rafts is required for biological activity. Nat Cell Biol 3:368–375
44. Prior IA, Muncke C, Parton RG, Hancock JF (2003) Direct visualization of Ras proteins in spatially distinct cell surface microdomains. J Cell Biol 160:165–170
45. Abankwa D, Gorfe AA, Inder K, Hancock JF (2010) Ras membrane orientation and nanodomain localization generate isoform diversity. Proc Natl Acad Sci USA 107:1130–1135
46. Lommerse PHM, Spaink HP, Schmidt T (2004) In vivo plasma membrane organization: results of biophysical approaches. Biochem et Biophys Acta 1664:119–131

47. Kusumi A, Suzuki K (2005) Toward understanding the dynamics of membrane-raft-based molecular interactions. Biochem et Biophys Acta 1746:234–251
48. Lommerse PH, Vastenhoud K, Pirinen NJ, Magee AI, Spaink HP, Schmidt T (2006) Single-molecule diffusion reveals similar mobility for the Lck, H-ras, and K-ras membrane anchors. Biophys J 91:1090–1097
49. Lommerse PHM, Snaar-Jagalska EB, Spaink HP, Schmidt T (2005) Single-molecule diffusion measurements of H-Ras at the plasma membrane of live cells reveal microdomain localization upon activation. J Cell Sci 118:1799–1809
50. Li SL, Miyata Y, Yahara I, Fujita-Yamaguchi Y (1993) Insulin- induced circular membrane ruffling on rat 1 cells expressing a high number of human insulin receptors: circular ruffles caused by rapid actin reorganization exhibit high density of insulin receptors and phosphotyrosines. Exp Cell Res 205:353–360
51. Tsakiridis T, Bergman A, Somwar R, Taha C, Aktories K, Cruz TF, Klip A, Downey GP (1998) Actin filaments facilitate insulin activation of the src and collagen homologous/mitogen-activated protein kinase pathway leading to DNA synthesis and c-fos expression. J Biol Chem 273:28322–28331
52. Khayat ZA, Tong P, Yaworsky K, Bloch RJ, Klip A (2000) Insulin-induced actin filament remodeling colocalizes actin with phosphatidylinositol 3-kinase and GLUT4 in L6 myotubes. J Cell Sci 113:279–290
53. Dadke S, Chernoff J (2003) Protein-tyrosine phosphatase 1B mediates the effects of insulin on the actin cytoskeleton in immortalized fibroblasts. J Biol Chem 278:40607–40611
54. Murakoshi H, Iino R, Kobayashi T, Fujiwara T, Ohshima C, Yoshimura A, Kusumi A (2004) Single-molecule imaging analysis of Ras activation in living cells. Proc Natl Acad Sci USA 101:7317–7322
55. Rotblat B, Prior IA, Muncke C, Parton RG, Kloog Y, Henis YI, Hancock JF (2004) Three separable domains regulate GTP-dependent association of H-ras with the plasma membrane. Mol Cell Biol 24:6799–6810
56. Belanis L, Plowman SJ, Rotblat B, Hancock JF, Kloog Y (2008) Galectin-1 is a novel structural component and a major regulator of HRas nanoclusters. Mol Biol Cell 19:1404–1414
57. Ries J, Yu SR, Burkhardt M, Brand M, Schwille P (2009) Modular scanning FCS quantifies receptor-ligand interactions in living multicellular organisms. Nat Methods 6:643–645
58. Shi X, Foo YH, Sudhaharan T, Chong SW, Korzh V, Ahmed S, Wohland T (2009) Determination of dissociation constants in living zebrafish embryos with single wavelength fluorescence cross-correlation spectroscopy. Biophys J 97:678–686
59. Shi X, Teo LS, Pan X, Chong SW, Kraut R, Korzh V, Wohland T (2009) Probing events with single molecule sensitivity in zebrafish and Drosophila embryos by fluorescence correlation spectroscopy. Dev Dyn 238:3156–3167
60. Friedrich M, Nozadze R, Gan Q, Zelman-Femiak M, Ermolayev V, Wagner TU, Harms GS (2009) Detection of single quantum dots in model organisms with sheet illumination microscopy. Biochem Biophys Res Commun 390:722–727
61. Lieschke GJ, Currie PD (2007) Animal models of human disease: zebrafish swim into view. Nat Rev Genet 8:353–367
62. Hsu CH, Wen ZH, Lin CS, Chakraborty C (2007) The zebrafish model: use in studying cellular mechanisms for a spectrum of clinical disease entities. Curr Neurovasc Res 4:111–120
63. Trede NS, Langenau DM, Traver D, Look AT, Zon LI (2004) The use of zebrafish to understand immunity. Immunity 20:367–379
64. Schaaf MJM, Koopmans WJA, Meckel T, van Noort J, Snaar-Jagalska EB, Schmidt T, Spaink HP (2009) Single-molecule microscopy reveals membrane microdomain organization of cells in a living vertebrate. Biophys J 97:1206–1214
65. Eichinger L et al (2005) The genome of the social amoeba *Dictyostelium disciodeum*. Nature 435:43–57

66. de Keijzer S, Sergé A, van Hemert F, Lommerse PHM, Lamers GEM, Spaink HP, Schmidt T, Snaar-Jagalska E (2008) A spatially restricted increase in receptor mobility is involved in directional sensing during *Dictyostelium disciodeum* chemotaxis. J Cell Sci 121:1750–1757
67. van Hemert F, Lazova MD, Snaar-Jagalska E, Schmidt T (2010) Mobility of G proteins is heterogenous and polarized during chemotaxis. J Cell Sci 123:2922–2930
68. Elzie CA, Colby J, Sammons MA, Janetopoulos C (2009) Dynamic localization of G proteins in *Dictyostelium disciodeum*. J Cell Sci 122:2597–2603
69. Ueda M, Sako Y, Tanaka T, Devreotes P, Yanagida T (2001) Single molecule analysis of chemotactic signaling in *Dictyostelium* cells. Science 294:864–867
70. Caterina MJ, Devreotes PN (1991) Molecular insights into eukaryotic chmotaxis. Faseb J 5:3078–3085
71. Devreotes P, Janetopoulos C (2003) Eukryotic chemotaxis: distinctions between directional sensing and polarization. J Biol Chem 278:20445–20448
72. Rappel WJ, Thomas PJ, Levine H, Loomis WF (2002) Establishing direction during chemotaxis in eukaryotic cells. Biophys J 83:1361–1367
73. Meinhardt H (1999) Orientation of chemotactic cells and growth cones: models and mechanisms. J Cell Sci 112:2867–2874
74. Narang A, Subramaniam KK, Lauffenburger DA (2001) A mathematical model for chemoattractant gradient sensing based on receptor-regulated membrane phospholipid signaling dynamics. Ann Biomed Eng 29:677–691
75. Potsma M, van Haastert PJM (2001) A diffusion-translocation model for gradient sensing by chemotacic cells. Biophys J 81:1314–1323
76. Ma L, Janetopoulos C, Yang L, Devreotes PN, Iglesias PA (2004) Two complementary, local excitation, global inhibition mechanisms acting in parallel can explain the chemoattractant-induced regulation of PI(3,4,5)P-3 response in *Dictyostelium* cells. Biophys J 87:3764–3774
77. Gamba A, de Candia A, Talia S, Coniglio A, Bussolino F, Serini G (2005) Diffusion-limited phase separation in eukaryotic chemotaxis. Proc Natl Acad Sci USA 103:16927–16932
78. Levine H, Kessler DA, Rappel WJ (2006) Directional sensing in eukaryotic chemotaxis: a balanced inactivation model. Proc Natl Acad Sci USA 103:9761–9766
79. Andrew N, Insall RH (2007) Chemotaxis in shallow gradients is mediated independently of PtdIns 3-kinase by biased choices between random protrusion. Nat Cell Biol 9:193–200
80. Bosgraaf L, Haastert PJV (2009) Navigation of chemotactic cells by parallel signalling to pseudopod persistance and orientation. PLoS One 4:e6842
81. Sasaki AT, Janetopoulos C, Lee S, Charest PG, Takeda K, Sundgeimer LW, Meili R, Devreotes PN, Firtel RA (2007) G protein-independent Ras/PI3K/F-actin circuit regulates basic cell motility. J Cell Biol 178:185–191
82. Palczewski K et al (2000) Crystal structure of Rhodopsin: a G-protein coupled receptor. Science 289:739–745
83. Rasmussen SGF et al (2007) Crystal structure of the human [bgr]2 adrenergic G-protein-coupled receptor. Nature 450:383–387
84. Jaakola VP et al (2008) The 2.6 Angstrom crystal structure of a human A2A adenosine receptor bound to an antagonist. Science 322:1211–1217
85. Nobles M, Benians A, Tinker A (2005) Heterotrimeric G proteins precouple with G protein-coupled receptors in living cells. Proc Natl Acad Sci USA 102:18706–18711
86. Parent CA, Devreotes PN (1999) A cell's sense of direction. Science 284:765–770
87. Jin T, Zhang N, Long Y, Parent CA, Devreotes PN (2000) Localization of the G protein$\beta\gamma$ complex in living cells during chemotaxis. Science 287:1034–1036
88. Janetopoulos C, Jin T, Devreotes P (2001) Receptor-mediated activation of heterotrimeric G-proteins in living cells. Science 291:2408–2411
89. Janssens PM, van Haastert PJ (1987) Molecular basis of transmembrane signal transduction in *Dictyostelium Disciodeum*. Microbiol Rev 51:396–418

90. Franca-Koh J, Kamimura Y, Devreotes P (2006) Navigating signaling networks: chemotaxis in *Dictyostelium Disciodeum*. Curr Opin Genet Dev 16:333–338
91. Weisswange I, Bretschneider T, Anderson KI (2005) The leading egde is a lipid diffusion barrier. J Cell Sci 118:4375–4380
92. Pawson T, Scott JD (1997) Signaling through scaffold, anchoring, and adaptor proteins. Science 278:2075–2080
93. Schutz GJ, Schindler H, Schmidt T (1997) Single molecule microscopy on model membranes reveals anomalous diffusion. Biophys J 73:1073–1080

Fluorescence Correlation and Cross-Correlation Spectroscopy Using Fluorescent Proteins for Measurements of Biomolecular Processes in Living Organisms

Yong Hwee Foo, Vladimir Korzh, and Thorsten Wohland

Abstract Fluorescence imaging and spectroscopy have played an important part in the advancement of the modern life sciences. Especially single molecule sensitive fluorescence techniques have allowed researchers to quantify biomolecular processes on the molecular level. However, these techniques were applied mainly in solution or in cell cultures that do not represent a physiologically relevant 3D environment. Results thus obtained are often not predictive of events in biomedically more interesting organisms or tissues. Therefore, many efforts are dedicated to bring single molecule fluorescence techniques into living model organisms. In this context, genetically encoded fluorescent proteins have played a major role. They exist in a wide variety (color, lifetime, Stokes shift, photoactivation, and environmental sensitivity) and ensure a 1:1 stoichiometry of labeling, important for the quantitation of data. In particular, two techniques, fluorescence correlation and cross-correlation spectroscopy (FCS and FCCS), profited from these labels and were successfully modified for applications in organisms. In this chapter, we will review the basics of FCS and FCCS, the challenges of applying FCS in organisms, and the importance of fluorescent proteins in this process. We provide a discussion of the different applications to date and give a perspective of new developments with promising potential for applications within organisms. The application of FCS and other single molecule techniques within living organisms is an important step in the quantitation of biomolecular events in physiologically relevant 3D environments.

Keywords Developmental biology · Fluorescence correlation spectroscopy · Fluorescence cross-correlation spectroscopy · Imaging FCS

Y.H. Foo and T. Wohland (✉)
Department of Chemistry, National University of Singapore, 3 Science Drive 3, Singapore 117543, Singapore
e-mail: chmwt@nus.edu.sg

V. Korzh
Institute of Molecular and Cell Biology, 61 Biopolis Drive, Proteos, Singapore 138673, Singapore
and
Department of Biological Sciences, National University of Singapore, 14 Science Drive 4, Singapore 117543, Singapore

G. Jung (ed.), *Fluorescent Proteins II*, Springer Ser Fluoresc (2012) 12: 213–248,
DOI 10.1007/4243_2011_16,
Published online: 31 March 2011

Contents

1 Introduction

Fluorescence imaging and spectroscopy have been defining techniques in the area of modern biology. Today, fluorescent tagging of molecules allows researchers to monitor even single molecules of interest in organisms (i.e., in vivo).[1] This advance has been possible because of the advent of green fluorescence protein (GFP), which allowed genetic labeling of proteins within cell cultures or in vivo in a selective and specific manner. In addition, mutants of GFP can be found throughout the visible spectrum [1, 2], which facilitates the monitoring of multiple proteins simultaneously. Most of them are nontoxic to the cell or organisms compared to organic dyes or quantum dots. Modern single molecule tools include Förster resonance

[1]In vivo, according to the common definition in biochemistry or molecular biology, refers to in living samples such as cell cultures and organisms, while in vitro refers to in tube or solution. However, in developmental biology, in vivo refers only to organism, while in vitro refers to in tube and in 2D cell culture. The term ex vivo is used when cells are directly taken from the organism and studied outside of the organism within a limited time. For the purpose of this chapter, we use the definitions of developmental biology, and therefore in vivo refers only to experiments in organism.

energy transfer (FRET), fluorescence lifetime imaging microscopy (FLIM), fluorescence recovery after photobleaching (FRAP), fluorescence correlation spectroscopy (FCS), and single particle tracking (SPT). These techniques are quantitative in nature and provide data about reaction rates, affinity of molecular interactions, fluorescence lifetime, diffusion characteristics, and photophysical properties. However, many of these fluorescence studies are conducted mainly in vitro (i.e., in a tube or 2D cell culture). The physiological relevance of 2D cell cultures has been recently questioned. Improvement is provided by 3D cell cultures which are said to bridge the gap between 2D cell cultures and physiological tissues [3], but nevertheless remains an in vitro approach. Therefore, there is a need to apply single molecule techniques within living organisms.

This chapter concentrates on FCS as one of the promising tools for quantitative measurements within living organisms. FCS is a single molecule sensitive technique commonly used to study molecular processes. It is based on extracting information that is embedded within the fluorescence signal fluctuations which are generated by the tagged molecules as they pass through an illuminated observation volume. The processes creating the fluctuations in the fluorescence signal can be due to photophysical properties of the label or movement of the labeled molecule. FCS extracts this information by transforming the fluorescence signal by a mathematical process known as autocorrelation to produce an autocorrelation function (ACF). With a typical resolution of nanoseconds and measurement times of seconds, processes happening in this time range contribute to the shape of the ACF. It is the high temporal resolution of FCS which allows diffusion processes (microseconds to milliseconds) such as Brownian diffusion [4], anomalous diffusion [5–7], and flow [8, 9] to be monitored. The time taken for these molecules to diffuse through the observation volume, the diffusion time, depends on the size of the molecules. In combination with the ability of the ACF to determine the number or concentration of molecules undergoing a certain diffusion process, one can determine and quantitate the kinetics or affinity of molecular interactions based on the change in diffusion time [10–18]. In addition, fast processes (nanoseconds to microseconds) such as triplet state dynamics [19–21], chemical reactions [22], rotational diffusion [23–28], and photophysics of fluorescent proteins [29–31] can be monitored. The typical observation volume in a confocal FCS instrument is ~0.5 femtoliter (fL), small enough to allow resolution at the subcellular level. FCS measurements have been performed in the cytoplasm, nucleus, and membranes of many common cell lines, bacteria, and yeast. A further advantage is that FCS works only in the concentration range from sub-nanomolar (nM) to a few micromolar (μM), close to typical physiological concentration ranges and well below what is commonly used in imaging. Therefore, overexpression of proteins is usually not required.

Fluorescence cross-correlation spectroscopy (FCCS) is a modality of FCS which uses two different fluorophores to measure biomolecular interactions. Signals from differently tagged molecules are detected in two separated detectors and are cross-correlated. The cross-correlation function (CCF) contains the information about interaction of the two molecules. This can be applied to quantitate the fraction of molecules in complexes or the K_d of molecular interactions. FCCS has been applied to study enzyme activities [32–34], polymerase chain reaction [35], DNA–protein or RNA–protein interactions [36, 37], protein–protein interactions [38–44], and receptor dimerization or oligomerization [45, 46]. The advantage of FCCS is that, unlike FRET, it is independent of the relative position, orientation, and spectral overlap of the two fluorophores, and thus is less likely to produce false-negative results.

The same advantages that make FCS and FCCS valuable tools for cell observations, namely the small observation volume, its sensitivity, its noninvasive character, and the possibility to quantify molecular interactions and dynamics make FCS also an ideal tool for observations in living organisms. Consequently, in recent years, FCS and FCCS have been applied to popular animal models such as zebrafish (*Danio rerio*), medaka (*Oryzias latipes*), nematodes (*Caenorhabditis elegans*), and fruit flies (*Drosophila melanogaster*). In this chapter, we will review the basics of FCS and FCCS, the challenges of applying FCS in vivo using fluorescence proteins (FPs) and modalities of FCS and FCCS which overcome some of the difficulties encountered by conventional FCS. We will then present a discussion of the different applications of FCS and FCCS in vivo to date and conclude by giving a perspective of new developments with promising potential for in vivo applications.

2 Principles of FCS and FCCS

The history of FCS dates back to 1972, when D. Magde, E. L. Elson, and W. W. Webb applied FCS to study the binding of ethidium bromide to DNA [22]. This was followed by a detailed discussion of the theory and setup of FCS [47, 48]. Although limited by the technology at that time resulting in poor temporal resolution, signal-to-background, and signal-to-noise ratio, FCS measurements represented a first step in a right direction. The decisive breakthrough came in the early 1990s when Rigler, Mets, and colleagues used a strongly focused laser and smaller pinhole size in addition to better technology to produce a diffraction-limited observation volume of less than 1 fL [4, 49]. After this advance, FCS became increasingly popular.

2.1 *Principles of FCS*

The principle of FCS is based on detecting fluorescent particles diffusing in and out of an observation volume. The observation volume can take on different sizes and shapes depending on the illumination and detection setup. In a typical confocal microscope, the observation volume is created using one-photon excitation (OPE) and using a pinhole to achieve axial sectioning of the observation volume. This confocal volume is an oblong shaped laser focal volume on the order of femtoliters. As fluorescent particles transit the observation volume (Fig. 1a), fluorescent fluctuations caused by the movement of the particle or any other process are recorded

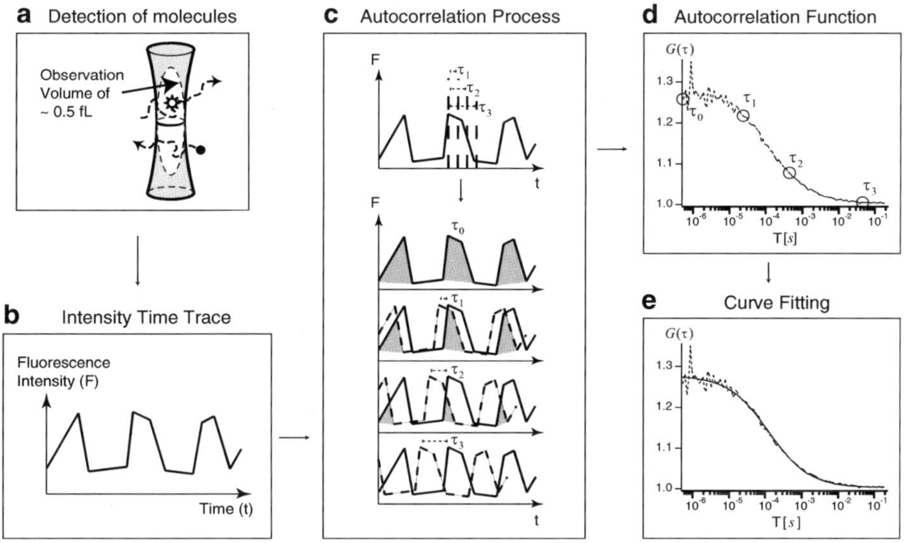

Fig. 1 Principles of FCS: (**a**) Detection of fluorophores in a confocal observation volume. (**b**) Fluorescence intensity time trace of particles moving through the observation volume. $\langle F(t) \rangle$ is the average intensity. (**c**) Illustration of the autocorrelation process. The ACF is a measure of the self-similarity of a signal at a delay time τ. (**d**) The autocorrelation process generates a decaying autocorrelation function. (**e**) Curve fitting of the ACF by theoretical ACF models

(Fig. 1b). In order to extract the information contained in the intensity time trace, its normalized ACF is calculated [50]:

$$G(\tau) = \frac{\langle F(t)F(t+\tau) \rangle}{\langle F(t) \rangle^2}. \quad (1)$$

During the ACF process, the fluorescent signal at time t is compared to the signal at time t + τ. The graphical representation is as if the signal is shifted in time and the "similarity" between the shifted and original signal is calculated (Fig. 1c). As the delay time τ increases, the signal correlates less (is less and less similar with the original signal) and hence the ACF decays to 1 at infinite τ (Fig. 1d). Finally, to extract information about the processes causing the fluorescence fluctuations, a theoretical ACF model is used to fit the experimental ACF (Fig. 1e).

Figure 2a shows the general shape of a typical ACF. The theoretical ACF expression is given by:

$$G(\tau) = G(0)P(\tau)D(\tau) + G(\infty), \quad (2)$$

where $G(0)$ is the amplitude of the ACF at $\tau = 0$ in the absence of any photodynamic processes $P(\tau)$. $P(\tau)$ is any process that changes the fluorescence characteristic of the fluorophore, e.g., triplet state populations or conformational changes

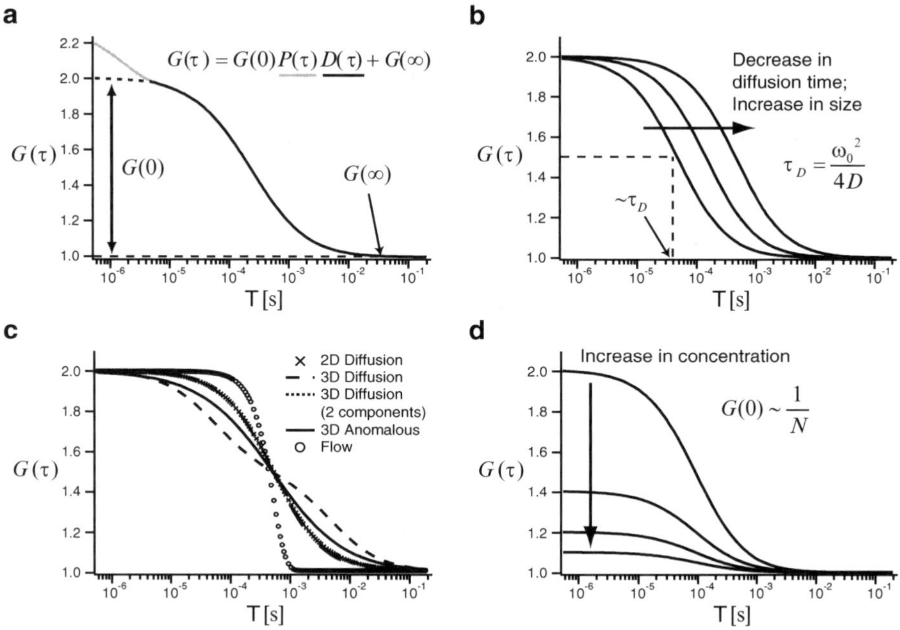

Fig. 2 The influence of different processes on the ACF: (**a**) Parts of an ACF. *Gray portion* of the ACF is contributed by photodynamic processes $P(\tau)$, while the *black portion* stems from diffusion processes $D(\tau)$. $G(0)$ is the amplitude of the ACF at $\tau = 0$. $G(\infty)$ is a convergence value for the ACF at long delay times. (**b**) The width of the ACF increases with the decreasing diffusion coefficient which is related to the size of the particles. (**c**) Shape of the ACF changes with different transport or diffusion processes. (**d**) The amplitude of the ACF decreases with increasing concentration

leading to quenching. These processes typically happen much faster than the diffusion process $D(\tau)$ of the particles. Hence, they show up as an additional shoulder at small τ values of the ACF. However, in some FPs, such processes can be on a similar timescale as diffusion (refer to Sect. 3.3). $D(\tau)$ is the characteristic diffusion process undertaken by the particles. $G(\infty)$ is a convergence value for the ACF at long (infinite) delay times. While in general it should be 1, it is usually added as a fit parameter and as an additional check for the quality of the data.[2] If $G(\infty)$ deviates significantly from the convergence value of 1, it possibly indicates photobleaching, sample movement, or other systematic deviations of the measurement. In the following subsections, $G(0)$, $D(\tau)$, and $P(\tau)$ are discussed in detail.

[2]In some cases, the ACF is defined slightly differently in which the ACF has a convergence value of 0 instead of 1. This different definition is only a change in offset. It does not influence the function of the ACF and does not alter any of the conclusions here.

2.1.1 Diffusion and Width of the ACF

The extraction of parameters from the experimental ACF requires the determination of ACF models based on the theory of the assumed process. For instance, to determine the ACF for particles diffusing through a confocal volume, one has to calculate the fluorescence signal arising from fluorescent particles undergoing Brownian motion and passing through a confocal observation volume. Theoretical ACF models have been determined for a wide variety of processes including translational and rotational diffusion, flow, chemical reactions, and photodynamics [50, 51]. Since diffusion typically dominates experimental ACFs in biological systems, we give here the theoretical model for particles diffusing in a 3D environment and discuss the different parameters on that example. Assuming a Gaussian laser profile, the 3D Brownian or free diffusion ACF with a photodynamic process is given by [49, 52]:

$$G(\tau) = G(0)P(\tau)\left(1 + \frac{\tau}{\tau_D}\right)^{-1}\left(1 + \frac{\tau}{K^2\tau_D}\right)^{-1/2} + G(\infty), \tag{3}$$

where the 3D diffusion model $D(\tau)$ is given by:

$$D_{3D}(\tau) = \left(1 + \frac{\tau}{\tau_D}\right)^{-1}\left(1 + \frac{\tau}{K^2\tau_D}\right)^{-1/2}. \tag{4}$$

K is the structural parameter defining a 3D observation volume in the case of a confocal laser spot:

$$K = \frac{z_0}{\omega_0}. \tag{5}$$

The average time (diffusion time) taken to diffuse through the laser focus is given by:

$$\tau_D = \frac{{\omega_0}^2}{4D}. \tag{6}$$

ω_0 and z_0 are the radial and axial distance of the laser focus at which the intensity profile decreases to e^{-2} of its maximal value, respectively. D is the diffusion coefficient of the particles. The width of the ACF is determined by the timescales of the processes. Since photophysical processes are usually taking place on a much faster timescale compared to diffusion, the overall width of the correlation function is determined solely by diffusion. Hence, the time the ACF needs to decay from $G(0)$ to half of its amplitude $G(0)/2$ is an estimate of the diffusion time τ_D (Fig. 2b). The hydrodynamic radius of a particle can be estimated by FCS using the Stokes–Einstein relation:

$$D = \frac{kT}{6\pi\eta r}, \tag{7}$$

where k is Boltzmann's constant; T is the absolute temperature; η is the viscosity of the medium; r is the hydrodynamic radius of the particle. For a spherical particle with radius r and molecular mass M, τ_D is proportional to $\sqrt[3]{M}$. Although τ_D can be estimated from the value of $G(\tau)$ at half maximum, a diffusion model is required to extract the value of τ_D.

Other common $D(\tau)$ models include:

$$D_{2D}(\tau) = \left(1 + \frac{\tau}{\tau_D}\right)^{-1} \tag{8}$$

for 2D or planar free diffusion [47] which is commonly used for membrane measurements.

$$D_{3D,Ano}(\tau) = \left(1 + \left(\frac{\tau}{\tau_D}\right)^{\alpha}\right)^{-1} \left(1 + \frac{1}{K^2}\left(\frac{\tau}{\tau_D}\right)^{\alpha}\right)^{-1/2} \tag{9}$$

for 3D anomalous diffusion [53] with the anomalous factor α. Values of $\alpha < 1$ indicate hindered or sub-diffusive diffusion, while $\alpha > 1$ indicates direct transport or super-diffusion.

$$\begin{aligned} D_{3D,2C}(\tau) = {} & F_1\left(1 + \frac{\tau}{\tau_{D1}}\right)^{-1} \left(1 + \frac{\tau}{K^2\tau_{D1}}\right)^{-1/2} \\ & + (1 - F_1)\left(1 + \frac{\tau}{\tau_{D2}}\right)^{-1} \left(1 + \frac{\tau}{K^2\tau_{D2}}\right)^{-1/2} \end{aligned} \tag{10}$$

for 2 components diffusion [10, 50]. This describes the detection of two different species of particles with two different τ_{Di}, where F_1 is the fraction of particles with τ_{D1}. It assumes that the two species have the same molecular brightness (covered in Sect. 2.1.2). This is particularly useful to determine the binding of a small fluorescent tagged molecule to a larger one. However, according to simulations, the two τ_D can be well resolved with statistical significance only if the slower molecule diffuses 1.6 times slower [54]. This corresponds to a mass ratio between the molecules of about 4. This ratio can be lower if one or both of the τ_D is known beforehand. Despite this, quantitating the binding of two species of molecules with similar mass is challenging. A more comprehensive list of ACF models has been reported [55]. The different processes give the ACF distinct shapes (Fig. 2c).

2.1.2 Concentration and Amplitude of the ACF

The amplitude of the ACF for pure diffusion at $\tau = 0$ is given by:

$$G(0) \sim \frac{1}{N}, \tag{11}$$

which shows that the amplitude of the ACF at $\tau = 0$ is inversely proportional to the average number of particles N (Fig. 2d). Therefore, it is desirable to keep the number of observed particles low. This can be done using a very small observation volume or keeping the concentration of the sample low. In general, FCS can measure samples with about 0.1–1,000 particles per observation volume which translates into sub-nanomolar to a low micromolar concentration range in the case of confocal FCS. The effective confocal volume V_{eff} in which fluorescent particles can be observed is given by [52]:

$$V_{eff} = \pi^{3/2} {\omega_0}^2 z_0. \tag{12}$$

Typically V_{eff} ranges from ~0.3 to 1 fL. V_{eff} can be measured in three different ways [56]. The first method is to do a dilution series using dye solutions of known concentrations. When the experimental N is plotted against the different concentrations of the dye, the slope gives the V_{eff}. The second method is to use a dye with a known diffusion coefficient D to determine τ_D and subsequently ω_0 and z_0 from experimental ACFs. The last method is to use a fluorescent bead to determine the point spread function and calculate the V_{eff}. Once the V_{eff} is known, (11) becomes:

$$G(0) = \frac{1}{N_A V_{eff} C}, \tag{13}$$

where C is the average concentration of the sample and N_A is Avogadro's constant. This method is only accurate if all the particles have the same molecular brightness, η or counts per particle per second (*cps*). The *cps*, experimentally determined by dividing the average intensity $\langle F(t) \rangle$ by $\langle N \rangle$, is influenced by the laser power, the detection efficiency of the optical system, and the quantum yield of the dye used. A higher *cps* gives a better signal-to-noise ratio [57]. If the observation volume contains different particles with different *cps*, (11) and (13) become a summation of all the i species with their particular η_i:

$$G(0) = \frac{\sum {\eta_i}^2 N_i}{\left(\sum \eta_i N_i\right)^2} = \frac{\sum {\eta_i}^2 C_i}{N_A V_{eff} \left(\sum \eta_i C_i\right)^2}. \tag{14}$$

The amplitude of the ACF is biased toward the species with higher η. A typical situation in which this can happen is in binding processes in which the fluorescent yield of a particle changes upon binding [14, 18].

2.1.3 Photodynamics

During the course of translating through the observation volume, any phenomenon that changes the fluorescence characteristics of the fluorophore will be detected. The most common process is the transition to the triplet state of the fluorophore. A fluorophore can flicker or blink as it cycles between a long-lived triplet and a short-lived singlet state. As this process is very fast, it is detected at small τ values (typically at ~10^{-6} s) of the ACF as an additional shoulder. It can be accounted for by the expression [19, 21]:

$$P(\tau) = 1 + \frac{F_{trip}}{1 - F_{trip}} \exp\left(-\frac{\tau}{\tau_{trip}}\right), \tag{15}$$

where F_{trip} is the fraction of the particles that reside in the triplet state and τ_{trip} is the triplet state relaxation time. For other processes that cycle between a dark and bright state while diffusing through the observation volume such as a loss of fluorescence due to a conformational change [58, 59] or a loss of fluorescence in different pH due to protonation [29], F_{trip} becomes the fraction in the dark state and τ_{trip} becomes the relaxation time for the process.

In summary, when the sample has different species with different *cps* undergoing different diffusion processes, (2) takes the general form [60]:

$$G(\tau) = \frac{1}{N_{tot}} P(\tau) \frac{\sum \eta_i^2 F_i D_i(\tau)}{(\sum \eta_i F_i)^2} + G(\infty), \tag{16}$$

where F_i is the fraction of the total particles N_{tot} having the corresponding η_i and $D_i(\tau)$. For in-depth discussions of the principles, equations and applications of FCS refer to these reviews [50, 51, 61].

2.2 Principles of FCCS

In contrast to an ACF, a CCF correlates signals from two different sources. In FCCS, photons collected from usually two separated detection channels each detecting particles tagged with fluorophores emitting at two different wavelengths are cross-correlated (Fig. 3). For example, protein X is tagged with GFP while protein Y is tagged with a red FP. If X binds to Y, whenever the "green" detector detects a signal, so will the "red" detector and the signals are correlated. The CCF contains information about the complexes formed between X and Y. The normalized CCF for the green (G) and red (R) channels is given by:

$$G_{GR}(\tau) = \frac{\langle F_G(t) F_R(t+\tau) \rangle}{\langle F_G(t) \rangle \langle F_R(t) \rangle}. \tag{17}$$

The amplitude of the CCF in a simple 1:1 stoichiometry binding case, assuming no changes in the brightness upon binding, no spectral crosstalk and negligible background is given by:

$$G_{GR}(0) = \frac{N_{gr}}{(N_g + N_{gr})(N_r + N_{gr})}, \tag{18}$$

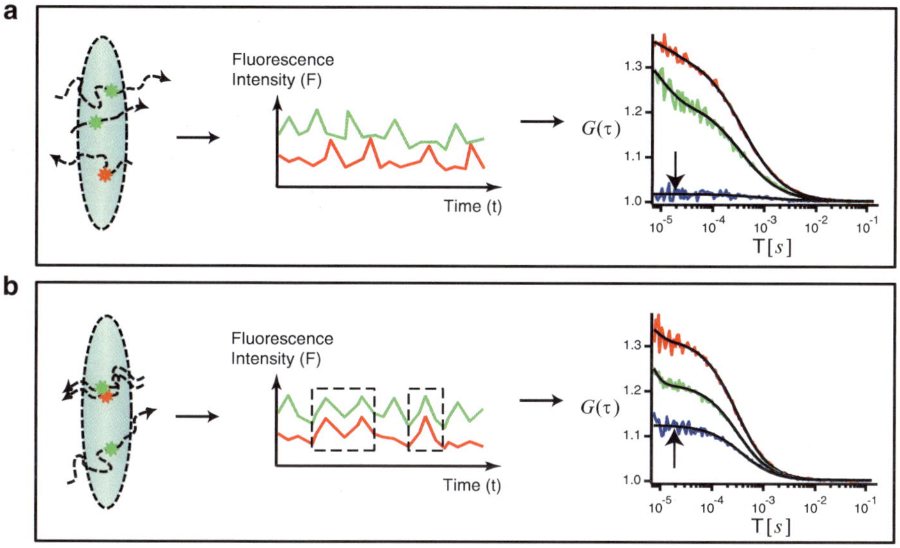

Fig. 3 Principle of FCCS: (**a**) Differently tagged particles move independently through the observation volume. Therefore, the signals are not correlated and their CCF is flat (*blue curve*). (**b**) If particles form complexes and move together through the observation volume, the signals of the differently tagged particles will correlate resulting in an elevated CCF with respect to the *green* and *red ACF*

where N_i and N_{gr} are the number of molecules of the free species and complex observed, respectively. As before, the two ACF amplitudes from each detection channel are inversely proportional to the total number of respectively molecules:

$$G_i(0) = \frac{1}{N_i + N_{gr}}. \tag{19}$$

The ratio of the amplitude of CCF to the ACFs indicates the amount of the molecules in complexes:

$$\frac{G_{GR}(0)}{G_G(0)} = \frac{N_{gr}}{N_r + N_{gr}}, \tag{20}$$

$$\frac{G_{GR}(0)}{G_R(0)} = \frac{N_{gr}}{N_g + N_{gr}}. \tag{21}$$

For example, the ratio of the CCF to the green ACF indicates the ratio of red tagged particles in complexes. Typically, if $N_r < N_g$, $G_{gr}(0)/G_g(0)$, which is limited by the amount of N_r and is of a higher value, is reported over the other ratio as a quantitative approach of reporting interaction.

However, this approach only provides an estimate because in practice crosstalk increases the CCF amplitude. Crosstalk is the leakage of signal from the green fluorophore into the red detection channel and vice versa due to the wide emission bands of most fluorophores. However, the crosstalk from the red fluorophore to the green channel is usually negligible. Hence, signals from a green fluorophore will cross-correlate with its own signals in the red channel resulting in an elevated CCF amplitude. This can lead to false-positive results. Crosstalk can be accounted for by treating it as another contribution of different *cps* (14). Assuming only crosstalk from the green fluorophore into the red channel, the ACF and CCF become [35]:

$$G_G(0) = \frac{\eta_{g,G}{}^2 N_g + \eta_{g,G}{}^2 N_{gr}}{\left(\eta_{g,G} N_g + \eta_{g,G} N_{gr}\right)^2} = \frac{1}{N_g + N_{gr}}, \tag{22}$$

$$\begin{aligned} G_R(0) &= \frac{\eta_{r,R}{}^2 N_r + \eta_{g,R}{}^2 N_g + \left(\eta_{r,R} + \eta_{g,R}\right)^2 N_{gr}}{\left[\eta_{r,R} N_r + \eta_{g,R} N_g + \left(\eta_{r,R} + \eta_{g,R}\right) N_{gr}\right]^2} \\ &= \frac{N_r + \left(\frac{\eta_{g,R}}{\eta_{r,R}}\right)^2 N_g + \left(1 + \frac{\eta_{g,R}}{\eta_{r,R}}\right)^2 N_{gr}}{\left[N_r + \frac{\eta_{g,R}}{\eta_{r,R}} N_g + \left(1 + \frac{\eta_{g,R}}{\eta_{r,R}}\right) N_{gr}\right]^2}, \end{aligned} \tag{23}$$

$$\begin{aligned} G_{GR}(0) &= \frac{\eta_{g,G}\left(\eta_{r,R} + \eta_{g,R}\right) N_{gr} + \eta_{g,G}\eta_{g,R} N_g}{\left(\eta_{g,G} N_g + \eta_{g,G} N_{gr}\right)\left[\eta_{r,R} N_r + \eta_{g,R} N_g + \left(\eta_{r,R} + \eta_{g,R}\right) N_{gr}\right]} \\ &= \frac{\left(1 + \frac{\eta_{g,R}}{\eta_{r,R}}\right) N_{gr} + \frac{\eta_{g,R}}{\eta_{r,R}} N_g}{\left(N_g + N_{gr}\right)\left[N_r + \frac{\eta_{g,R}}{\eta_{r,R}} N_g + \left(1 + \frac{\eta_{g,R}}{\eta_{r,R}}\right) N_{gr}\right]}, \end{aligned} \tag{24}$$

where $\eta_{i,j}$ is the *cps* of fluorescent species *i* detected in channel *j*. The individual *cps* can be determined using samples with only one of the fluorophores present. Once the ratio $\eta_{g,R}/\eta_{r,R}$ is known, the three equations can be solved for the values *of* N_g, N_r, and N_{gr}. Usually crosstalk can be avoided using fluorophores with well-separated emission spectra. However, in the case of FPs, the selection is limited. Crosstalk can also be avoided by alternating excitation schemes where only one type of fluorophore is excited at a time and only the corresponding detection channel is recorded. Alternating excitation is discussed in detail in Sect. 4.

In recent years, FCCS was developed into a quantitative method for the determination of dissociation constants (K_d) of biomolecular interactions [41–44, 62].

In order to determine concentrations of free and bound molecules, the individual effective volumes V_{eff} for the two detection channels and the cross-correlation need to be known. In dual-color FCCS (DC-FCCS) [63], excitation of the two fluorophores is achieved by two lasers. The two laser foci will not have the same size due to different wavelengths and might be shifted with respect to each other. Hence, the cross-correlation effective volume, which is where the binding complex is detected, needs to be calculated resulting in three different V_{eff} [63]. To avoid the alignment of two lasers, single-wavelength FCCS (SW-FCCS) [64] or two-photon excitation FCCS (TPE-FCCS) [65] can be used resulting in only a single V_{eff}. Reviews covering theoretical and practical aspect as well as recent advances in FCCS are available [66–68].

2.3 *Experimental Realization*

FCCS was first shown in 1989 by Rička and colleagues [69] who used a single laser wavelength to obtain a "green" fluorescence signal from fluorescent polystyrene lattices and a scattered "blue" signal from nonfluorescent latex particles. Both signals were cross-correlated. In 1997, Schwille and colleagues [63] used two lasers to excite two different dyes with different emission spectra and cross-correlating their signals. With the help of better technology, FCCS became a powerful tool with the ability for the determination of molecular interactions. A typical confocal setup of DC-FCCS is shown in Fig. 4. In an FCS setup, only one laser and one of the detectors are used. Emission filters select the range of wavelengths to be detected by high quantum efficiency avalanche photodiode (APD) detectors in photon counting mode. A hardware correlator performs the auto- and cross-correlation. Correlation can also be done by a software correlator [70, 167]. A scanning laser microscope is usually coupled to an FCS/FCCS module to allow the user to obtain an image of the sample before selecting the position in the sample to perform FCS measurements [71].

3 Challenges of Applying FCS and FCCS In Vivo

The discussion and equations in the previous section in general can only be applied in solution. When applied to cell cultures or organisms, especially using FPs as the probe, special attention has to be paid. In this section, we discuss issues facing measurements in cell cultures and organisms, and common solutions.

3.1 *Optimizing Measurement Conditions*

A typical FCS or FCCS biochemical experiment in vitro involves tagging the molecule of interest with an organic dye in a solution with low background. Photons

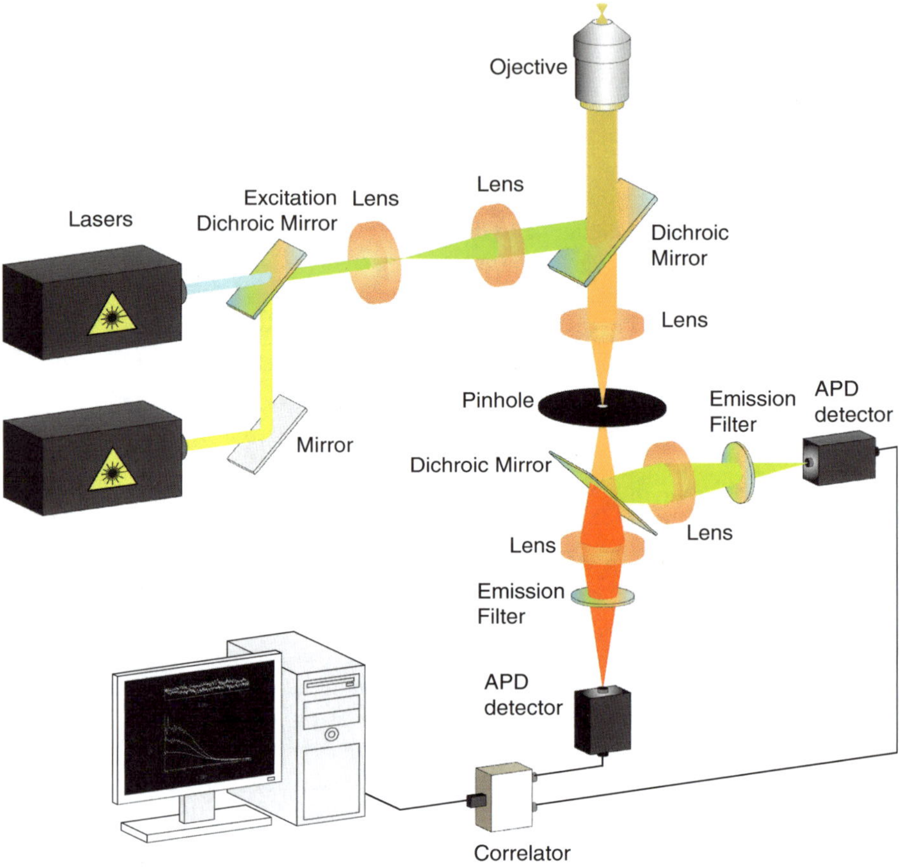

Fig. 4 Dual-color FCCS (DC-FCCS) setup. APD: Avalanche photodiode. For the performance of SW-FCCS, only one of the lasers is used. For TP-FCCS, one pulsed laser is used

generated from the dyes are easily detected achieving high signal-to-noise ratio. In contrast, the fluorophores preferred in organisms are fluorescent proteins (FPs) due to the ease, selectivity, and control of stoichiometry and expression levels of genetic labeling. In general, FPs have lower quantum efficiency than the commonly used organic dyes. Working in organisms further reduces the detection efficiency of the fluorescent proteins due to scattering and absorption in the tissue. This limits the working distance of the technique. Furthermore, bulk photobleaching of the fluorescent proteins can pose a serious problem in the confined volume of a cell. Hence, there is a fine balance between reduction of the laser power to reduce photobleaching and keeping it sufficient for producing and recording enough fluorescent photons to calculate an ACF. Hence, the first and foremost concern for experiment in vivo with FCS or FCCS is to establish the range of *cps*, which gives a good ACF with minimal photobleaching. A *cps* of 800–1,500 Hz is usually sufficient for a single 30 s measurement in organisms [43]. Even a *cps* as low as 500 Hz was reported to be sufficient [72].

3.2 Correcting for Background Autofluorescence

In measurements in vitro, background is usually low and the organic fluorophores used have high *cps*. Hence, background contribution from the buffer or dark counts of the detector is insignificant. Due to the low *cps* of FPs obtained in vivo, the native fluorescence background (autofluorescence) in cells or biological tissues often significantly contributes to the ACF. If the background autofluorescence has elevated *cps*, the autofluorescence has to be considered as another fluorescent species and has to be account for by (14). The background fluorescence can also be removed using lifetime gating [73] or fluorescence lifetime correlation spectroscopy (FLCS) [74], where contribution from unwanted background fluorescence can be separated based on the fluorescence lifetime. In the case where the background is noncorrelated (background which does not yield an ACF), a simple addition of a correction factor to the theoretical ACF is sufficient. Noncorrelated background can be due to noise from the detector or scattered light from the sample which are random in nature. This simply decreases the amplitude of the ACF, resulting in the apparent detection of more particles. The expression for the apparent ACF amplitude with noncorrelative background contribution B is:

$$G_{ACF,app}(0) = \frac{\sum \eta_i{}^2 N_i}{\left(\sum \eta_i N_i + B\right)^2}. \tag{25}$$

This can be rewritten as a factor of the total average fluorescence intensity F_{meas} measured:

$$G_{ACF,app}(0) = \frac{(F_{meas} - B)^2}{F_{meas}{}^2} G_{ACF}(0). \tag{26}$$

Similarly, the apparent CCF is given by:

$$G_{CCF,app}(0) = \left(1 - \frac{B_G}{F_{meas,G}}\right)\left(1 - \frac{B_R}{F_{meas,R}}\right) G_{CCF}(0), \tag{27}$$

where $F_{\mathrm{meas},i}$ is the total average fluorescence intensity in the corresponding channels. Background can be corrected using these background correlation factors into the theoretical models to fit the experimental ACF/CCF [41, 75] or by solving (25) [43–45]. Equation (27) assumes that both B_G and B_R are not correlated, and if a correlated background signal is present it is treated as a separate species.

3.3 Selecting Fluorescent Proteins for FCS/FCCS

Although GFP and its many variants provide sufficient coverage of the visible spectrum, for the FCS/FCCS experiment the selection of FPs is still limited. As a low expression level is usually required in an FCS/FCCS experiment, it is crucial that the FPs used have high molecular brightness and are photostable. However, in general, FPs are not as photostable and as bright compared to organic dyes or quantum dots. Many of them have complex photodynamic processes which are not well understood. All these factors have to be considered when deciding on an FP to use in FCS/FCCS studies.

The enhanced GFP (EGFP) from Clontech is one of the most stable and brightest GFP mutants and commonly used in FCS/FCCS studies but has a weak tendency to dimerize [168]. Monomeric forms of GFP are available [76, 77]. For FCCS, a red FP is needed to pair up with EGFP to reduce crosstalk. Often mRFP1 [78] has been used to partner with EGFP [33, 38, 39, 43–45]. In recent years, mCherry [79], reported to be more photostable than mRFP1, is also increasingly used [41, 42, 80, 81]. However, red FPs have disadvantages because of their complex photodynamics and poor photostability. Just like organic dyes, FPs can transit to triplet states and flicker/blink depending on the environment and measurement conditions. These photodynamic processes can be much slower and more complex than in typical organic dyes. These, thus, show up as an additional shoulder in the medium correlation time region and can influence the shoulder of the diffusion process in the ACF. Extracting the diffusion time and actual amplitude of the curve becomes complicated. In fact, the first few FCS experiments using GFP looked at its photodynamics [21, 29, 82]. Investigation into the photodynamics of YFP [30] and the tetramer DsRed [31] soon followed. Because YFP has a strong blinking behavior and is less photostable than GFP, it is less often used. The red FPs, mRFP1, mCherry, and mStrawberry also have complex photodynamic processes that can complicate data analysis [83]. Moreover, not all of the expressed proteins are fluorescent; 60% of mRFP1 has been reported to be in a dark state [84], while 50% of mCherry is not matured, at least in yeast [41]. A recently developed FP, mKeima, with a large Stokes shift can be coupled with ECFP to perform SW-FCCS [85]. Limited by the selection of monomeric RFPs suitable for FCS/FCCS, mCherry is the best available choice as a partner for EGFP at the moment.

3.4 Working with Organisms

The challenges mentioned so far are common in cell cultures and in organisms. When measuring in vivo, there are some additional issues that require particular attention. The most common problem is the working distance of the technique.

Even for a transparent tissue, for example the zebrafish embryo, tissue will still cause scattering and absorption which prevents the detection of sufficient photons. TPE, which will be discussed in the next section, has been proposed to overcome this. However, the increase in photobleaching at the laser focus and saturation effects have restricted its application in FCS [86–88]. In some cases, excision of tissue or dissection of the animal, i.e., ex vivo conditions are required to overcome this problem [89, 90].

Distortion of the focal volume due to variations in refractive index in thick tissue is another problem. Since the size of the effective volume is typically determined by a calibration measurement in solution, it is uncertain how accurate this calibration is for in vivo measurements. The reliability of experimentally determined diffusion times and concentrations depends largely on knowing the exact size of the volume in the actual sample. The distortion of excitation volumes extends to DC-FCCS. Since two different wavelengths are used, the distortion and chromatic aberration for each volume are different. SW-FCCS and TPE-FCCS are two methods that use a single wavelength for excitation to partially avoid this problem. The issue of distortion of the focal volume in vivo can be addressed using two-foci cross-correlation spectroscopy (TFCCS) covered in the next section.

Natural movements of the organism during measurement introduces artifacts or complicates data analysis as it distorts the ACF; hence, measurement near the heart should be avoided [90]. Depending on the organism, anesthetics are available to immobilize the organism (see, for instance, the study of tricaine in zebrafish [91]).

Figure 5 illustrates the issue of protein expression levels in organisms. Figure 5a shows a confocal image of zebrafish embryo expressing a membrane receptor, Cxcr4b, tagged with EGFP. Different regions of the embryo express different amounts of the protein. FCS can only be performed in those that are low in expression. Hence, cells suitable for FCS may not be visible under the normal settings for imaging. Figures 5b and c shows a *Drosophila* embryo expressing

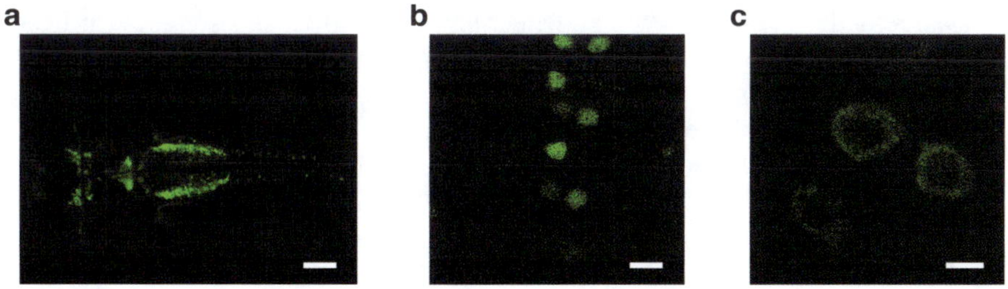

Fig. 5 (**a**) Zebrafish embryo expressing a membrane receptor Cxcr4b-EGFP. The scale bar represents a size of 100 μm. (**b**) *Drosophila* embryo showing motor neurons expressing cytosolic EGFP. The scale bar represents 10 μm. (**c**) *Drosophila* embryo showing motor neurons expressing mCD8-EGFP membrane-targeted protein. The scale bar represents 5 μm

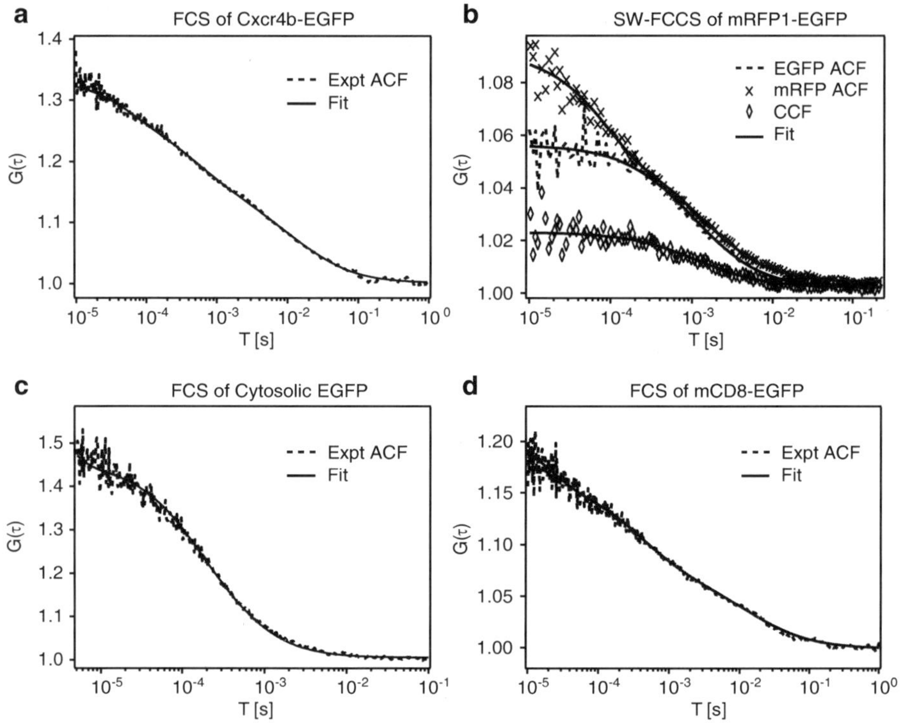

Fig. 6 (**a**) ACF of the membrane receptor Cxcr4b measured in a zebrafish embryo. (**b**) ACFs (*cross* and *dash*) and CCF (*diamond*) of a tandem mRFP1-EGFP positive control, where both mRFP1 and EGFP are excited simultaneously using a 515 nm laser (SW-FCCS) in a zebrafish embryo. (**c**) ACF of cytosolic EGFP expressed in a *Drosophila* embryo. (**d**) ACF of membrane-targeted EGFP-mCD8 in a *Drosophila* embryo

cytosolic EGFP and membrane-targeted EGFP-mCD8 mouse protein in the motor neurons, respectively. The background noise in Fig. 5b and c is due to the tuning of the detector on the laser scanning microscope to very high sensitivity to detect the low expressed proteins. In general, detecting expression levels suitable for FCS is different from normal confocal imaging conditions. If the tagged proteins are introduced exogenously instead of a transgenic line, suitable amount of the injected mRNA or plasmid and the time in which FCS is to be performed when the protein is expressed at a suitable level are critical.

Figure 6 contains some example of FCS/FCCS measurements in vivo. Figure 6a is an ACF of the membrane receptor Cxcr4b measured in zebrafish embryo, while Fig. 6b is an FCCS experiment of a tandem mRFP1-EGFP positive control. These tandem FPs are commonly used to define the upper limit of an FCCS experiment. The ACFs of cytosolic EGFP and membrane-targeted EGFP-mCD8 expressed in *Drosophila* embryo are shown in Fig. 6c and d, respectively.

4 FCS and FCCS Modalities

4.1 Two-Photon Excitation

In TPE, the fluorophore absorbs two photons within a very short time ($<10^{-15}$ s). The sum of the two photons' energies is the energy required for excitation. This means that light with a twice longer wavelength can be used to excite the same dye compared to OPE. However, the probability of TPE is very low, and the photon emission probability increases quadratically with the excitation intensity. Therefore, a more expensive high energy femto- or pico-second pulsed laser is needed to generate the high flux of excitation photons required. Only a small localized region at the laser focus (~1 fL) has the intensity high enough for TPE. This gives TPE real 3D sectioning without the need of any spatial filters such as a pinhole.

TPE microscopy was introduced [92] to take advantage of the 3D sectioning. The localized excitation reduces photobleaching in the bulk of the sample as well as reduction of photodamage to a living cell. The longer wavelength used, typically in the infrared region, provided deeper tissue penetration. Hence, TPE is suitable for thick tissue imaging. TPE was combined with FCS on live cells [93, 94]. A unique feature of TPE is that the excitation spectra of many dyes overlap making it possible to excite more than one dye using TPE at one particular wavelength. Using the dyes Rhodamine Green and Texas Red, it was shown that it is possible to perform TPE-FCCS using a single wavelength [65]. This removes the need for the alignment of two lasers in DC-FCCS, making the experimental setup and concentration calculations simpler. TPE-FCCS was later applied to a fusion protein of red-shifted GFP and DsRed [95]. Recent TPE-FCCS studies have been reported [96–101]. In these publications, only Rosales et al. (using ECFP and EYFP) and Savatier et al. (using mCerulean and mCherry) perform TPE-FCCS using FPs. The challenge in TPE-FCCS for FPs is finding two suitable FPs to be optimally excited with a single wavelength giving similar *cps* (to minimize emission crosstalk) and low photobleaching.

There are few articles comparing TPE-FCS with OPE-FCS [86, 88, 94, 102]. Despite the advantages of TPE-FCS, it suffers from poor signal-to-noise ratio (low *cps*) which places the FPs at a disadvantage when compared to the higher quantum yield organic dyes. This low *cps* is mainly attributed to increased photobleaching at the laser focus and saturation effects [86–88].

4.2 Alternating Excitation

In a typical alternating excitation experiment, two dyes are alternatively excited by two lasers at a rate much faster than their diffusion times. This requires time-correlated single photon counting (TCSPC) to synchronize the pulsing lasers and the photons generated by the respective laser. The main advantage of using alternating excitation is to avoid false positives caused by spectral crosstalk in FCCS as well as to avoid the hassle of mathematically accounting for crosstalk as shown in Sect. 2.2. This is helpful in the case of FPs which usually have a long emission tail. As was first shown by Thews et al., this approach removes crosstalk from the CFP and YFP pair in FCCS [103]. This method of performing FCCS was then expanded by Müller et al. in the same year [104] naming the method pulsed interleaved excitation (PIE). A similar alternating excitation scheme is alternating laser excitation (ALEX) [105]. ALEX's pulse is typically in the microsecond timescale, while PIE's pulse is in the nanosecond time scale. Hence, PIE is capable of producing correlation function of sub-microsecond resolution.

4.3 Two-Focus Cross-Correlation Spectroscopy

Two-focus or two-beam cross-correlation uses two focal spots side by side to record fluorescence fluctuations. It was originally applied to a system with flow. These fluctuations can be cross-correlated to determine direction and velocity of the flow [106–108]. The main advantage of TFCCS with regards to measurement in organisms is the accurate determination of the focal volume width directly within the tissue during measurement. Typically, the width of the observation volume is predetermined in solution using a dye of a known diffusion coefficient. However, when going from solution to cell cultures or in vivo, due to the refractive index mismatch of tissue and water (typically the immersion and mounting medium used in biological samples), the observation volume is distorted. Calibration of the focal volume width by a dye in solution may be a poor estimate. In TFCCS, the laser width and diffusion coefficient of the molecule of interest can be directly extracted from the same biological sample in which the experiment is performed without the need of any dye calibration once the distance between the two foci is known [62].

4.4 Multiplexed FCS

Multiplexed correlation spectroscopy uses more than two detection volumes to simultaneously perform FCS in different spots. As different parts of a living cell or organism are not homogenous, there is a need to simultaneously probe different parts of the sample. Blom et al. used diffractive optical elements (DOEs) to create multiple focal spots of 2×2 [109] and 4×1 [110] to perform parallel FCS in microarrays. DC-FCCS with the 2×2 DOEs was later reported [111]. Takahashi and colleagues sequentially used four foci to perform FCS either at different spots in the same cell or in four different cells [112]. The same group later used total internal reflection FCS (TIR-FCS) coupled with seven photomultiplier tube detectors to perform FCS simultaneously on seven different spots in living cells [113]. Confocal FCS was also coupled to an electron multiplying charged coupled device (EMCCD) camera, which is an array detector, with the intention of multiplexing

[114]. They use this EMCCD array detection method with TIR illumination to acquire 1,600 point measurements simultaneously [115]. An array of pinholes, based on a stationary Nipkow disk used in a spinning disk confocal microscope, was coupled to an EMCCD camera to achieve tens to hundreds of measurements simultaneously [116]. Recently, the use of single plane illumination microscopy (SPIM) in combination with fast EMCCD camera-based FCS lead to the simultaneous measurement of more than 4,000 contiguous ACFs in zebrafish [117].

4.5 Scanning FCS

Scanning FCS (SFCS) in general refers to scanning the laser in the sample repeatedly either in a line [53, 62, 118–122] or in a circular pattern [123–127] while recording the fluorescence fluctuations. A circular scan works similarly to TFCCS in the sense that the determination of the diffusion coefficient of the fluorophore is achieved once the scanning radius is known without any additional calibrations [127]. The information obtained from scanning contains spatial and temporal components of the scan. The photons can then be correlated spatially and/or temporally. Scanning also allows the measurement of very slow or immobile molecules [125]. Very slow diffusing molecules require longer measurement times to acquire enough photons for statistical accuracy; hence, scanning overcomes this problem. Many proteins especially membrane localized proteins diffuse slowly or are immobile. Hence, SFCS allows the study of such molecules where conventional FCS fails. Scanning also reduces the time fluorophores spend in the laser focus thus reducing photobleaching.

Using a scanning mirror, Berland and colleagues combined TPE and circular scanning to detect molecular aggregation [123]. Petrášek and colleagues compared the photobleaching effects of TPE with and without scanning [128]. They have shown that scanning improves the signal-to-noise ratio of TPE. They applied this circular SFCS with TPE to study protein dynamics in *C. elegans* before asymmetric cell division [126]. A series of line scanning FCS/FCCS techniques, which combined TFCCS and alternating excitation, for membrane studies with the capability to correct for membrane movements was recently developed [121, 129] and applied in zebrafish embryos [62, 130]. Spinning disk confocal microscopy, which uses an array of ~10,000 pinholes to quickly scan through the sample, has recently been coupled with FCS [131].

4.6 Raster Image Correlation Spectroscopy

Raster image correlation spectroscopy (RICS) uses the images acquired from a confocal laser scanning microscope to perform correlations instead of using a separate FCS module. The fluorescence intensity in one of the pixels which is considered to be a single point measurement which can be spatially and/or temporally correlated. A review of the different types of image correlation spectroscopy methods has been recently published [132]. Aggregation, number densities, and the mobility of molecules can be determined from these methods. However, as the stacks of images are acquired at different time points, the methods are suitable only for immobile or very slow moving processes, or chemically fixed cells. RICS, which is a type of image correlation spectroscopy developed recently, exploits the time structure hidden during the acquisition of an image to allow one to monitor much faster processes in the microseconds to seconds timescale [133, 134]. It has also been extended to perform two color cross-correlation [135]. Although not yet implemented in organisms, the results in 2D cell cultures are promising. Since confocal imaging is routinely done in organisms, it is a matter of time before RICS is applied in living organisms.

4.7 Imaging Total Internal Reflection FCS

Total internal reflection fluorescence microscopy (TIRFM) uses the evanescent electromagnetic field generated when the incident laser is total internally reflected while traveling from an optically denser to a less dense medium. This evanescent field can project ~70–300 nm into the sample. This enables high sensitivity for regions at the cover slide, allowing processes close to the cover slide, such as membrane dynamics, to be studied. In addition, bulk photobleaching and phototoxicity are reduced. TIR has been coupled to FCS (TIR-FCS) in the late 1970s and early 1980s to study molecular dynamics near to the glass surface [136–138]. However, those measurements are limited to a single spot in the sample. Kannan and colleagues coupled TIR with EMCCD camera-based FCS allowing the measurement of whole FCS images on cell membranes [115]. As the diffusion dynamics on a cell membrane are slower than in the cytoplasm, the milliseconds time resolution of the EMCCD still allows such dynamics to be studied. This method, imaging TIR-FCS (ITIR-FCS), was later extended to perform spatiotemporal cross-correlation studies in cells [139]. ITIR-FCS has not been reported in organisms yet, but a TIR-based single molecule study using YFP as a probe in zebrafish embryos has already been reported [140].

4.8 Single Plane Illumination Microscopy-FCS

SPIM uses a thin, diffraction-limited light sheet to illuminate the specimen [141]. A second objective is placed at a right angle to the light sheet to collect the fluorescence. This allows imaging of a whole single section of a sample. Performing a z-scan using the light sheet generates a 3D image. In contrast to confocal imaging, only the observed section is illuminated; hence, photobleaching and phototoxicity are dramatically reduced [142]. SPIM can be coupled to an EMCCD camera to achieve multiplexed FCS at a single section of a specimen. SPIM-FCS has recently been demonstrated in zebrafish embryo using injected fluorescence microspheres [117]. Similar to ITIR-FCS, it is limited by the time resolution of the imaging device, hence restricting measurements to processes at the millisecond timescale at the moment.

5 Application of FCS/FCCS *In Vivo*

At the time of writing, the number of publications of FCS/FCCS being applied in organisms is just a handful, illustrating the fact that application of this technique in vivo is still at its infancy. To our knowledge, FCS/FCCS has been applied to animal models such as nematodes (*C. elegans*), fruit flies (*D. melanogaster*), medaka (*O. latipes*), and zebrafish (*D. rerio*). Some of these reports are still exploring and optimizing the ways FCS can be applied in living tissue. Not all the reports covered here deal with intact organism as manipulation of the organism such as excision or dissection is required to reach regions deep inside the organism for measurement. As FCS/FCCS are now regularly been applied in 2D cell cultures, it is a matter of time before the technique will be used regularly in vivo.

5.1 Nematodes (C. elegans)

C. elegans is a well-studied, easy to maintain multicellular worm with a short life cycle. Its genome is completely sequenced and many genetic tools are available. This invertebrate animal is transparent even as an adult. Its development pattern is well understood down to a single cell. For example, *C. elegans* is a good model to study cell polarization and asymmetric cell division. *C. elegans* also happened to be the first eukaryotic organism in which the GFP gene was encoded and expressed [143].

Using circular TPE-SFCS, Petrášek and colleagues investigated the diffusion characteristics of proteins PAR-2 and NMY-2, which are related to cell division, in the cytosol and on the cortex of the *C. elegans* embryo before asymmetric cell division [126]. First, they used static FCS to measure the diffusion distribution of these proteins in the cytosol. However, they noted that while measurement in the cytosol was successful, measurements on the cortex was complicated due to the failure to separate the movement of the organism and the slow diffusion time of these proteins in the data analysis. Circular SFCS was then used to overcome this problem as well as photobleaching and the low statistical accuracy due to the slow motion of the molecules on the cortex. In addition to the display of the usual temporal correlation for the whole duration of the scan, the scans can be plotted as the number of period of the scan against the position of the scan within a single period. This 2D image map can be correlated spatiotemporally. This spatiotemporal correlation can distinguish different mode of transport better compared to the usual temporal correlation. The authors came to the conclusion that NMY-2 has the characteristic of directed motion, while PAR-2 displays multicomponent diffusive or even sub-diffusive behavior in the cortex.

5.2 Fruit Fly (D. melanogaster)

Fruit fly is another favorite animal model. It is small, easy to handle, has a relatively low maintenance cost, short lifecycle, and its embryo develops outside the body. It has many genetic methodologies, many of which have been transferred to other animal model.

Although as early as 2004, in a conference paper, it was reported that TPE-FCS can be applied in *Drosophila* imaginal wing disks to monitor the molecular dynamics of the morphogen Dpp-EGFP [144], it was not until 2009 that FCS was again applied in *Drosophila*. Bhattacharya and colleagues being interested in chromatin organization in mouse embryonic stem (ES) cell differentiation investigated the molecular dynamics of some nuclear proteins in cell cultures as well as in the *Drosophila* embryo [145]. FRAP, FCS, fluorescence loss in photobleaching (FLIP), and fluorescence anisotropy were the main techniques used in their study. As the dynamics of nuclear proteins is linked to genome regulation, they started by comparing the molecular dynamics of nuclear-related proteins in mouse ES cells and primary mouse embryonic fibroblasts (PMEF). FCS in general was used to compare the diffusion time as well as the detection of anomalous diffusion of these proteins in both cell types, while FRAP was used to determine the amount of fluorescence that can be recovered. In the second part of the study, a *Drosophila* transgenic line with a core histone protein H2B-EGFP was used to monitor the changes of the chromatin states during development. The core histone proteins within the developing embryo are highly mobile before cellularization, but their mobility decreases as cellularization and differentiation programs set in, revealing a transition from plastic to frozen chromatin assembly during cellular differentiation.

Shi and colleagues demonstrated that by dissecting the *Drosophila* embryo at stage 16, in which the transparency of the embryo is less than the earlier stages, and mounting it on a cover glass, the exposed central nervous system can be accessed by FCS in an ex vivo experiment [90]. A simple measurement of diffusion time between flies expressing cytosolic EGFP and membrane-bound protein mCD8-EGFP in the aCC/RP2 motor neurons were reported. They showed that FCS differentiates the diffusion characteristics of the membrane-bound mCD8-EGFP from the cytosolic EGFP in conditions very close to those existent in vivo.

5.3 *Zebrafish (D. rerio)*

Zebrafish is a popular animal model for many areas of scientific research such as developmental biology, genetics, neurobiology, drug screening, and cancer research. A recent review on the achievements and potential of zebrafish for modeling human diseases, drug discovery, and development is available [146]. Reviews focusing on the area of cancer research [147, 148], drug studies [149, 150], and transgenic lines for neurodegenerative diseases have been published. Zebrafish is easily maintained, which makes it a low-cost vertebrate that breeds regularly producing a large number of offspring. Its genetics is well known and hence many genetic techniques and transgenic lines are available. The main advantage with respect to optical studies is that its embryo is optically transparent and develops outside the mother. This facilitates the applications of many fluorescence techniques and in this case FCS/FCCS to study blood flow, protein mobility, and protein–protein interactions in zebrafish embryos.

Fluid flow dynamics in organisms are thought to influence the development of tissues and organs. Techniques such as confocal scanning microscopy and particle image velocimetry have been used to quantify blood flow in zebrafish [151]. However, these techniques either have low spatial resolution or require the addition of tracer particles into the blood. Pan et al. applied line SFCS at different angle with respect to the flow channel to determine the 2D flow direction in microchannel and applied it in zebrafish embryo blood vessel [122]. The autofluorescence in the serum allowed determination of flow direction and velocity without any external tagging or injection of fluorescent particles. Using a model with two flows, the alternating slow and fast blood flows in the zebrafish cardiac cycle stemming from systolic (contraction) and diastolic (relaxation) periods were distinguished. Extension to 3D flow was reported and tested in zebrafish embryo brain blood vessels [152]. Later on, static FCS was used to detect the presence of blood circulation in sinusoids of the developing zebrafish liver [153]. In another publication, the same group compared the penetration depth of OPE and TPE-FCS, characterized flow across blood vessels, and the mobility of a membrane receptor, Cxcr4b-EGFP, in living zebrafish embryos [90].

Shi and colleague used single-wavelength excitation to perform static SW-FCCS in zebrafish embryos and CHO cells to measure the apparent K_d for a Rho-GTPase and its effector [43]. The K_d was considered to be apparent because competitive binding from the endogenous protein, unknown interaction of the proteins of interest with other proteins and photobleaching or dark states from the FPs influence the amount of detected complexes. The expression vectors encoding the mRFP1-tagged constitutively active G12V form of the Rho-GTPase Cdc42 and EGFP-IQGAP1, which is an actin-binding scaffolding protein, were co-injected into 1 of 16 cells of early zebrafish embryos or co-transfected into CHO cells. The T17N dominantly inactive form of Cdc42 was used as a negative control. A single 514-nm laser line was used to excite the mRFP and EGFP, simultaneously avoiding the uneven distortion and alignment of two lasers in the case of DC-FCCS.

The apparent K_d was reported to be ~100 nM in zebrafish embryo, while a value of ~1,000 nM was obtained in 2D cell culture, indicating the difference of performing the experiment in vivo and in vitro and emphasizing a need to make a radical turn from in vitro studies toward in vivo-based applications.

In a parallel effort, Ries and colleagues developed a series of line scanning FCS/FCCS to measure slow membrane dynamics on artificial and yeast cell membranes [121]. The line scans were performed by scanning the laser beam horizontally in *xy* direction perpendicularly through a vertical membrane at a rate much faster than the diffusion time of the fluorophore. The linear scan path is chosen in such a way that the detection volume spends a short time inside the membrane to minimize photobleaching. The line scan containing the spatial information of the position of the membrane was aligned and the intensity trace from the membrane extracted and correlated. This corrects for the membrane movements which can cause data analysis to be difficult. This approach was applied in living zebrafish embryo to study the mobility of mRFP-labeled fibroblast growth factor receptors Fgfr1 and Fgfr4 in cell membranes of living zebrafish embryos as well as to determine their binding affinities to their EGFP-labeled ligand Fgf8 [62]. First, scanning TFCCS was used to study the mobility of the receptors and at the same time determine the size of the observation volume required for the quantitation of concentrations. Next, using the extracellular concentration determined by conventional FCS and the concentration of receptor–ligand complexes determined by dual-color scanning FCCS, the K_d of the receptor–ligand binding was calculated. The same group also investigated the concentration gradients of the morphogen Fgf8 in zebrafish embryos [130]. A morphogen provides positional information by generating a protein concentration gradient. mRNA of Fgf8-EGFP was injected into a single cell at the 32-cell stage, hence restricting the source of ligand production. TFCCS was used to investigate whether the movement of Fgf8-EGFP is by diffusion or directed transport. The former was found to be the case. In addition, readings taken at different distances from the source indicated that the concentration of Fgf8-EGFP decreased with increasing distance from the source, confirming that Fgf8-EGFP establishes a gradient. Changing the level of receptor-mediated endocytosis influenced the shape of the concentration gradient. It was concluded that the mechanism of Fgf8 morphogen involved Brownian diffusion from the source through the extracellular space in the presence of a sink, which is generated by endocytosis to form and maintain a morphogen gradient.

5.4 Medaka (O. latipes)

Similar to zebrafish, medaka is a well-studied fish model laying transparent embryos. Unlike the tropical zebrafish, the medaka embryo can tolerate a wide range of temperature from 4 to 35°C [154], and its embryos are equally, if not more, transparent.

Nagao et al. applied FCS, FRAP, and time-lapse LSM ex vivo using excised portion of medaka embryo body to investigate the molecular dynamics of proteins in primordial germ cells (PGCs) [89]. After excision of the medaka embryo tail region, the exposed section was placed on a glass bottom dish. FCS was used ex vivo to measure the diffusion coefficient of a granule-related protein Olvas-EGFP expressed in PGCs in the early embryonic stages. The protein is localized in granule-like structures in the cytoplasm of the PGC which are thought to be nuages. Diffusion in the granules were measured by FRAP, probably due to the higher concentration and lower mobility of Olvas-EGFP in the granules. Due to the movement of the specimen, FCS measurements were restricted to only 3 s. Results suggested that Olvas-EGFP shuttles between the cytosol and the granules. Upon expression in HeLa cells, Olvas-EGFP was localized to granules similar to nuages in medaka PGCs. It behaved similar to two other proteins of PGC granules – Nanos and Tudor. Despite the fact that most of the study was conducted in cell culture, the authors have shown that by manipulating the medaka sample, FCS and other fluorescence techniques can potentially be applied to study protein dynamics in vivo.

6 Outlook

Currently, most FCS applications are still conducted in vitro despite the fact that present technology already allows the application of FCS/FCCS in vivo. There are a number of limitations hampering the transition to in vivo studies. On one side, there is the problem that despite their simplicity to be used efficiently, the model animals are much more productive in hands of trained personnel and often special holding facilities are required. On the other side, single molecule techniques, including FCS, remain largely inaccessible to most biological laboratories. Although there are commercial systems available, they are not as ubiquitously available as other biophysical techniques. And the most advanced systems, which are often the system necessary for in vivo research, are not yet commercialized. Many modalities of FCS/FCCS are still being developed and are usually used by only a few groups around the world. These new FCS modalities often need self-written software packages, require newly developed data treatment algorithms, and need handling and maintenance by specialists. Therefore, most of the in vivo single molecule research requires close collaborations between developmental biology and biophysical laboratories.

Luckily, these days the *C. elegans*, zebrafish, and other developmental biology laboratories are quite numerous, and the number of biophysical groups engaged in single molecule research has risen considerably over the years. It is understood that some biological problems may not be feasible to study in fish embryos or adult worms, but many projects will gain from being transplanted into in vivo systems. But we expect that in the future, as FCS/FCCS becomes more popular and accessible to more laboratories, we should see more studies done in vivo.

At the time of writing, some new FCS modalities have emerged that could advance in vivo FCS research. The combination of stimulated emission depletion (STED) with FCS (STED-FCS) can reduce the diffraction-limited volume allowing higher concentration and smaller region to be measured [155]. FCS has also been used with nearfield scanning optical microscopy [156], nano-apertures [157], and supercritical angle illumination [158] to manipulate the observation volume. FLCS, where contributions due to different fluorescent species can be separated based on their lifetime and not emission wavelengths, is able to generate ACF free from unwanted background fluorescence or individual ACFs from spectrally similar fluorescent species as long as their fluorescence lifetime can be resolved [74, 159]. It can also separate particles with similar diffusion time, something which the conventional FCS cannot achieve. FLCS has been applied to living cells very recently [160]. Optical fiber-based FCS has also been developed [161–164]. Although these studies were performed in solution, optical fiber-based FCS has the potential of measuring in remote areas of an organism. To apply such new modalities to living organisms requires additional research and creative innovations. The initial motivation of applying FCS/FCCS or single molecule techniques in organisms is its physiological relevance. Techniques that induce high amount of laser radiation thus inducing photodamage and phototoxicity, or require extensive physical manipulation of the organism to the point that they are no longer physiological relevant should be avoided. Ideally, the measurement should be noninvasive and use the lowest amount of radiation per photon detected as possible. Light sheet-based illumination as used in SPIM-FCS greatly enhances the multiplexing capabilities of FCS, allowing the measurements of whole areas in an organism simultaneously with greatly reduced phototoxicity [117, 142]. This allows capturing many more measurements per organism and makes it possible to establish FCS/FCCS images in which each pixel reports not on the fluorescence intensity but on molecular parameters, including molecular mobility, concentration, and degree of biomolecular interactions.

As GFPs and its variants are probably going to be the preferred choice of reporters for studies in organisms in the near future, applications going from in vitro to in vivo have to deal with the constraint of working with relatively lower molecular brightness and photostability. Although in recent years, there has been many advances in generating new FPs [165, 166], and the photophysical properties of GFPs are largely understood, there are still important questions remaining. Are there complex photodynamic processes or multiple fluorescent states in the FP used that hinder the accuracy of monitoring the process of interest? Are all of the FPs tagged to the protein of interest fluorescent as a result of mis-folding or non-maturing FPs? These questions are important in FCS/FCCS as well as in any quantitative technique which uses FPs as reporters. Since single molecule techniques are now commonly used with FPs, some of these studies may help answer these questions.

Going hand in hand with the advances in FPs, advances in technologies such as better detectors and optical systems, more efficient data treatment and computing power are constantly being developed. Studies with higher spatial and temporally resolution can be expected in the future. With the wealth of quantitative information FCS and FCCS can generate, they are powerful techniques to be applied in living organisms.

7 Summary

In this chapter, we introduced the basic principles of FCS and FCCS, and the type of quantitative information they can provide. We discussed the challenges and solutions in applying FCS/FCCS in vivo using FPs as probes and introduced the different modalities of FCS/FCCS. We reviewed the current applications of FCS and FCCS in vivo and offer some perspective on the potential of this technique. Despite its nearly 40-year history, FCS is steadily developing, improving, and gaining popularity and is now successfully applied in organisms. Recent advances in FCS aim especially at improving FCS performance in vivo. High resolution STED-FCS makes new length scales below 100 nm and thus higher concentrations accessible. Scanning FCS resolves problems of sample movements, protein mobility, and photobleaching, and camera-based SPIM-FCS increases the number of simultaneous measurements per sample by several orders of magnitude, while reducing significantly the phototoxicity of the measurements. The future for FCS is going to be exciting indeed.

Acknowledgments We would like to thank Lin Shin Teo and Xianke Shi for the *Drosophila* and zebrafish embryo pictures. Yong Hwee Foo received a scholarship from the National University of Singapore. Thorsten Wohland and Vladimir Korzh gratefully acknowledge funding by a grant from the Biomedical Research Council in Singapore (07/1/21/19/488, R-143 000-351-305) and the IMCB institutional grant from the Agency for Science, Technology and Research (A-STAR) of Singapore, respectively.

References

1. Matz MV, Lukyanov KA, Lukyanov SA (2002) Family of the green fluorescent protein: journey to the end of the rainbow. Bioessays 24(10):953–959
2. Muller-Taubenberger A, Anderson KI (2007) Recent advances using green and red fluorescent protein variants. Appl Microbiol Biotechnol 77(1):1–12
3. Pampaloni F, Reynaud EG, Stelzer EH (2007) The third dimension bridges the gap between cell culture and live tissue. Nat Rev Mol Cell Biol 8(10):839–845
4. Mets Ü, Rigler R (1994) Submillisecond detection of single rhodamine molecules in water. J Fluoresc 4(3):259–264
5. Schwille P, Korlach J, Webb WW (1999) Fluorescence correlation spectroscopy with single-molecule sensitivity on cell and model membranes. Cytometry 36(3):176–182
6. Weiss M, Hashimoto H, Nilsson T (2003) Anomalous protein diffusion in living cells as seen by fluorescence correlation spectroscopy. Biophys J 84(6):4043–4052
7. Banks DS, Fradin C (2005) Anomalous diffusion of proteins due to molecular crowding. Biophys J 89(5):2960–2971

8. Magde D, Webb WW, Elson EL (1978) Fluorescence correlation spectroscopy. III. Uniform translation and laminar flow. Biopolymers 17(2):361–376
9. Foquet M et al (2004) Focal volume confinement by submicrometer-sized fluidic channels. Anal Chem 76(6):1618–1626
10. Kinjo M, Rigler R (1995) Ultrasensitive hybridization analysis using fluorescence correlation spectroscopy. Nucleic Acids Res 23(10):1795–1799
11. Schwille P, Oehlenschlager F, Walter NG (1996) Quantitative hybridization kinetics of DNA probes to RNA in solution followed by diffusional fluorescence correlation analysis. Biochemistry 35(31):10182–10193
12. Auer M et al (1998) Fluorescence correlation spectroscopy: lead discovery by miniaturized HTS. Drug Discovery Today 3(10):457–465
13. Pack CG et al (1999) Analysis of interaction between chaperonin GroEL and its substrate using fluorescence correlation spectroscopy. Cytometry 36(3):247–253
14. Wohland T et al (1999) Study of ligand-receptor interactions by fluorescence correlation spectroscopy with different fluorophores: evidence that the homopentameric 5-hydroxytryptamine type 3As receptor binds only one ligand. Biochemistry 38(27):8671–8681
15. Van Craenenbroeck E, Engelborghs Y (1999) Quantitative characterization of the binding of fluorescently labeled colchicine to tubulin in vitro using fluorescence correlation spectroscopy. Biochemistry 38(16):5082–5088
16. Octobre G et al (2005) Monitoring the interaction between DNA and a transcription factor (MEF2A) using fluorescence correlation spectroscopy. C R Biol 328(12):1033–1040
17. Kobayashi T et al (2004) Detection of protein-DNA interactions in crude cellular extracts by fluorescence correlation spectroscopy. Anal Biochem 332(1):58–66
18. Rauer B et al (1996) Fluorescence correlation spectrometry of the interaction kinetics of tetramethylrhodamin alpha-bungarotoxin with Torpedo californica acetylcholine receptor. Biophys Chem 58(1–2):3–12
19. Widengren J, Rigler R, Mets Ü (1994) Triplet-state monitoring by fluorescence correlation spectroscopy. J Fluoresc 4(3):255–258
20. Widengren J, Mets U, Rigler R (1995) Fluorescence correlation spectroscopy of triplet states in solution: a theoretical and experimental study. J Phys Chem 99(36):13368–13379
21. Widengren J (1999) Photodynamic properties of green fluorescent proteins investigated by fluorescence correlation spectroscopy. Chem Phys 250:171–186
22. Magde D, Elson E, Webb WW (1972) Thermodynamic fluctuations in a reacting system-measurement by fluorescence correlation spectroscopy. Phys Rev Lett 29(11): 705–708
23. Ehrenberg M, Rigler R (1974) Rotational brownian motion and fluorescence intensify fluctuations. Chem Phys 4(3):390–401
24. Aragón SR, Pecora R (1975) Fluorescence correlation spectroscopy and Brownian rotational diffusion. Biopolymers 14(1):119–137
25. Kask P et al (1987) Fluorescence correlation spectroscopy in the nanosecond time range: rotational diffusion of bovine carbonic anhydrase B. Eur Biophys J 14(4):257–261
26. Kask P et al (1989) Separation of the rotational contribution in fluorescence correlation experiments. Biophys J 55(2):213–220
27. Tsay JM, Doose S, Weiss S (2006) Rotational and translational diffusion of peptide-coated CdSe/CdS/ZnS nanorods studied by fluorescence correlation spectroscopy. J Am Chem Soc 128(5):1639–1647
28. Loman A et al (2010) Measuring rotational diffusion of macromolecules by fluorescence correlation spectroscopy. Photochem Photobiol Sci 9(5):627–636
29. Haupts U et al (1998) Dynamics of fluorescence fluctuations in green fluorescent protein observed by fluorescence correlation spectroscopy. Proc Natl Acad Sci USA 95(23): 13573–13578

30. Schwille P et al (2000) Fluorescence correlation spectroscopy reveals fast optical excitation-driven intramolecular dynamics of yellow fluorescent proteins. Proc Natl Acad Sci USA 97(1):151–156
31. Malvezzi-Campeggi F et al (2001) Light-induced flickering of DsRed provides evidence for distinct and interconvertible fluorescent states. Biophys J 81(3):1776–1785
32. Kettling U et al (1998) Real-time enzyme kinetics monitored by dual-color fluorescence cross-correlation spectroscopy. Proc Natl Acad Sci USA 95(4):1416–1420
33. Saito K et al (2004) Direct detection of caspase-3 activation in single live cells by cross-correlation analysis. Biochem Biophys Res Commun 324(2):849–854
34. Kohl T, Haustein E, Schwille P (2005) Determining protease activity in vivo by fluorescence cross-correlation analysis. Biophys J 89(4):2770–2782
35. Rigler R et al (1998) Fluorescence cross-correlation: a new concept for polymerase chain reaction. J Biotechnol 63(2):97–109
36. Rippe K (2000) Simultaneous binding of two DNA duplexes to the NtrC-enhancer complex studied by two-color fluorescence cross-correlation spectroscopy. Biochemistry 39(9): 2131–2139
37. Ohrt T et al (2008) Fluorescence correlation spectroscopy and fluorescence cross-correlation spectroscopy reveal the cytoplasmic origination of loaded nuclear RISC in vivo in human cells. Nucleic Acids Res 36(20):6439–6449
38. Baudendistel N et al (2005) Two-hybrid fluorescence cross-correlation spectroscopy detects protein-protein interactions in vivo. ChemPhysChem 6(5):984–990
39. Muto H et al (2006) Fluorescence cross-correlation analyses of the molecular interaction between an Aux/IAA protein, MSG2/IAA19, and protein-protein interaction domains of auxin response factors of Arabidopsis expressed in HeLa cells. Plant Cell Physiol 47(8):1095–1101
40. Oyama R et al (2006) Protein-protein interaction analysis by C-terminally specific fluorescence labeling and fluorescence cross-correlation spectroscopy. Nucleic Acids Res 34(14): e102
41. Maeder CI et al (2007) Spatial regulation of Fus3 MAP kinase activity through a reaction-diffusion mechanism in yeast pheromone signalling. Nat Cell Biol 9(11):1319–1326
42. Slaughter BD, Schwartz JW, Li R (2007) Mapping dynamic protein interactions in MAP kinase signaling using live-cell fluorescence fluctuation spectroscopy and imaging. Proc Natl Acad Sci USA 104(51):20320–20325
43. Shi X et al (2009) Determination of dissociation constants in living zebrafish embryos with single wavelength fluorescence cross-correlation spectroscopy. Biophys J 97(2): 678–686
44. Sudhaharan T et al (2009) Determination of in vivo dissociation constant, KD, of Cdc42-effector complexes in live mammalian cells using single wavelength fluorescence cross-correlation spectroscopy. J Biol Chem 284(20):13602–13609
45. Liu P et al (2007) Investigation of the dimerization of proteins from the epidermal growth factor receptor family by single wavelength fluorescence cross-correlation spectroscopy. Biophys J 93(2):684–698
46. Neugart F et al (2009) Detection of ligand-induced CNTF receptor dimers in living cells by fluorescence cross correlation spectroscopy. Biochim Biophys Acta 1788(9):1890–1900
47. Elson EL, Magde D (1974) Fluorescence correlation spectroscopy. I. Conceptual basis and theory. Biopolymers 13(1):1–27
48. Magde D, Elson EL, Webb WW (1974) Fluorescence correlation spectroscopy. II. An experimental realization. Biopolymers 13(1):29–61
49. Rigler R et al (1993) Fluorescence correlation spectroscopy with high count rate and low background: analysis of translational diffusion. Eur Biophys J 22(3):169–175
50. Thompson NL (1991) Fluorescence correlation spectroscopy. In: Lakowicz JR (ed) Topics in fluorescence spectroscopy, 1st edn. Plenum Press, New York

51. Krichevsky O, Bonnet G (2002) Fluorescence correlation spectroscopy: the technique and its applications. Rep Prog Phys 65:251–297
52. Aragon SR, Pecora R (1976) Fluorescence correlation spectroscopy as a probe of molecular dynamics. J Chem Phys 64:1791
53. Wachsmuth M, Waldeck W, Langowski J (2000) Anomalous diffusion of fluorescent probes inside living cell nuclei investigated by spatially-resolved fluorescence correlation spectroscopy. J Mol Biol 298(4):677–689
54. Meseth U et al (1999) Resolution of fluorescence correlation measurements. Biophys J 76 (3):1619–1631
55. Shi X, Wohland T (2009) Fluorescence correlation spectroscopy. In: Diaspro A (ed) Nanoscopy and multidimensional optical fluorescence microscopy. Taylor & Francis, Boca Raton
56. Ruttinger S et al (2008) Comparison and accuracy of methods to determine the confocal volume for quantitative fluorescence correlation spectroscopy. J Microsc 232(2):343–352
57. Koppel DE (1974) Statistical accuracy in fluorescence correlation spectroscopy. Phys Rev A 10(6):1938–1945
58. Bonnet G, Krichevsky O, Libchaber A (1998) Kinetics of conformational fluctuations in DNA hairpin-loops. Proc Natl Acad Sci USA 95(15):8602–8606
59. Widengren J et al (2001) Two new concepts to measure fluorescence resonance energy transfer via fluorescence correlation spectroscopy: theory and experimental realizations. J Phys Chem A 105(28):6851–6866
60. Haustein E, Schwille P (2003) Ultrasensitive investigations of biological systems by fluorescence correlation spectroscopy. Methods 29(2):153–166
61. Lakowicz JR (2006) Fluorescence correlation spectroscopy. In: Lakowicz JR (ed) Principles of fluorescence spectroscopy, 3rd edn. Springer, New York
62. Ries J et al (2009) Modular scanning FCS quantifies receptor-ligand interactions in living multicellular organisms. Nat Methods 6:643–645
63. Schwille P, Meyer-Almes FJ, Rigler R (1997) Dual-color fluorescence cross-correlation spectroscopy for multicomponent diffusional analysis in solution. Biophys J 72(4): 1878–1886
64. Hwang LC, Wohland T (2004) Dual-color fluorescence cross-correlation spectroscopy using single laser wavelength excitation. ChemPhysChem 5(4):549–551
65. Heinze KG, Koltermann A, Schwille P (2000) Simultaneous two-photon excitation of distinct labels for dual-color fluorescence crosscorrelation analysis. Proc Natl Acad Sci USA 97(19):10377–10382
66. Weidemann T et al (2002) Analysis of ligand binding by two-colour fluorescence cross-correlation spectroscopy. Single Mol 3(1):49–61
67. Bacia K, Kim SA, Schwille P (2006) Fluorescence cross-correlation spectroscopy in living cells. Nat Methods 3(2):83–89
68. Hwang LC, Wohland T (2007) Recent advances in fluorescence cross-correlation spectroscopy. Cell Biochem Biophys 49(1):1–13
69. Rička J, Binkert T (1989) Direct measurement of a distinct correlation function by fluorescence cross correlation. Phys Rev A 39(5):2646–2652
70. Magatti D, Ferri F (2003) 25 ns software correlator for photon and fluorescence correlation spectroscopy. Rev Sci Instrum 74(2):1135–1144
71. Pan X et al (2007) Multifunctional fluorescence correlation microscope for intracellular and microfluidic measurements. Rev Sci Instrum 78(5):053711
72. Mütze J, Ohrt T, Schwille P (2009) Fluorescence correlation spectroscopy in vivo. Laser Photon Rev. doi:10.1002/lpor.200910041
73. Lamb DC et al (2000) Sensitivity enhancement in fluorescence correlation spectroscopy of multiple species using time-gated detection. Biophys J 79(2):1129–1138
74. Böhmer M et al (2002) Time-resolved fluorescence correlation spectroscopy. Chem Phys Lett 353(5–6):439–445

75. Wenger J et al (2006) Dual-color fluorescence cross-correlation spectroscopy in a single nanoaperture: towards rapid multicomponent screening at high concentrations. Opt Express 14(25):12206–12216
76. Snapp EL et al (2003) Formation of stacked ER cisternae by low affinity protein interactions. J Cell Biol 163(2):257–269
77. Zacharias DA et al (2002) Partitioning of lipid-modified monomeric GFPs into membrane microdomains of live cells. Science 296(5569):913–916
78. Campbell RE et al (2002) A monomeric red fluorescent protein. Proc Natl Acad Sci USA 99 (12):7877–7882
79. Shaner NC et al (2004) Improved monomeric red, orange and yellow fluorescent proteins derived from Discosoma sp. red fluorescent protein. Nat Biotechnol 22(12):1567–1572
80. Huet S et al (2010) Nuclear import and assembly of influenza A virus RNA polymerase studied in live cells by fluorescence cross-correlation spectroscopy. J Virol 84(3):1254–1264
81. Lillemeier BF et al (2010) TCR and Lat are expressed on separate protein islands on T cell membranes and concatenate during activation. Nat Immunol 11(1):90–96
82. Jung G et al (2000) The role of dark states in the photodynamics of the green fluorescent protein examined with two-color fluorescence excitation spectroscopy. J Phys Chem A 104 (5):873–877
83. Hendrix J et al (2008) Dark states in monomeric red fluorescent proteins studied by fluorescence correlation and single molecule spectroscopy. Biophys J 94(10):4103–4113
84. Hillesheim LN, Chen Y, Muller JD (2006) Dual-color photon counting histogram analysis of mRFP1 and EGFP in living cells. Biophys J 91(11):4273–4284
85. Kogure T et al (2006) A fluorescent variant of a protein from the stony coral Montipora facilitates dual-color single-laser fluorescence cross-correlation spectroscopy. Nat Biotechnol 24(5):577–581
86. Dittrich PS, Schwille P (2001) Photobleaching and stabilization of fluorophores used for single-molecule analysis with one- and two-photon excitation. Appl Phys B 73(8):829–837
87. Berland K, Shen G (2003) Excitation saturation in two-photon fluorescence correlation spectroscopy. Appl Opt 42(27):5566–5576
88. Eggeling C, Volkmer A, Seidel CA (2005) Molecular photobleaching kinetics of Rhodamine 6G by one- and two-photon induced confocal fluorescence microscopy. ChemPhysChem 6 (5):791–804
89. Nagao I et al (2008) Analysis of the molecular dynamics of medaka nuage proteins by fluorescence correlation spectroscopy and fluorescence recovery after photobleaching. FEBS J 275(2):341–349
90. Shi X et al (2009) Probing events with single molecule sensitivity in zebrafish and Drosophila embryos by fluorescence correlation spectroscopy. Dev Dyn 238(12):3156–3167
91. Malone MH et al (2007) Laser-scanning velocimetry: a confocal microscopy method for quantitative measurement of cardiovascular performance in zebrafish embryos and larvae. BMC Biotechnol 7:40
92. Denk W, Strickler JH, Webb WW (1990) Two-photon laser scanning fluorescence microscopy. Science 248(4951):73–76
93. Berland KM, So PT, Gratton E (1995) Two-photon fluorescence correlation spectroscopy: method and application to the intracellular environment. Biophys J 68(2):694–701
94. Schwille P et al (1999) Molecular dynamics in living cells observed by fluorescence correlation spectroscopy with one- and two-photon excitation. Biophys J 77(4):2251–2265
95. Kohl T et al (2002) A protease assay for two-photon crosscorrelation and FRET analysis based solely on fluorescent proteins. Proc Natl Acad Sci USA 99(19):12161–12166
96. Rosales T et al (2007) Quantitative detection of the ligand-dependent interaction between the androgen receptor and the co-activator, Tif2, in live cells using two color, two photon fluorescence cross-correlation spectroscopy. Eur Biophys J 36(2):153–161

97. Swift JL et al (2007) Two-photon excitation fluorescence cross-correlation assay for ligand-receptor binding: cell membrane nanopatches containing the human micro-opioid receptor. Anal Chem 79(17):6783–6791
98. Ruan Q, Tetin SY (2008) Applications of dual-color fluorescence cross-correlation spectroscopy in antibody binding studies. Anal Biochem 374(1):182–195
99. Nguyen TT et al (2009) Effects of various small-molecule anesthetics on vesicle fusion: a study using two-photon fluorescence cross-correlation spectroscopy. J Phys Chem B 113:10357–10366
100. Li N et al (2010) Multiple Escherichia coli RecQ Helicase monomers cooperate to unwind long DNA substrates: a fluorescence cross-correlation spectroscopy study. J Biol Chem 285(10):6922–6936
101. Savatier J et al (2010) Estrogen receptor interactions and dynamics monitored in live cells by fluorescence cross-correlation spectroscopy. Biochemistry 49(4):772–781
102. Brand L et al (1997) Single-molecule identification of coumarin-120 by time-resolved fluorescence detection: comparison of one- and two-photon excitation in solution. J Phys Chem A 101(24):4313–4321
103. Thews E et al (2005) Cross talk free fluorescence cross correlation spectroscopy in live cells. Biophys J 89(3):2069–2076
104. Müller BK et al (2005) Pulsed interleaved excitation. Biophys J 89(5):3508–3522
105. Kapanidis AN et al (2004) Fluorescence-aided molecule sorting: analysis of structure and interactions by alternating-laser excitation of single molecules. Proc Natl Acad Sci USA 101(24):8936–8941
106. Brinkmeier M, Rigler R (1995) Flow analysis by means of fluorescence correlation spectroscopy. Exp Tech Phys 41:205–210
107. Brinkmeier M et al (1999) Two-beam cross-correlation: a method to characterize transport phenomena in micrometer-sized structures. Anal Chem 71(3):609–616
108. Dittrich PS, Schwille P (2002) Spatial two-photon fluorescence cross-correlation spectroscopy for controlling molecular transport in microfluidic structures. Anal Chem 74(17): 4472–4479
109. Blom H et al (2002) Parallel fluorescence detection of single biomolecules in microarrays by a diffractive-optical-designed 2 x 2 fan-out element. Appl Opt 41(16):3336–3342
110. Blom H et al (2002) Parallel flow measurements in microstructures by use of a multifocal 4 x 1 diffractive optical fan-out element. Appl Opt 41(31):6614–6620
111. Gösch M et al (2005) Parallel dual-color fluorescence cross-correlation spectroscopy using diffractive optical elements. J Biomed Opt 10(5):054008
112. Takahashi Y et al (2005) Analysis of cellular functions by multipoint fluorescence correlation spectroscopy. Curr Pharm Biotechnol 6(2):159–165
113. Ohsugi Y, Kinjo M (2009) Multipoint fluorescence correlation spectroscopy with total internal reflection fluorescence microscope. J Biomed Opt 14(1):014030
114. Kannan B et al (2006) Electron multiplying charge-coupled device camera based fluorescence correlation spectroscopy. Anal Chem 78(10):3444–3451
115. Kannan B et al (2007) Spatially resolved total internal reflection fluorescence correlation microscopy using an electron multiplying charge-coupled device camera. Anal Chem 79(12):4463–4470
116. Needleman DJ, Xu Y, Mitchison TJ (2009) Pin-hole array correlation imaging: highly parallel fluorescence correlation spectroscopy. Biophys J 96(12):5050–5059
117. Wohland T et al (2010) Single plane illumination fluorescence correlation spectroscopy (SPIM-FCS) probes inhomogeneous three-dimensional environments. Opt Express 18(10):10627–10641
118. Petersen NO (1986) Scanning fluorescence correlation spectroscopy. I. Theory and simulation of aggregation measurements. Biophys J 49(4):809–815
119. Petersen NO, Johnson DC, Schlesinger MJ (1986) Scanning fluorescence correlation spectroscopy. II. Application to virus glycoprotein aggregation. Biophys J 49(4):817–820

120. Koppel DE et al (1994) Scanning concentration correlation spectroscopy using the confocal laser microscope. Biophys J 66(2 Pt 1):502–507
121. Ries J, Schwille P (2006) Studying slow membrane dynamics with continuous wave scanning fluorescence correlation spectroscopy. Biophys J 91(5):1915–1924
122. Pan X et al (2007) Characterization of flow direction in microchannels and zebrafish blood vessels by scanning fluorescence correlation spectroscopy. J Biomed Opt 12(1):014034
123. Berland KM et al (1996) Scanning two-photon fluctuation correlation spectroscopy: particle counting measurements for detection of molecular aggregation. Biophys J 71(1):410–420
124. Ruan Q et al (2004) Spatial-temporal studies of membrane dynamics: scanning fluorescence correlation spectroscopy (SFCS). Biophys J 87(2):1260–1267
125. Skinner JP, Chen Y, Muller JD (2005) Position-sensitive scanning fluorescence correlation spectroscopy. Biophys J 89(2):1288–1301
126. Petrášek Z et al (2008) Characterization of protein dynamics in asymmetric cell division by scanning fluorescence correlation spectroscopy. Biophys J 95(11):5476–5486
127. Petrášek Z, Schwille P (2008) Precise measurement of diffusion coefficients using scanning fluorescence correlation spectroscopy. Biophys J 94(4):1437–1448
128. Petrášek Z, Schwille P (2008) Photobleaching in two-photon scanning fluorescence correlation spectroscopy. ChemPhysChem 9(1):147–158
129. Ries J, Schwille P (2008) New concepts for fluorescence correlation spectroscopy on membranes. Phys Chem Chem Phys 10(24):3487–3497
130. Yu SR et al (2009) Fgf8 morphogen gradient forms by a source-sink mechanism with freely diffusing molecules. Nature 461(7263):533–536
131. Sisan DR et al (2006) Spatially resolved fluorescence correlation spectroscopy using a spinning disk confocal microscope. Biophys J 91(11):4241–4252
132. Kolin DL, Wiseman PW (2007) Advances in image correlation spectroscopy: measuring number densities, aggregation states, and dynamics of fluorescently labeled macromolecules in cells. Cell Biochem Biophys 49(3):141–164
133. Digman MA et al (2005) Measuring fast dynamics in solutions and cells with a laser scanning microscope. Biophys J 89(2):1317–1327
134. Digman MA et al (2005) Fluctuation correlation spectroscopy with a laser-scanning microscope: exploiting the hidden time structure. Biophys J 88(5):L33–L36
135. Digman MA et al (2009) Detecting protein complexes in living cells from laser scanning confocal image sequences by the cross correlation raster image spectroscopy method. Biophys J 96(2):707–716
136. Hirschfeld T, Block MJ, Mueller W (1977) Virometer: an optical instrument for visual observation, measurement and classification of free viruses. J Histochem Cytochem 25 (7):719–723
137. Thompson NL, Burghardt TP, Axelrod D (1981) Measuring surface dynamics of biomolecules by total internal reflection fluorescence with photobleaching recovery or correlation spectroscopy. Biophys J 33(3):435–454
138. Thompson NL, Axelrod D (1983) Immunoglobulin surface-binding kinetics studied by total internal reflection with fluorescence correlation spectroscopy. Biophys J 43(1): 103–114
139. Sankaran J et al (2009) Diffusion, transport, and cell membrane organization investigated by imaging fluorescence cross-correlation spectroscopy. Biophys J 97(9):2630–2639
140. Schaaf MJ et al (2009) Single-molecule microscopy reveals membrane microdomain organization of cells in a living vertebrate. Biophys J 97(4):1206–1214
141. Huisken J et al (2004) Optical sectioning deep inside live embryos by selective plane illumination microscopy. Science 305(5686):1007–1009
142. Reynaud EG et al (2008) Light sheet-based fluorescence microscopy: more dimensions, more photons, and less photodamage. HFSP J 2(5):266–275
143. Chalfie M et al (1994) Green fluorescent protein as a marker for gene expression. Science 263(5148):802–805

144. Wang Z et al (2004) In vivo FCS measurements of ligand diffusion in intact tissues. SPIE 5323:177–183
145. Bhattacharya D et al (2009) Spatio-temporal plasticity in chromatin organization in mouse cell differentiation and during Drosophila embryogenesis. Biophys J 96(9):3832–3839
146. Lieschke GJ, Currie PD (2007) Animal models of human disease: zebrafish swim into view. Nat Rev Genet 8(5):353–367
147. Stoletov K, Klemke R (2008) Catch of the day: zebrafish as a human cancer model. Oncogene 27(33):4509–4520
148. Feitsma H, Cuppen E (2008) Zebrafish as a cancer model. Mol Cancer Res 6(5):685–694
149. Bowman TV, Zon LI (2010) Swimming into the future of drug discovery: in vivo chemical screens in zebrafish. ACS Chem Biol 5(2):159–161
150. Chakraborty C et al (2009) Zebrafish: a complete animal model for in vivo drug discovery and development. Curr Drug Metab 10(2):116–124
151. Hove JR (2006) Quantifying cardiovascular flow dynamics during early development. Pediatr Res 60(1):6–13
152. Pan X et al (2009) Line scan fluorescence correlation spectroscopy for three-dimensional microfluidic flow velocity measurements. J Biomed Opt 14(2):024049
153. Korzh S et al (2008) Requirement of vasculogenesis and blood circulation in late stages of liver growth in zebrafish. BMC Dev Biol 8:84
154. Furutani-Seiki M, Wittbrodt J (2004) Medaka and zebrafish, an evolutionary twin study. Mech Dev 121(7–8):629–637
155. Kastrup L et al (2005) Fluorescence fluctuation spectroscopy in subdiffraction focal volumes. Phys Rev Lett 94(17):178104
156. Vobornik D et al (2008) Fluorescence correlation spectroscopy with sub-diffraction-limited resolution using near-field optical probes. Appl Phys Lett 93(16):163904-3
157. Wenger J et al (2005) Single molecule fluorescence in rectangular nano-apertures. Opt Express 13(18):7035–7044
158. Ries J et al (2008) Supercritical angle fluorescence correlation spectroscopy. Biophys J 94 (1):221–229
159. Kapusta P et al (2007) Fluorescence lifetime correlation spectroscopy. J Fluoresc 17(1): 43–48
160. Chen J, Irudayaraj J (2010) Fluorescence lifetime cross correlation spectroscopy resolves EGFR and antagonist interaction in live cells. Anal Chem 82(15):6415–6421
161. Garai K, Muralidhar M, Maiti S (2006) Fiber-optic fluorescence correlation spectrometer. Appl Opt 45(28):7538–7542
162. Garai K, Sureka R, Maiti S (2007) Detecting amyloid-beta aggregation with fiber-based fluorescence correlation spectroscopy. Biophys J 92(7):L55–L57
163. Chang YC et al (2008) Two-photon fluorescence correlation spectroscopy through a dual-clad optical fiber. Opt Express 16(17):12640–12649
164. Aouani H et al (2009) Optical-fiber-microsphere for remote fluorescence correlation spectroscopy. Opt Express 17(21):19085–19092
165. Shaner NC, Patterson GH, Davidson MW (2007) Advances in fluorescent protein technology. J Cell Sci 120(Pt 24):4247–4260
166. Nienhaus GU, Wiedenmann J (2009) Structure, dynamics and optical properties of fluorescent proteins: perspectives for marker development. Chemphyschem 10(9–10):1369–1379
167. Wahl M, et al (2003) Fast calculation of fluorescence correlation data with asynchronous time-correlated single-photon counting. Opt Express 11(26):3583–3591
168. Yang F, et al (1996) The molecular structure of green fluorescent protein. Nat Biotech 14(10):1246–1251

Investigating the Life Cycle of HIV with Fluorescent Proteins

Viola Baumgärtel, Sergey Ivanchenko, Barbara Müller, and Don C. Lamb

Abstract Sensitive fluorescence methods based on fluorescent proteins (FPs) are being widely applied in the life sciences. Virology is no exception and the entry, assembly, and release of individual viruses are being investigated in living cells with unprecedented temporal and spatial resolution. In this chapter, we will focus on how FPs are used to explore the interaction of the human immunodeficiency virus (HIV) with its host cell. The first challenge is to fluorescently label the virion without interfering with virus functionality. Once components of the virus are labeled, processes such as entry, assembly, and budding can be directly followed optically in live cells in real time. Using single virus tracing, the interaction of individual viruses with the plasma membrane, virion uptake, as well as the kinetics of HIV assembly and release have been investigated. Furthermore, a variety of fluorescence microscopy techniques, e.g., Förster resonance energy transfer or image correlation microscopy, as well as FPs with special properties, e.g., photoconvertible proteins, offer new means of gaining mechanistic insight into processes involved in HIV replication.

Keywords ESCRT · HIV-1 · Live-cell-imaging · Single virus tracing

---

V. Baumgärtel and S. Ivanchenko
Physical Chemistry, Department of Chemistry and Biochemistry, Ludwig-Maximilians University, Butenandtstraße 11, 81377 Munich, Germany

B. Müller (✉)
Department of Infectious Diseases, Virology, University Hospital Heidelberg, Im Neuenheimer Feld 324, 69120 Heidelberg, Germany
e-mail: Barbara_Mueller@med.uni-heidelberg.de

D.C. Lamb (✉)
Physical Chemistry, Department of Chemistry and Biochemistry, Ludwig-Maximilians University, Butenandtstraße 11, 81377 Munich, Germany

Department of Physics, University of Illinois at Urbana-Champaign, Urbana, IL 61801, USA
e-mail: D.Lamb@lmu.de

G. Jung (ed.), *Fluorescent Proteins II*, Springer Ser Fluoresc (2012) 12: 249–278, DOI 10.1007/4243_2011_32
Published online: 7 September 2011

Contents

1 Introduction

Viruses are obligatory intracellular parasites, which enter suitable host cells and rely on exploiting cellular machineries and pathways to ensure their own replication. Accordingly, close interactions of viral components with host cell factors are involved in all stages of the viral life cycle. These interactions have to be precisely regulated and each interaction represents a potential target for novel antiviral strategies. During the past decade, the use of fluorescent proteins (FPs) in virology has become an invaluable tool for elucidating the dynamics of virus–cell interactions.

1.1 The Use of Fluorescent Proteins in Virology

Viruses strongly depend on exploiting the biochemical mechanisms and pathways of their host cell for their own replication. Therefore, all steps of the viral life cycle are intimately coupled to intracellular processes and the visualization of these steps requires the identification of virions or subviral structures within the context of an infected cell. The small size of virus particles, with capsid diameters ranging from ~20 nm for parvoviruses to ~500 nm for mimiviruses, are near or beyond the resolution limits of traditional optical microscopy; therefore, viruses cannot be visualized directly by conventional light microscopy. Furthermore, the viral replication cycle comprises an ecliptic phase: viral genome replication and protein production is preceded or accompanied by disintegration of the physical particle. During this period, the virion does not exist as a morphologically identifiable structure in the infected cell until virus progeny is formed in the assembly process. While electron microscopy (EM), immuno-EM, cryo-EM, and more recently cryo-electron tomography have been instrumental for detailed analyzes of virion morphogenesis and architecture as well as for visualization of individual steps in virus–cell interaction, these techniques provide only static images that do not convey information regarding the dynamics of the observed processes and are restricted to the analysis of free virions or morphologically identifiable subviral structures within the cell.

The discovery of FPs and their adaptation to cell biology applications has greatly expanded our possibilities to study individual events in the viral life cycle. Many virological applications of FPs do not differ, in principle, from their use in other areas of cell biology, e.g., the investigation of the intracellular trafficking of an individual viral protein or its colocalization or interaction with a particular host cell factor. Beyond that, fluorescence-labeling strategies together with advanced live-cell imaging techniques make it possible to visualize the dynamic interactions of individual, complete virions with a host cell. The transition from ensemble measurements to single virus tracing (SVT) has allowed us to unravel processes that are difficult to investigate by traditional virological techniques. Since many individual particles may interact with one host cell in a nonsynchronized manner, it is difficult to dissect events occurring in parallel and on a fast time scale using bulk biochemical or virological methods. For instance, some viruses have been found to use different pathways for cytoplasmic entry. Direct observation of individual virions in combination with fluorescent labeling of interacting cellular proteins can dissect different individual entry events occurring in parallel in the same host cell and thereby distinguish between different entry pathways. Likewise, the real-time observation of intracellular transport processes or the interaction of the virus with a specific host factor yields information on the sequence of events and on dynamics that is not easily obtained by any other method.

1.2 Human Immunodeficiency Virus Type I

Although FPs are being used to investigate many different types of viruses, we will focus here on their applications to elucidate the life cycle of Human Immunodeficiency Virus Type I (HIV-1). The retrovirus HIV-1 was identified in 1983 as the etiological agent of the acquired immunodeficiency syndrome AIDS. Its genome consists of a +strand ssRNA of approximately 9,800 nucleotides and encodes 15 mature protein products, most of which are components of the viral particle (Fig. 1a, c) [4]. The spherical virion (Fig. 1c) with a diameter of ~145 nm [5, 6] is surrounded by a lipid envelope derived from the host cell plasma membrane, which carries the viral Env glycoprotein. Approximately 15 Env spikes, each consisting of three dimers composed of the transmembrane protein gp41 and the

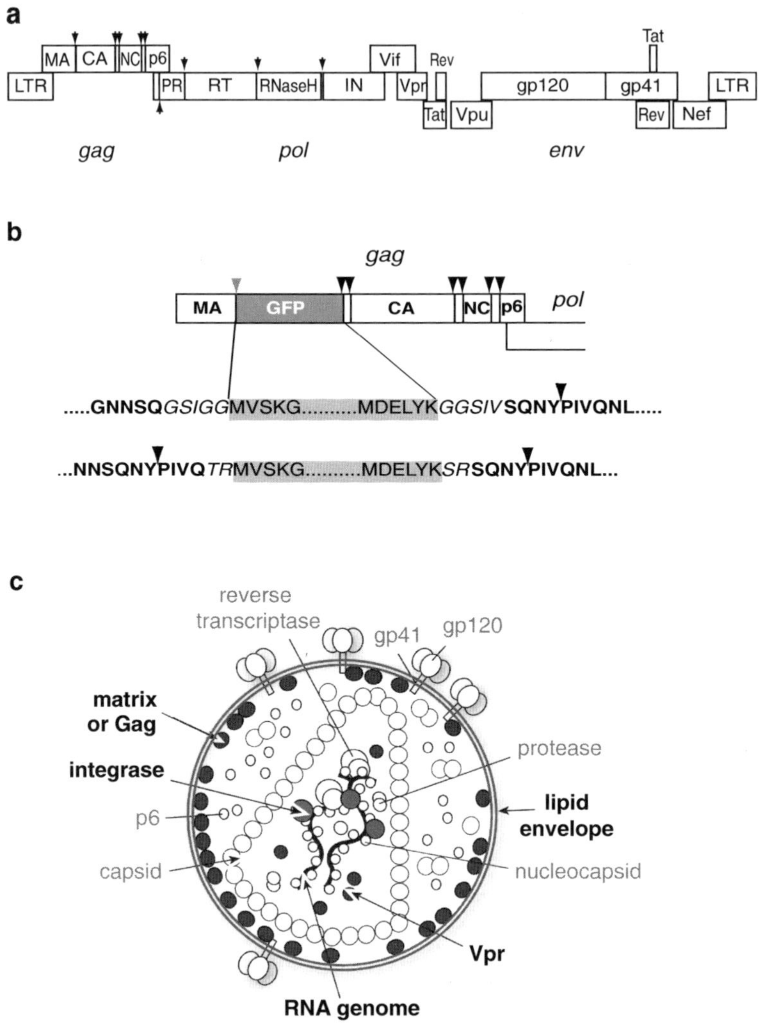

Fig. 1 HIV-1 genome organization and possible sites of fluorescent protein insertion (**a**) Schematic representation of the HIV genome organization. *Arrows* indicate cleavage sites of HIV protease. (**b**) Insertion of a fluorescent protein (GFP, *gray box*) within the HIV-1 Gag polyprotein. Constructs with either one [1] or two protease cleavage sites [2] flanking the FP moiety have been described. *Bold* case, HIV Gag derived amino acid sequence; *italics*, linker sequences; *arrowheads*, protease cleavage site. (**c**) Schematic structure of the mature HI virion. Viral components that have been labeled by fusion to an FP moiety (or by expression of a membrane-targeted FP derivative in the case of the lipid envelope [3]) are highlighted in *bold* case. See main text for references

surface protein gp120, are found on the viral surface [7, 8]. The inner protein shell of the virion is constructed from ~2,500 copies of the main structural polyprotein Gag, which makes up roughly 50% of the particle mass [9]. The Gag polyprotein consists of four domains – matrix, capsid, nucleocapsid and p6 – that after proteolytic separation by the virus encoded protease form the structural components of the mature virion [10]. In the mature particle, the inside of the lipid envelope is lined with a layer of matrix, attached to the membrane through an N-terminal myristoyl anchor and a patch of basic surface charges. A conical capsid built from capsid molecules encases the inner nucleoprotein core, which consists of two copies of the RNA genome, tightly condensed by interaction with the nucleocapsid domain. The virion also contains the viral $p6^{Gag}$ protein, the virus encoded enzymes protease, reverse transcriptase, and integrase present in approximately 125 copies per virion and several hundred molecules of the viral accessory protein Vpr as well as small amounts of the accessory proteins Vif and Nef (summarized in [9]). A number of cellular proteins, most prominently actin, cyclophilin A, and ubiquitin, have also been detected in purified viral particles [11]. The functional role of cellular proteins incorporated into HIV particles – if any – is currently unclear in most cases.

1.3 The Life Cycle of HIV

A general overview of the life cycle of HIV is shown in Fig. 2. HIV replicates in cells of the human immune system, the most important host cells being CD4 positive T-cells and macrophages. While several factors on the cell surface, e.g., heparan sulfate proteoglycans [12, 13], can mediate unspecific cell attachment, productive HIV entry into these cells requires the specific interaction of Env with the cellular receptor CD4 and a coreceptor molecule, most importantly one of the cellular chemokine receptors CCR5 or CXCR4 [14]. The sequential interaction of Env with CD4 and a coreceptor molecule triggers a series of conformational changes within gp41 that result in the fusion of the viral envelope with the host cell membrane and release of the viral core into the cytoplasm [15]. While it was initially assumed that productive HIV fusion occurs exclusively at the plasma membrane, this view has more recently been challenged (see Sect. 4.1). The structural transformations of the core following cytoplasmatic entry – a process termed uncoating – remain currently enigmatic. HIV RT carried within the core converts the ssRNA genome into a dsDNA form, which is subsequently transported into the nucleus, where integrase catalyzes the covalent insertion of the proviral DNA into the host cell genome. The finding that many mutations and conditions that affect viral capsid stability also impair reverse transcription suggests that uncoating and the subsequent steps are coupled, but the coordination of uncoating, reverse transcription, nuclear import and integration, as well as the functional roles of the viral proteins and host cell factors involved in the process are currently not well understood. The expression of HIV structural proteins from the integrated proviral genome is mediated by the cellular transcription and translation machineries and regulated by the viral transcriptional activator Tat as well as the viral protein Rev, which mediates the nuclear export of unspliced or partially spliced viral mRNA. The late stages of assembly and release of progeny virions

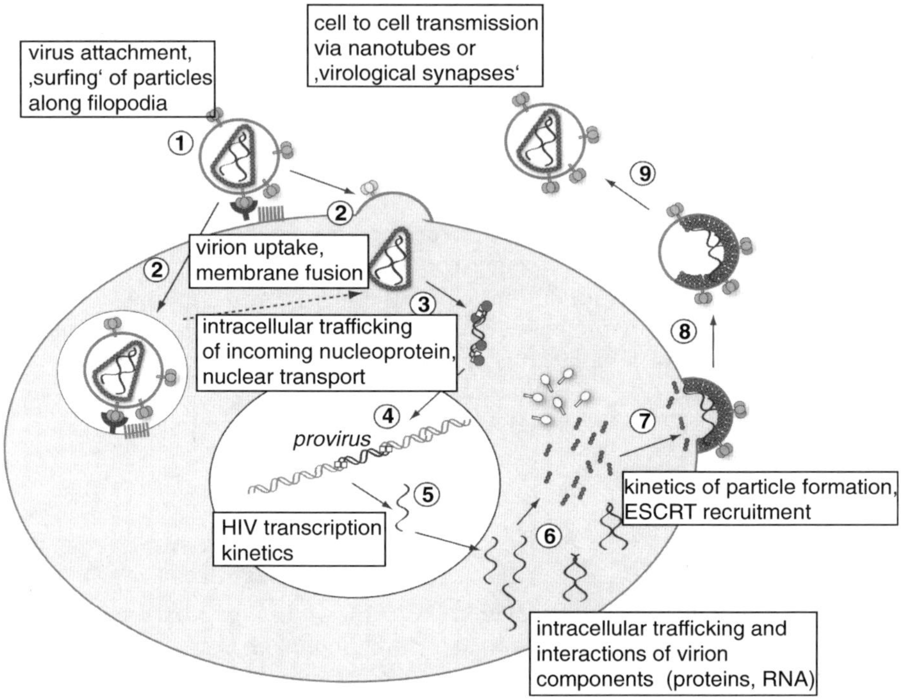

Fig. 2 Invesitgation of individual steps in HIV replication using fluorescent proteins The HIV replication cycle can be schematically divided into the following steps (*1*) Binding to the cellular membrane surface receptor CD4 and a coreceptor molecule (CXCR4 or CCR5, respectively). (*2*) Cytoplasmic entry of the viral core by fusion of the envelope with a cellular membrane (plasma membrane or endosomal membrane). (*3*) Uncoating, reverse transcription and formation of the pre-integration complex. (*4*) Nuclear import of the pre-integration complex and integration of the proviral genome into the host genome. (*5*) Transcription and nuclear export of viral mRNA and genomic RNA. (*6*) Translation of viral proteins by the host cell machinery. (*7*) Trafficking of virion components to the plasma membrane and assembly of the immature virion. (*8*) Budding and release of virus progeny. (*9*) Proteolytic cleavage of Gag followed by morphological maturation of the virus. *Boxes* highlight events in virus replication that have been investigated using fluorescent fusion proteins in conjunction with modern live cell imaging techniques

are orchestrated by the main structural polyprotein Gag. This protein is translated in the cytoplasm and traffics to the plasma membrane, where it assembles into spherical immature particles. Gag expressed alone is sufficient for assembly and release of virus like particles [16]. In the viral context, Gag is also responsible for the incorporation of the viral genomic RNA and other viral and cellular components into the virion, as well as for the recruitment of host cell factors required for virus budding from the plasma membrane (see Sect. 4.2). Concomitant with release, proteolytic processing of the Gag polyprotein by protease into its functional subunits matrix, capsid, nucleocapsid, and p6 triggers extensive morphological rearrangements within the particle, resulting in the characteristic cone-shaped core structure. This converts the immature, noninfectious virion into its mature, infectious form capable of infecting a new host cell.

2 Preparation of Fluorescently Labeled HIV-1 Particles

2.1 Labeling of HIV-1 Proteins by Fusion with Fluorescent Proteins

A major challenge in the process of generating fluorescently labeled virus derivatives suitable for SVT is the introduction of one or more fluorophores into the virus without affecting virus functionality. Live-cell imaging experiments over tens of minutes with high time resolution require the use of suitably bright objects. However, the attachment of a large number of fluorophores can interfere with the functionality of the virus and needs to be tested. One possibility is to use chemical labeling strategies with synthetic, photostable fluorophores. For example, covalent modification of viral surface proteins by amino- or thiol-reactive fluorophores, labeling of the virus envelope by lipophilic dyes or the insertion of fluorescent intercalators into the viral genome have been successfully applied for labeling of different viruses [17]. However, these strategies are difficult to control and may lead to virus preparations of varying infectivity. Genetic labeling of viruses by insertion of a gene encoding for an FP, on the other hand, results in defined virus derivatives whose functionality with respect to individual replication steps can be analyzed. However, the context for insertion of an FP-encoding gene has to be carefully selected (rationally or by random insertion followed by screening of variants), and the resulting virus derivative needs to be well characterized with respect to function.

The possibility that the fusion to a 27-kDa FP moiety may affect intracellular localization, properties, or functionality of the protein of interest is a general concern for all live-cell imaging studies and needs to be considered when analyzing fluorescently labeled HIV-1 proteins. Beyond that, the molecular biology and compact organization of viruses presents additional pitfalls for attempts to label proteins in the viral context. As for other viruses, HIV-1 uses strategies of genetic economy in order to encode as much information as possible within the limitations of a small genome. Specifically, HIV-1 employs (1) overlapping open reading frames (see Fig. 1a) as well as overlap of open reading frames with regulatory RNA sequences, (2) differential RNA splicing, (3) translational frameshifting, (4) expression of polyproteins that need to be processed in a highly organized manner in order to produce infectious virions, and (5) assembly of the virion structure from arrangements of multiple identical subunits. All of these mechanisms present obstacles for the introduction of *fp* genes into the viral genome. The insertion of the *egfp* gene into the HIV genome increases the length of the genomic RNA to be replicated and packaged in viral particles by approximately 7%. Beyond that, tagging viral structural proteins with the relatively bulky FP domain may interfere with the assembly of regular structures required for virion formation and/or lead to molecular crowding within the restricted volume of the virus particle. It has been shown that the architecture of HIV is more flexible than that of many other viruses, which is reflected in a variable particle diameter and a wide range of the number of Gag molecules per virion [5, 6]. Nevertheless, extensive homo- and heteromeric interactions between HI virion components require a careful, structure-guided selection of potential labeling sites. Furthermore, at many positions within the HIV genome, the introduction of a foreign gene will not only change the HIV open reading frame of interest but also at the same time introduce unwanted modifications of other viral gene products encoded by overlapping open reading frames and/or interfere with regulatory sequences important for virus replication. This limits the number of potential *fp* introduction sites in the viral context.

A final concern for the study of human pathogens by live-cell imaging is that unfixed samples of pathogenic viruses must be handled in certified facilities with the appropriate biosafety level (BSL3 for wild-type HIV-1). If the required equipment is not available under the appropriate biosafety level conditions, experiments have to be performed using either fixed samples or subviral constructs containing deletions that are officially classified for use in lower level biosafety laboratories. In the case of HIV-1, variants with deletion of both long terminal repeat regions express most viral proteins when put under the control of an exogeneous promoter and form entry competent particles, but cannot undergo reverse transcription. Such variants may be classified for use under relaxed biosafety conditions.

It is not always necessary to work with the full viral construct. Depending on the questions to be addressed, the expression of an individual FP-tagged viral protein may suffice. HIV-1 Gag by itself can directly assemble and be released as virus-like particles, and CMV promoter-driven expression constructs encoding codon-derivatized versions of Gag that allow the expression of HIV-1 Gag independent of the viral Rev protein [18, 19]. In this context, a FP of choice can be fused to the Gag C-terminus [20, 21]. It should be noted, however, that codon-optimized, Rev-independent versions of Gag are expressed in very high levels in transfected cells and, along with codon optimization or the addition of heterologous RNA export signals, can influence the intracellular localization of Gag (e.g., [22, 23]). Furthermore, other viral components involved in the process of particle formation as well as the generation of mature viruses cannot be investigated using this approach. This problem can partly be circumvented by exogeneous expression of a tagged version of the protein of interest together with the wild-type virus, thereby leaving the sequence and stoichiometry of the endogenous viral gene products unaltered. In the case of HIV-1, it has been shown that the viral Gag and Gag-Pol polyproteins as well as the viral Env glycoprotein are incorporated into the virion upon exogeneous co-expression. Furthermore, the accessory protein Vpr, which is present in several hundred copies in the mature virion [24], is targeted to the budding site through specific interaction with HIV-1 Gag [25]. This property has been exploited to package various fusion proteins into HIV particles. FP-tagged versions of Vpr itself [26, 27] as well as a fusion protein that mediates the incorporation of integrase-GFP into the virus through transient, HIV protease-sensitive fusion to Vpr [28] have been successfully used to label the inner core of HIV particles and follow core components subsequent to HIV cell entry. Pseudotyped particles carrying a labeled version of the viral Env protein would also be of interest for the analysis of HIV entry or assembly. However, while functional FP tagged variants of Env have been successfully developed for other retroviruses [29], this is not the case for HIV-1.

Despite the numerous pitfalls and difficulties in labeling viruses with FPs, many successful fluorescently labeled constructs of different proteins in HIV-1 have been developed. In the case of HIV-1 Gag, it is possible to introduce a fluorescent label to the polyprotein in the authentic viral context (Fig. 1b). Structural and cryo-EM analyzes revealed, that the Gag polyprotein displays a modular architecture in which the folded domains of its functional subunits are connected by extended flexible linker regions [10]. In the assembly process, Gag molecules are aligned in a parallel fashion in patches of local order, resulting in an area of low protein density between the layers of the folded domains of matrix and capsid in the immature particle [30]. Further, the immature Gag shell is not a completely closed sphere, but the Gag layer encompasses only ~2/3 of the lipid envelope, leaving a large gap of low protein density [9, 31]. This structural flexibility allows the introduction of additional proteins within the extended and flexible C-terminus of matrix. Small insertions directly upstream of the matrix-capsid cleavage site are tolerated without affecting viral infectivity in tissue culture [1, 32], but even the introduction of the complete GFP domain, which adds more than 50% to the molecular mass of the Gag protein, is possible. The FP-labeled virus is infectious, but infectivity is significantly reduced. However, wild type morphology and infectivity is restored upon cotransfection of the modified construct with wild-type provirus in equimolar amounts [1]. While this strategy does not allow us to follow the virus through a complete replication cycle or multiple rounds of replication, labeled particles for the study of virus entry can be produced, or the release of virions from transfected cells can be observed. A derivative of this construct that carries an additional protease cleavage site between matrix and eGFP (HIV^{iGFP}) has been reported to display increased infectivity and sustain virus replication in MT-4T-cells [2].

Finally, labeling strategies have been developed that allow detection of specific RNA molecules by fluorescence microscopy [33]. In order to follow HIV RNA in a live cell, the viral genome – or a deleted variant thereof retaining the structural element required for virion incorporation (psi) – can be tagged with multiple copies of RNA stem-loop sequences acting as specific and high affinity binding sites for the bacteriophage MS2 coat protein or the *E. coli* antitermination protein BglG, respectively. Coexpression of these binding partners as FP-tagged derivatives then results in specific attachment of fluorescent labels to viral RNA molecules, which made it possible to study the dynamics of HIV RNA trafficking and Gag-RNA interactions by live cell imaging [34].

The selection of an optimal labeling strategy strongly depends on the question to be addressed. If the aim of the study is to investigate the dynamics of an individual viral protein or its interactions with cellular factors, the optimal strategy may be to express an FP-tagged version of this protein of interest by itself. This has for instance been exploited to study intracellular localization and interactions of the accessory proteins Tat [35] or Rev [36]. For the investigation of virus particle assembly, FP-labeled Gag will in most cases be a good choice, since this protein is centrally involved in the processes of particle formation and release. Interactions between other viral components, e.g., RNA, with viral budding sites can then be followed by introducing a differently colored label to this other component of interest. Due to the high number of molecules incorporated per virion, FP-tagged Gag expressed in the viral context is also the best strategy if the experiment requires a bright labeling of individual particles with a single color (for example, for following virus cell attachment or cell to cell transfer). For experiments that require a distinction between complete virions and subviral nucleoprotein complexes resulting from virus cell fusion, however, a combination between a labeled component of the inner core of the virion (integrase or Vpr) and a differently colored lipid membrane marker (e.g., S15-mCherry, [3]) is required. This principle may be expanded to triple labeling strategies (e.g., including a label at matrix) in order to separately follow virion components in the process of entry.

2.2 *Preparation of Fluorescent Virus Particles for Studies of Virus–Cell Entry*

HIV entry studies can be performed by addition of purified and characterized fluorescently labeled particles to host cells. The most convenient method providing high particle yields is the transfection of 293T cells with a suitable labeled virus derivative. By cotransfection of several different constructs, virus particles carrying multiple differently colored labels attached to different viral proteins (e.g., Matrix. mCherry or a myristoylated FP lining the envelope and eGFP.Vpr in the viral core [3, 26] can be generated. Labeled virions can be purified from the supernatant of transfected cells by ultracentrifugation through a 20% (w/w) sucrose cushion or by use of size exclusion chromatography spin columns. It should be noted that both are rather crude purification methods and the resulting virus preparations may contain significant amounts cellular vesicles and pelletable cell debris – though invisible in the live cell imaging analysis if they do not contain the fluorescently labeled viral protein(s). More rigorous purification methods, in particular separation through iodixanol gradient centrifugation [37], will remove membrane vesicles and other impurities. However, this is not recommendable for virions intended for SVT studies because such treatment also affects virus infectivity. Rather, the amount of impurities can be limited by using appropriate transfection procedures (the standard $CaPO_4$ transfection method is preferable to using polyethyleneimine or most lipofection reagents) and by harvesting after relatively short incubation times posttransfection (24–30 h). Purified virus particles should be resuspended in PBS supplemented with 10 mM Hepes pH 7.2 and 10% FCS and stored in aliquots at −80°C; repeated freeze-thaw cycles must be avoided. Necessary quality control for particle preparations includes (1) the determination of virus concentration by enzyme-linked immunosorbent assay or quantitative immunoblot using antiserum raised against HIV-1 capsid, (2) recording of the appropriate fluorescence spectrum under standardized conditions, and (3) immunoblotting using various antisera to detect the labeled viral protein of interest, to monitor overall particle composition as well as to ensure proper proteolytic maturation of Gag. In our work with double labeled HIV derivatives, we had observed that Gag maturation was impaired upon incorporation of larger amounts of eGFP.Vpr [26]; this may also be the case upon incorporation of other labeled protein derivatives. Since proper Gag maturation is a prerequisite for HIV infectivity, it should be controlled to ensure the quality of virion preparations. (4) In the case of multilabeled particles, colabeling efficiency needs to be determined from images of particles bound to glass cover slips recorded under standardized conditions and automatically evaluated using e.g., an ImageJ plugin [26].

2.3 *Host Cells*

The selection of the appropriate host cell is crucial for SVT analyzes. Both, the questions to be addressed as well as potential limitations of the experimental setup need to be considered. For conceptual reasons, it is always preferable to employ natural target cells of the virus – e.g., T-cell lines or primary immune cells in the case of HIV. The use of such cells may be mandatory to address certain questions, for example, intercellular transmission of virions through a virological synapse. However, some properties of these cells can present insurmountable difficulties for sensitive and high-resolution live-cell analyzes. Primary T-cells and T-cell lines grow in suspension and need to be attached to a surface for continuous observation. They display relatively high mobility and the cytoplasm is confined to a narrow space between plasma membrane and nucleus. Thus, it may be necessary to revert to adherent cell lines as a model for imaging experiments. The HeLa cell line is not derived from natural target cells of HIV, but transfected HeLa cells (for studying virus production) or HeLa derivatives carrying the HIV receptor and coreceptor (for analysis of virus entry) have been widely used as model HIV host cell lines in biochemical and virological studies. They may thus be employed for SVT, provided that the model character of this setup is acknowledged.

3 Fluorescence Microscopy in Virology

Using FPs, the dynamics of the viral life cycle can be monitored in real time using live-cell imaging. Fluorescence microscopy has the advantage over EM or bulk biochemical or virological approaches that it is minimally invasive and the dynamics of the process can be directly observed in real time. Current detectors such as avalanche photodiodes, hybrid detectors, and back-illuminated CCD cameras have quantum efficiencies over 50% (up to 95%) at visible wavelengths, providing high sensitivity to fluorescently labeled molecules. A few precautions need to be taken to minimize autofluorescence of the cells. Even when working with viruses that are labeled with hundreds of FPs, it is advantageous to use ultrasensitive detection (with high-transmission optics and sensitive detectors) as the excitation power used for the experiments can be reduced, minimizing phototoxicity effects.

Numerous microscopy methods are used to investigate the life cycle of viruses in live cells. Wide-field (WF) microscopy has the advantage that it is fast and has the highest sensitivity. It is often used for investigating the uptake of labeled viruses as there is little background fluorescence from the cells. In contrast, confocal microscopy is the method of choice for experiments in which cells express labeled viral proteins. The confocal geometry provides the best contrast, especially when the fluorescence background of the cell is high. However, confocal systems are typically slow as the diffraction-limited confocal volume has to be scanned over the region of interest. This disadvantage can be overcome with a spinning disk confocal microscope (SDCM), where several confocal volumes are simultaneously scanned over the area of interest, making data collection rates similar to that of WF systems. For the observation of events occurring at the plasma membrane, total internal reflection fluorescence (TIRF) microscopy can be used. In TIRF microscopy, the evanescent field excites only a thin layer of ~100–200 nm above the cover slip, keeping the background low even in samples where fluorescent molecules are present in the cytosol. It has therefore been used to study the dynamics of HIV budding site formation at the plasma membrane ([38, 39], see below). However, TIRF microscopy has the drawback that only the basal membrane of the cell can be observed. This membrane is not freely accessible to diffusing particles from the outside medium for studying virus entry; in experiments addressing the late stages of HIV replication, many of the released particles become trapped between the plasma membrane and the cover slip.

There are a number of fluorescence techniques that have been combined with microscopy to gain detailed information over viral processes in the cell. We briefly discuss a few of them below.

3.1 Förster Resonance Energy Transfer

Förster Resonance Energy Transfer (FRET) is an excellent method for visualizing the interaction of molecules within the intracellular environment. For details of FRET and its use in imaging, we refer the reader to the review of Jares-Erijman and Jovin [40]. Using CFP and YFP attached to Gag, the oligomerization of retroviral Gag was investigated using FRET in live cells [41]. Spearman and coworkers showed that the majority of Gag oligomerizes at the plasma membrane. Larson et al. used a similar FRET assay in Rous sarcoma virus to demonstrate that there is some oligomerization of Gag also in the cytosol [42]. Further work from

Poole and coworkers used FRET in fixed cells to show that Gag interacts with viral RNA early after expression in the perinuclear region of the cell [43]. These few examples serve only to illustrate that FRET is an important tool for studying molecular interactions in virology and, when coupled with microscopy, can be used to investigate the location and dynamics of these interactions in living cells in real time.

3.2 *Fluorescence Fluctuation Spectroscopy*

In the early 1970s, Fluorescence Correlation Spectroscopy (FCS) was developed at Cornell University [44, 45, 46]. Through a temporal correlation of the measured fluorescence intensity, the timescale and amplitude of the fluctuations are analyzed and information regarding the concentration, reaction rates, and mobility of molecules determined. Since the development of FCS, a number of methods have been developed based on analysis of the fluctuations in fluorescence intensity and are referred to as Fluorescence Fluctuation Spectroscopy (FFS). A more detailed description of these methods can be found in the chapter in this book on Fluorescence Fluctuation Methods by Wohland. The group of Webb used FCS to measure the mobility of Rous sarcoma virus Gag in the cytosol [42]. Gag was found to interact with molecules within the cytosol, showing a much slower diffusion coefficient than measured for monomeric GFP alone.

In addition to investigating the temporal fluctuations in fluorescence intensity, the fluctuations in the amplitude of the fluorescence intensity can also be analyzed [47–49]. From the amplitude of the fluorescence intensity, it is possible to extract information regarding the molecular brightness of the fluorescing molecules, which is well suited for investigating the oligomerization of molecules. The group of Mueller used brightness analysis to estimate the number of Gag molecules in individual virions [50]. They found that in particle preparations from transfected cells, the number of Gag molecules per particle depended on Gag expression levels, varying from 750 to ~2,500 molecules per virion. The higher value is consistent with what has been derived from electron tomography analysis of HIV [9]. Analysis of cytosolic HIV-1 Gag by brightness analysis revealed a concentration dependent oligomerization [51].

As the distribution of molecules in space follows the same physical principles as the distribution of molecules at a particular position with time, the number, and brightness of molecules can not only be determined from the temporal fluctuations in fluorescence intensity but also from spatial fluctuations. This analysis approach, called Image correlation spectroscopy (ICS), was developed in the group of Petersen [52–54] and has been applied to estimate the number of VPS4 monomers in a complex that interacts with HIV assembly sites [55] as discussed in more detail below.

3.3 Single Virus Tracing

In contrast to bulk virological and fluorescence experiments, SVT involves the direct microscopic observation of individual virions in their interaction with the host cell [17]. The pioneering work of Seisenberger and coworkers [56] has shown that it is possible to follow single viruses over extended periods. In combination with fast and sensitive EMCCD cameras and specific labeling techniques, fluorescence live-cell microscopy is able to acquire images at 30 ms per frame and localize particles with accuracy on the order of a few nanometers. Initial SVT experiments were performed using WF microscopy [56]. Later, TIRF microscopy was employed as it is particularly well suited for events that occur at the cell membrane such as HIV-1 genome recruitment to the nascent bud [34] and Gag assembly [38, 39]. More recently, three-dimensional SVT is also being performed using z-stacking with a scanning confocal microscope such as a SDCM [57–59].

In single-virus tracing experiments, individual viruses are tracked as they undergo particular steps in the virus life cycle such as cell entry, transport within cells, or assembly of virus at the plasma membrane. The concentration of virus must thus be low enough to allow a clear distinction of individual virus particles – typically, on the order of 100 virions per cell or less. In addition, data collection rates must be fast enough to recognize the same particles in adjacent frames. Typical data acquisition rates vary from a few seconds for a z-stack (10–21 planes) for three-dimensional measurements down to 30 ms per frame for fast moving particles in a single plane. Viruses can then be tracked manually by selecting each particle in each frame of the movie [for example, using the "ParticleAnalysis" plugin from ImageJ (http://rsb.info.nih.gov/ij/)]. However, meaningful conclusions from SVT experiments require the analysis of datasets comprising a significant number of individual events. This is greatly simplified by using software packages that automate or semiautomate the tracking process. The selection of the appropriate tracking software depends strongly on the preconditions in each recorded movie like particle density, signal-to-noise ratio, or particle movement. Automatic tracking can be performed using programs such as the ImageJ plugin ViroTracker [60] or commercial versions like the Velocity quantification module (Improvision, Perkin Elmer Inc.). The important steps in tracking particles are first the localization of the individual particles in each frame and then a correspondence algorithm that determines the position of the individual viruses in the following frame. Correspondence finding is challenging as signals from individual particles can merge, disappear from the focus, new particles enter or become visible in the field of view, or particles can change their motional behavior during the recording time. Therefore, a given software package may need to be adjusted according to the requirements of a particular experimental setup.

Once the trajectories of individual virions have been extracted from the movie, the interaction of viral proteins with each other or with components of the cells can be investigated. Interactions can be visualized through changes in the motional behavior of the particles as well as through multicolor experiments where potential interaction partners and/or cellular organelles have been labeled in a different color and monitored in different channels.

4 Elucidating the Life Cycle of HIV-1 with Fluorescence Microscopy

Different virus types use different strategies to infect and replicate within their respective host cells, but a common feature of all viral replication cycles is that they comprise numerous highly dynamic steps, ranging from cell attachment and entry of virions, the intracellular transport and dynamic interactions of viral components to the assembly and release of progeny virions. Fluorescence technologies are ideally suited to investigate the dynamic aspect of these processes. Even though FP derivatives are widely used in virology, we will mainly focus here on FP applications used for the investigation of the HIV-1 life cycle (schematically illustrated in Fig. 2). Different fluorescence microscopy techniques and protein derivatives are presented as possible strategies to elucidate dynamic or mechanistic details of each HIV-1 replication step from entry over assembly to egress.

4.1 Entry of HIV-1 Virions

The delivery of the viral genome into a suitable host cell involves attachment of the virion to the plasma membrane of the target cell, followed by viral entry mediated through specific virus–receptor interactions. Enveloped viruses, like HIV, enter by fusion of the envelope with a cellular membrane. Depending on the virus, this may either occur by direct fusion at the plasma membrane or through uptake by clathrin-dependent or -independent pathways [61, 62] followed by fusion from within the endosomal compartment. The latter pathway is mandatory if the viral fusion mechanism is triggered by low pH, as in the case of influenza virus. However, the mechanism of endocytic uptake may differ and the same virus may exploit different pathways in parallel or switch between them depending on cell type or in response to inhibition. Using SVT, it is possible to differentiate between entry pathways, even if they occur in parallel on the same cell. For example, SVT experiments revealed that clathrin-dependent as well as clathrin-independent pathways can be used for productive entry of influenza virus [63].

For HIV, initial attachment can be nonspecific, i.e., independent of the binding of the viral Env protein to the cellular receptor. Using HIV labeled with eGFP within Gag, we investigated the interaction of HIV with HeLaP4 cells stably expressing CD4 and the viral coreceptors [64]. Approximately 20% of the viruses were immobilized immediately upon contact with the plasma membrane of the cell and the fraction of immobilized particles was proportional to the concentration of heparan sulfate-linked proteoglycans (HSPG) on the cell surface. In other cell types with lower HSPG levels, HIV-1 may exploit other nonspecific interactions mediated by surface proteins like intracellular adhesion molecules incorporated into HI virions (ICAMs). Low-affinity unspecific attachment can lead to nonspecific endocytic uptake and degradation, but may also serve to concentrate virus at the plasma membrane and increase the probability for the virus to encounter receptor molecules at the cell surface followed by stronger specific binding. At least in the case of infection of cells in tissue culture by free virions, there appears to be a kinetic competition between association/dissociation of attached virions and virus uptake by fusion or endocytosis [65, 66].

Specific HIV–cell interactions are mediated by binding of HIV gp120 to the cellular receptor CD4, followed by interaction with a coreceptor molecule (CCR5 or CXCR4, respectively). Coreceptor binding triggers conformational rearrangements in gp41, which mediate fusion between the viral lipid envelope and the cell membrane, allowing cytoplasmatic entry of the viral core [15].

The observations that HIV-1 Env mediated membrane fusion is pH independent, does not require endocytosis of the viral receptor and can occur between the plasma membranes of adjacent cells displaying Env and CD4, respectively [97, 98, 99] led to the assumption that the plasma membrane represents the site of productive HIV-1 fusion. Efficient uptake of HIV-1 by endocytosis has been observed, but was initially considered to be nonproductive. However, evidence from virological analyzes [67] indicated that productive endocytic entry can occur at least in some cell types. More recently, endocytosis, followed by pH-independent, receptor-mediated fusion from the endosome, has even been proposed to be the only productive entry pathway based on biochemical evidence and results from SVT experiments [59]. The relative importance of different entry pathways for HIV-1 in different cell types still requires further exploration.

The lack of a functional FP-labeled derivative of HIV-1 Env prevents the analysis of the dynamics of the fusion event through direct microscopic observation of the separation of Env and viral core components. Studies using the lipophilic dye DiD as a membrane marker and nucleocapsid-GFP as a marker for soluble content of the virus indicated that lipid mixing occurred rapidly, while release of the viral content was delayed and occurred on a time scale of 10 min [59, 68]. Retroviral fusion at the plasma membrane was investigated using double labeled pseudotyped particles carrying matrix.mCherry as a marker for the viral interior and a fully functional YFP-tagged version of the murine leukemia virus Env protein [29]. Color separation was dependent on the fusogenicity of the Env protein and the analysis of color separation events revealed, that fusion occurred within seconds of virus-cell contact [69].

4.2 *Assembly of HIV-1 at the Plasma Membrane of Hela Cells*

As for entry, viruses can take multiple routes for exiting the cell. In the case of HIV-1, EM analyzes have shown that release occurs by budding from the plasma membrane of infected cells. This is not only the case for T-cells but also for macrophages, where an apparently intracellular budding compartment, initially proposed to be of endosomal origin [70] has been identified as a specialized invagination of the plasma membrane [71–73]. The plasma membrane is also the location of budding site formation in the model cell lines used for fluorescence analysis. Consistent with this, live cell imaging analyzes indicated that initiation of Gag assembly occurred at the plasma membrane [74], rather than at late endosomal membranes, as had been previously suggested. The viral genomic RNA is recruited by Gag and involved in the assembly process. Experiments using viral RNA tagged with 24 stem loops that bind eGFP-labeled MS2 protein of bacteriophage coexpressed with FP-tagged HIV-1 Gag [34] showed that RNA is immobilized at the plasma membrane in a Gag-dependent manner. Initially, Gag-FP fluorescence is undetectable, indicating that only a few molecules of Gag are involved in the early assembling structure. This then serves as a nucleation site for Gag shell recruitment as indicated by the onset of Gag.mCherry fluorescence a few minutes after arrival of RNA at the plasma membrane (Fig. 3).

The kinetics of the assembly of Gag alone [39] and Gag in the context of the entire viral genome [38] have been investigated using either TIRF microscopy or

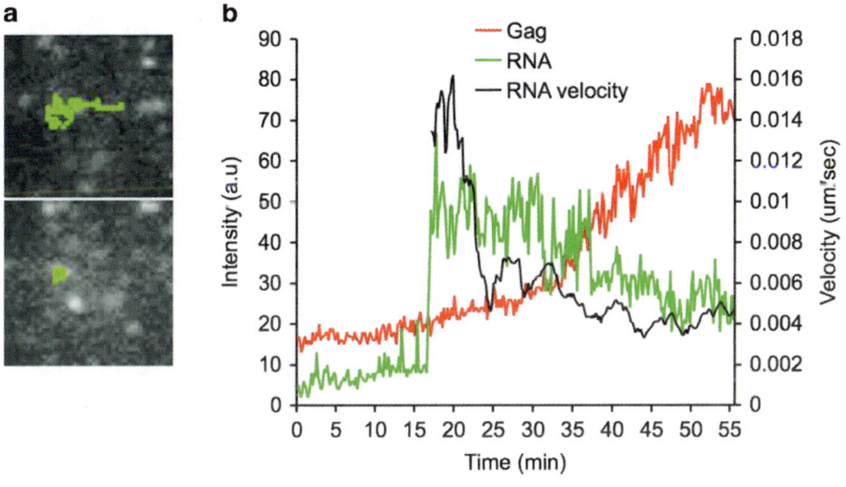

Fig. 3 HIV Gag polyprotein interaction with the viral RNA genome (**a**) Individual assembly sites in HeLa cells transfected with Gag.mCherry polyprotein and MS2-NLS-GFP viral RNA were imaged using TIRF microscopy. Images showing an membrane area with several assembly sites and the trajectories (*green line*) of one viral RNA punctum; before (*top*) and after (*bottom*) Gag. mCherry bocames visible colocalizing with the previously detected RNA punctum. Images are 5×5 μm. (**b**) Fluorescence intensity trace (in arbitrary units) of the Gag.mCherry polyprotein (*red line*) and the labeled MS2-NLS-GFP viral RNA (*green line*), as shown in (**a**). The increase in intensity of labeled RNA indicates trapping of the RNA at the membrane due to interaction with a low number of Gag molecules and is followed by gradual recruitment of further Gag.mCherry. The lateral velocity of the labeled RNA (*black line*, averaged using a sliding window of 21 frames) decreases due to association with the nascent Gag particle [34]

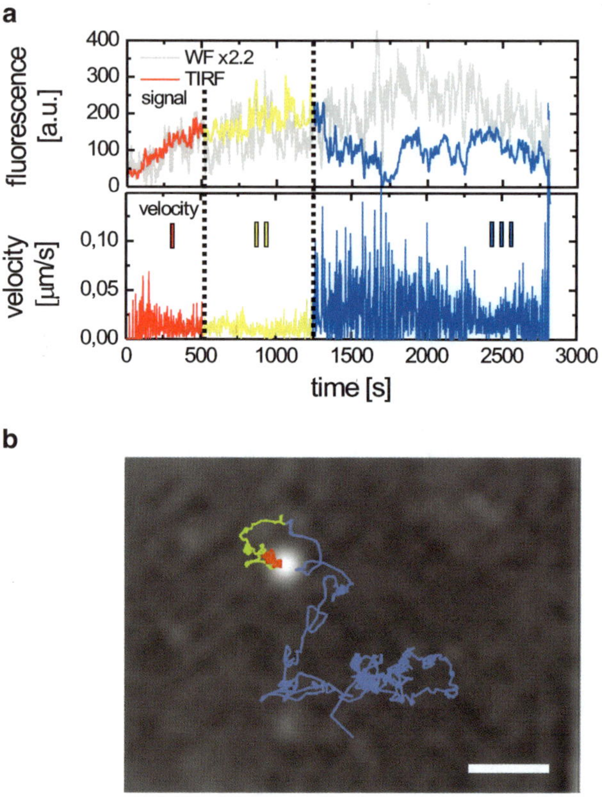

Fig. 4 Tracking assembly and release of HIV particles (**a**) *Top*, fluorescence intensity trace, showing assembly and release of an individual HIV^{eGFP}particle at the HeLa cell plasma membrane recorded using TIRF microscopy (*colored line*). Three phases are observed: phase I (*red*) showing the increase in fluorescence intensity due to the olimerization of Gag at the nascent assembly site, phase II (*yellow*) showing a plateau in fluorescence intensity when Gag assembly is complete and the virion presumably interacts with cellular components, and phase III (*blue*), where the fluorescence intensity decreases or the particle disappears. The corresponding intensity level in the wide field trace is shown in *gray*. *Bottom*, instantaneous velocity plot of the same HIV particle shows a strong increase in mobility of the particle in phase III (*blue*) in contrast to phase I (*red*) and II (*yellow*). (**b**) 2-D trajectory of the individual HIV particle shown in *panel a*, color coding illustrates the three different phases. Adapted from [38]

alternating TIRF/WF microscopy, respectively. The assembly process was found to proceed in three phases as exemplified in Fig. 4a. In phase I, Gag proteins are recruited to the plasma membrane, indicated by an exponential increase in fluorescence intensity. This phase was found to be completed in 5–6 min in cells expressing Gag.eGFP only [39] and on the order of 8–9 min when Gag.eGFP was expressed in the context of the complete HIV genome [38]. In phase II, the fluorescence reaches a plateau and no new Gag is delivered to the assembly site. In phase III, a decrease in fluorescence signal is observed, often accompanied an onset of rapid movement or disappearance of the particle.

Gag might either be recruited to the budding site directly from the cytosolic pool or arrive to the budding site by lateral diffusion along the plasma membrane. At first glance, the latter mechanism may look more appealing: confining diffusion to only two dimensions would significantly increase the rate of assembly driven by random diffusion. However, a high affinity of monomeric Gag to the plasma membrane might result in spontaneous formation of Gag patches lacking genomic RNA. Instead, HIV utilizes cooperative interactions to a high degree to form genome-containing nucleation sites. The Gag polyprotein comprises many interaction domains which mediate Gag-membrane, Gag–Gag, and Gag–RNA interactions [10]. For the individual interactions to work together, they should be weak enough to avoid premature nucleation. This explains why RNA with just a few Gag molecules attached serves as a nucleator and supports further growth of the Gag shell.

This theory is supported by the finding that a cytosolic pool of Gag molecules, rather than membrane-bound Gag molecules, contribute to the growing bud [38]. Here, we utilized a fusion construct of Gag labeled with the photoconvertible fluorescent protein mEosFP [75], which undergoes an irreversible transition from a green- to red-emitting state upon illumination at ~400 nm. Cells expressing the Gag.mEosFP protein in the viral context were illuminated with a 405-nm laser in TIRF mode, photoconverting plasma membrane bound Gag.mEosFP from a green- to red-emitting state while leaving the cytosolic fraction of molecules mostly unconverted. The newly formed viruses displayed a higher ratio of green (i.e., nonconverted Gag.mEosFP) to red fluorescence than the Gag.mEosFP pool available from the surrounding membrane, indicating that Gag is mostly recruited directly from the cytosol to the assembly site.

4.3 Egress of HIV-1 Virions

By tracking individual HIV-budding sites, we can follow the virus through the assembly process. Fig. 4b shows the trajectory of an HI virion from initiation of assembly to release. As we have observed for many individual budding sites, the transition from Phase II to Phase III is accompanied by a pronounced increase in the motility of the virion. In this particular case, the particle could be tracked for several frames before disappearing. From the mean-squared-displacement analysis of the trajectory of the virion, we calculated a hundred-fold increase in the diffusion coefficient of the particle upon the onset of Phase III. The characteristics of the mean squared displacement indicate Brownian motion, revealing that the particle diffuses freely outside of the cell. The observation that the diffusion coefficient is lower than measured for virions in free solution can be explained by trapping of the particle in a pocket between the cover slip and the plasma membrane.

To determine if the assembly and release rates were influenced by the geometry of the TIRF setup, control experiments were performed on the dorsal plasma membrane using a SDCM. Fig. 5 shows the fluorescence intensity and three-

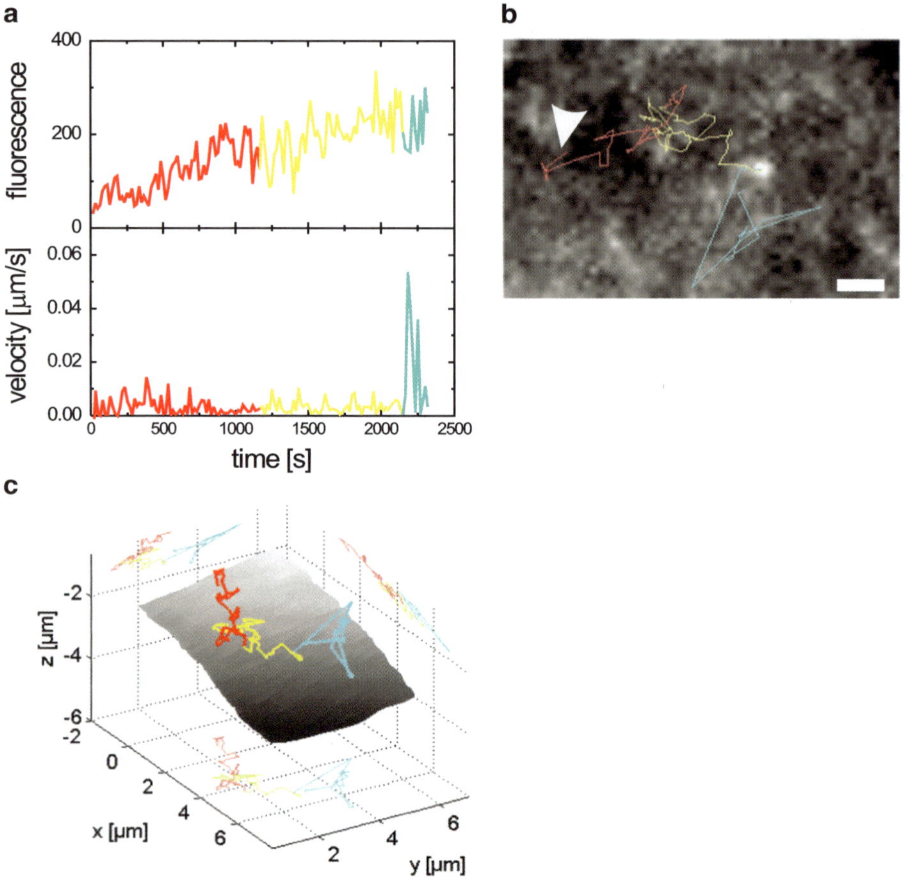

Fig. 5 HIV assembly and release using 3-D particle tracking (**a**) *Top*, fluorescence intensity trace of an individual assembling HIV^{eGFP} particle. *Bottom*, corresponding instantaneous velocity plot of the same particle; both datasets were obtained from the full 3D information from SDCM images, smoothed over 3 frames. (**b**) 2-D projection of the SDCM trajectory overlaid on top of the projected image, *arrowhead* marks the beginning of the trajectory, scale bar 1 µm. (**c**) 3-D representation of the trajectory, shown in *panel b*, that takes into account the z-position of the virus particle . The membrane surface is shown in *gray*. In all panels the color coding is according to the three assembly phases shown in Fig. 4. Adapted from [38]

dimensional trajectory of an HIV particle from initiation to release. No difference in the assembly rates or release rates was detected when comparing measurements from the ventral or dorsal membranes of the cell.

At present, particle release can only be detected based on changes in the motional behavior of the virion. Released particles that stay immobilized at the plasma membrane are therefore not classified as release events and this potential bias to the statistics needs to be considered. However, a sufficient number of release events were detected to gain significant statistics regarding the release process, although the exact time point of release is unclear. The development of more sophisticated fluorescence techniques utilizing special properties of FPs to detect release events with more accuracy is currently under way.

4.4 Interaction of HIV-1 Particles with ESCRT

During HIV-1 budding, the cellular ESCRT machinery (Endosomal Sorting Complex Required for Transport; [76]), consisting of four heteromeric complexes (ESCRT-0 to -III), promotes virion release from the infected host cell [100]. This machinery is involved in a number of cellular membrane fission events, most importantly the formation and abscission of vesicles at the multivesicular body and cytokinesis. While the interactions between components of the cellular ESCRT machinery have been well characterized by numerous studies and the involvement of a subset of ESCRT proteins in HIV-1 release is firmly established [76, 100], the mechanism of action of individual ESCRT proteins in the process of HIV release and the order of events at the viral budding site are less well understood. Fluorescently labeled variants of many ESCRT proteins are available, but analysis of HIV–ESCRT interactions by live cell imaging is complicated by the fact that most components of the ESCRT machinery, in particular ESCRT-III become nonfunctional or dominant negative when overexpressed as FP-tagged derivatives [77–80, 81]. Using stably transfected cells expressing low levels of individual FP-tagged ESCRT proteins, Jouvenet and coworkers [82] investigated their arrival time at the HIV-1 and Equine infectious anemia virus (EIAV) Gag assembly sites. They observed that the ESCRT-I related protein ALIX gradually accumulated at EIAV budding sites in parallel with Gag, whereas ESCRT-III and VPS4 were recruited to HIV-1 and EIAV budding sites only after the completion of Gag assembly. In our experiments, we focused in detail on the interaction of VPS4A with nascent assembly sites (Fig. 6). In cells coexpressing eGFP-VPS4A and $\text{HIV}^{\text{mCherry}}$ fusion proteins, we visualized distinct fluorescent bursts of eGFP-VPS4A transiently recruited to nascent $\text{HIV}^{\text{mCherry}}$ particles assembling at the plasma membrane [55]. Independent of the cellular expression level, the number of colocalizing eGFP-VPS4A bursts increased with the number of $\text{HIV}^{\text{mCherry}}$ assembly sites and saturated at around 75% of colocalization. The coexpression of a late domain defective mutant of $\text{HIV}^{\text{mCherry}}$ strongly reduced the number of eGFP-VPS4A bursts to a basal level that was also observed in cells not producing HIV-1. Using moment analysis [53], a type of ICS [52, 54], the number of VPS4A subunits per bursts could be estimated from fluctuations of the eGFP fluorescence signal, yielding a range of two to five dodecamers, which assemble from cytosolic monomers.

To gain more insight into the dynamics of the VPS4A-HIV interaction with respect to the overall kinetics of HIV assembly, we determined the durations of Phase I and Phase II and scaled them to correspond to the respective averaged duration. The resulting average kinetic scheme of HIV assembly is shown in the

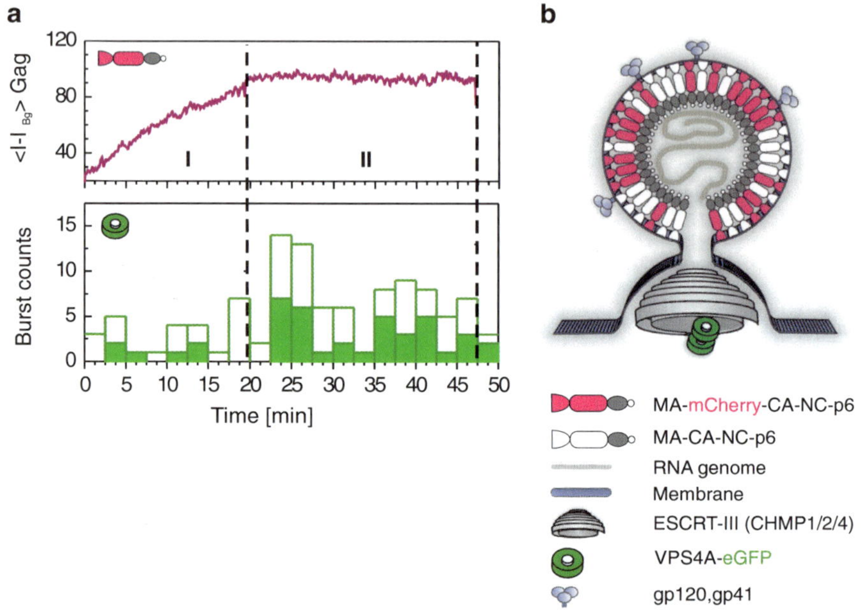

Fig. 6 Time-points of the VPS4-HIV interaction during HIV-1 budding (**a**) *Top*, mean Gag. mCherry intensity trace adapted to the average kinetic rate, phase I indicates the assembly phase, phase II the lag phase of the HIV-1 budding process. *Bottom*, histogram of all VPS4 burst positions representing the time of VPS4 activity relative to the start of HIV-1 assembly, all bursts colocalizing with Gag.mCherry (*open bars*) or single burst events (*filled bars*). Adapted from [55]. (**b**) A model of ESCRT components at the HIV-1 budding site. The immature virus shell is formed underneath the plasma membrane from Gag.mCherry and Gag(wt) proteins. A dome shaped high molecular assembly of ESCRT-III narrows the membrane neck. Several dodecamers of VPS4 transiently interact with this structure and are directly involved in the membrane scission event

upper panel of Fig. 6b. The time at which bursts were detected in the VPS4A channel were scaled onto the same kinetic scheme and shown in the lower panel of Fig. 6b. eGFP-VPS4A activity, as indicated by each fluorescent burst, mainly appeared during phase II and before the viral particles entered phase III. This suggests a more direct involvement of VPS4A in the budding process of HIV-1 [55] than previously assumed from in vitro studies of ESCRT mediated membrane scission within giant unilamellar vesicles [83, 84]. Involvement of VPS4 in the release process is also indicated by the observation that ESCRT-III components still localize to budding site in the presence of dominant negative VPS4, but fail to catalyze virus release [82]. The findings from live cell imaging are consistent with a model in which several VPS4 dodecamers transiently bind to and promote a dome-shaped assembly of ESCRT-III molecules [101] constricting the membrane neck (Fig. 6b). Promotion of dome formation as well as rapid resolution of the structure by concerted action of several VPS4 dodecamers might contribute to the process of membrane fission.

4.5 Cell-to-Cell Transmission of HIV-1

Most experiments involving HIV infection in tissue culture, including most SVT studies, have been carried out using cell-free virus. It has, however, been shown that the efficiency of virus transmission through direct contact between an infected and a noninfected T-cell is several orders of magnitude higher and is believed to be an important pathway in vivo. Several different mechanisms of directed cell-to-cell spread have been described [85, 86]. It should be noted that the transmission of HIV between T-cells through so-called "virological synapses" [87] as well as the transfer of virus from one cell to the other along filopodial structures [29] still involves virus budding and specific cytoplasmatic entry of virions; thus, the mechanisms of the actual release and entry events are presumably not fundamentally different from those studied using cell-free virus.

While it is difficult to demonstrate the occurrence of HIV cell-to-cell transmission by bulk virological or biochemical techniques, live cell imaging has the advantage that it allows the direct visualization of transfer events. Cell-to-cell transfer of HIV through virological synapses has been investigated using fluorescently labeled virus [58]; however, current studies lack the spatial and temporal sensitivity required for SVT analysis. Time-lapse fluorescence imaging of Jurkat T-cells expressing HIV^{iGFP} showed focal accumulation of Gag leading to a 1–3 μm button-shaped disk at the contact site between virus expressing and uninfected CD4+ T cells [58]. Cotransfection of Gag-iCerulean and Gag-iVenus constructs resulted in a high FRET signal at the synaptic button, confirming homo-oligomerization of Gag at these points. In 3D video microscopy using a SDCM, the complete cell-to-cell transfer could be tracked starting from Gag-iGFP movement toward the cell adhesion site and into the nascent synapse (0.10–0.25 μm/s). This was then followed by the release of HIV into and distal migration (0.12 μm/s) within the target cell. Virus transfer was coreceptor dependent as shown by inhibition by the CXCR4 antagonist AMD3100 and led to productive infection of target cells.

The cell-to-cell transmission of retrovirus along filopodial bridges has also been demonstrated by live-cell microscopy. Using fluorescently labeled murine leukemia virus (MLV), the group of Mothes demonstrated the directed "surfing" of individual particles along filopodia established from a noninfected cell toward an MLV producing cell [88]. Receptor-dependent transfer of HIV-1 along membrane nanotubes connecting infected and noninfected T-cells has also been described [89]. The importance of these processes for HIV transmission in vivo remains to be further characterized.

5 Summary and Outlook

As for many fields in the areas of biology, biophysics, biochemistry, and medicine, the development and application of FPs has a profound impact on virological research. Until recently, most of the advances in virology were derived from biochemistry and

EM experiments. Now, it is possible to perform quantitative fluorescence experiments on viruses and analyze protein interactions, kinetics, and transport processes in real time in living cells. While for the purpose of this chapter we have concentrated on a handful of experiments that demonstrate how FPs are being applied to elucidate the life cycle of HIV-1, there are many excellent investigations being performed on different viruses using FPs [17]. The methods developed and experience gained from HIV-1 imaging analyzes may serve to guide similar experiments with other viruses, provided that suitable labeling strategies can be devised.

Experiments summarized above have laid the foundation for detailed investigations of the kinetics of the processes in the life cycle of HIV-1. An obvious extension of this research is to characterize the effect of antiviral drugs on dynamic events in the life cycle of the virus. Beyond that, the continuous development of FPs with distinct physicochemical properties will expand our possibilities for investigating virus–cell interactions. Protein variants with different excitation properties, increased photostability, and higher quantum yield are being developed [90, 91]. The exchange of specific amino acids situated near the chromophore center lead to so-called photochromic FPs, which can be reversibly switched between fluorescent on and off states [92, 93]. The development of new, smaller, in vivo labeling systems in the future will be very important for virological studies where current fusion proteins are too bulky for labeling of some viral components. With added functionality, it will be possible to exploit multiple fluorescence parameters for multichannel measurements, determining particle release or monitoring the maturation process of the virion. In addition, specialized FP variants allow the monitoring of environmental conditions as pH [94], calcium [95], and redox potential [96] with high time resolution and may be employed to explore more complex processes. In summary, FPs have become a very powerful tool in virology, making it possible to visualize viruses and their interactions in real time during virtually all phases of the viral life cycle and promising even more exciting possibilities in the future.

References

1. Müller B, Daecke J, Fackler OT, Dittmar MT, Zentgraf H, Kräusslich HG (2004) Construction and characterization of a fluorescently labeled infectious human immunodeficiency virus type 1 derivative. J Virol 78:10803–10813
2. Hubner W, Chen P, Del Portillo A, Liu Y, Gordon RE, Chen BK (2007) Sequence of human immunodeficiency virus type 1 (HIV-1) Gag localization and oligomerization monitored with live confocal imaging of a replication-competent, fluorescently tagged HIV-1. J Virol 81:12596–12607
3. Campbell EM, Perez O, Melar M, Hope TJ (2007) Labeling HIV-1 virions with two fluorescent proteins allows identification of virions that have productively entered the target cell. Virology 360:286–293
4. Frankel A, Young J (1998) HIV-1: fifteen proteins and an RNA. Annu Rev Biochem 67:1–25
5. Briggs JA, Wilk T, Welker R, Krausslich HG, Fuller SD (2003) Structural organization of authentic, mature HIV-1 virions and cores. EMBO J 22:1707–1715

6. Wilk T, Gross I, Gowen BE, Rutten T, de Haas F, Welker R, Krausslich HG, Boulanger P, Fuller SD (2001) Organization of immature human immunodeficiency virus type 1. J Virol 75:759–771
7. Chertova E, Chertov O, Coren LV, Roser JD, Trubey CM, Bess JW Jr, Sowder RC 2nd, Barsov E, Hood BL, Fisher RJ, Nagashima K, Conrads TP, Veenstra TD, Lifson JD, Ott DE (2006) Proteomic and biochemical analysis of purified human immunodeficiency virus type 1 produced from infected monocyte-derived macrophages. J Virol 80:9039–9052
8. Zhu P, Chertova E, Bess J Jr, Lifson JD, Arthur LO, Liu J, Taylor KA, Roux KH (2003) Electron tomography analysis of envelope glycoprotein trimers on HIV and simian immunodeficiency virus virions. Proc Natl Acad Sci USA 100:15812–15817
9. Carlson LA, Briggs JA, Glass B, Riches JD, Simon MN, Johnson MC, Muller B, Grunewald K, Krausslich HG (2008) Three-dimensional analysis of budding sites and released virus suggests a revised model for HIV-1 morphogenesis. Cell Host Microbe 4:592–599
10. Ganser-Pornillos BK, Yeager M, Sundquist WI (2008) The structural biology of HIV assembly. Curr Opin Struct Biol 18:203–217
11. Ott DE (2008) Cellular proteins detected in HIV-1. Rev Med Virol 18:159–175
12. Mondor I, Ugolini S, Sattentau QJ (1998) Human immunodeficiency virus type 1 attachment to HeLa CD4 cells is CD4 independent and gp120 dependent and requires cell surface heparans. J Virol 72:3623–3634
13. Saphire AC, Bobardt MD, Zhang Z, David G, Gallay PA (2001) Syndecans serve as attachment receptors for human immunodeficiency virus type 1 on macrophages. J Virol 75:9187–9200
14. Alkhatib G, Berger EA (2007) HIV coreceptors: from discovery and designation to new paradigms and promise. Eur J Med Res 12:375–384
15. Pierson TC, Doms RW (2003) HIV-1 entry and its inhibition. Curr Top Microbiol Immunol 281:1–27
16. Gheysen D, Jacobs E, de Foresta F, Thiriart C, Francotte M, Thines D, De Wilde M (1989) Assembly and release of HIV-1 precursor Pr55gag virus-like particles from recombinant baculovirus-infected insect cells. Cell 59:103–112
17. Brandenburg B, Lee LY, Lakadamyali M, Rust MJ, Zhuang X, Hogle JM (2007) Imaging poliovirus entry in live cells. PLoS Biol 5(7):e183
18. Graf M, Deml L, Wagner R (2004) Codon-optimized genes that enable increased heterologous expression in mammalian cells and elicit efficient immune responses in mice after vaccination of naked DNA. Methods Mol Med 94:197–210
19. Schneider R, Campbell M, Nasioulas G, Felber BK, Pavlakis GN (1997) Inactivation of the human immunodeficiency virus type 1 inhibitory elements allows rev-independent expression of Gag and Gag/protease and particle formation. J Virol 71:4892–4903
20. Hermida-Matsumoto L, Resh MD (2000) Localization of human immunodeficiency virus type 1 Gag and Env at the plasma membrane by confocal imaging. J Virol 74:8670–8679
21. Perrin-Tricaud C, Davoust J, Jones IM (1999) Tagging the human immunodeficiency virus gag protein with green fluorescent protein. Minimal evidence for colocalisation with actin. Virology 255:20–25
22. Jin J, Sturgeon T, Weisz OA, Mothes W, Montelaro RC (2009) HIV-1 matrix dependent membrane targeting is regulated by Gag mRNA trafficking. PLoS One 4(8):e6551
23. Swanson CM, Puffer BA, Ahmad KM, Doms RW, Malim MH (2004) Retroviral mRNA nuclear export elements regulate protein function and virion assembly. EMBO J 23:2632–2640
24. Muller B, Tessmer U, Schubert U, Krausslich HG (2000) Human immunodeficiency virus type 1 Vpr protein is incorporated into the virion in significantly smaller amounts than gag and is phosphorylated in infected cells. J Virol 74:9727–9731
25. Kondo E, Mammano F, Cohen EA, Gottlinger HG (1995) The p6gag domain of human immunodeficiency virus type 1 is sufficient for the incorporation of Vpr into heterologous viral particles. J Virol 69:2759–2764
26. Lampe M, Briggs J, Endress T, Glass B, Riegelsberger S, Kräusslich H, Lamb D, Bräuchle C, Müller B (2007) Double-labelled HIV-1 particles for study of virus-cell interaction. Virology 360:92–104

27. McDonald D, Vodicka MA, Lucero G, Svitkina TM, Borisy GG, Emerman M, Hope TJ (2002) Visualization of the intracellular behavior of HIV in living cells. J Cell Biol 159:441–452
28. Albanese A, Arosio D, Terreni M, Cereseto A (2008) HIV-1 pre-integration complexes selectively target decondensed chromatin in the nuclear periphery. PLoS One 3(6):e2413
29. Sherer NM, Lehmann MJ, Jimenez-Soto LF, Horensavitz C, Pypaert M, Mothes W (2007) Retroviruses can establish filopodial bridges for efficient cell-to-cell transmission. Nat Cell Biol 9:310–315
30. Fuller SD, Wilk T, Gowen BE, Krausslich HG, Vogt VM (1997) Cryo-electron microscopy reveals ordered domains in the immature HIV-1 particle. Curr Biol 7:729–738
31. Wright ER, Schooler JB, Ding HJ, Kieffer C, Fillmore C, Sundquist WI, Jensen GJ (2007) Electron cryotomography of immature HIV-1 virions reveals the structure of the CA and SP1 Gag shells. EMBO J 26:2218–2226
32. Rudner L, Nydegger S, Coren LV, Nagashima K, Thali M, Ott DE (2005) Dynamic fluorescent imaging of human immunodeficiency virus type 1 gag in live cells by biarsenical labeling. J Virol 79:4055–4065
33. Bertrand E, Chartrand P, Schaefer M, Shenoy SM, Singer RH, Long RM (1998) Localization of ASH1 mRNA particles in living yeast. Mol Cell 2:437–445
34. Jouvenet N, Simon S, Bieniasz P (2009) Imaging the interaction of HIV-1 genomes and Gag during assembly of individual viral particles. Proc Natl Acad Sci USA 106: 19114–19119
35. Marcello A, Cinelli RA, Ferrari A, Signorelli A, Tyagi M, Pellegrini V, Beltram F, Giacca M (2001) Visualization of in vivo direct interaction between HIV-1 TAT and human cyclin T1 in specific subcellular compartments by fluorescence resonance energy transfer. J Biol Chem 276:39220–39225
36. Daelemans D, Costes SV, Cho EH, Erwin-Cohen RA, Lockett S, Pavlakis GN (2004) In vivo HIV-1 Rev multimerization in the nucleolus and cytoplasm identified by fluorescence resonance energy transfer. J Biol Chem 279:50167–50175
37. Dettenhofer M, Yu XF (1999) Highly purified human immunodeficiency virus type 1 reveals a virtual absence of Vif in virions. J Virol 73:1460–1467
38. Ivanchenko S, Godinez WJ, Lampe M, Kräusslich H-G, Eils R, Rohr K, Bräuchle C, Müller B, Lamb DC (2009) Dynamics of HIV-1 assembly and release. PLoS Pathog 5(11):e1000652
39. Jouvenet N, Bieniasz PD, Simon SM (2008) Imaging the biogenesis of individual HIV-1 virions in live cells. Nature 454:236–240
40. Jares-Erijman EA, Jovin TM (2003) FRET imaging. Nat Biotechnol 21:1387–1395
41. Derdowski A, Ding L, Spearman P (2004) A novel fluorescence resonance energy transfer assay demonstrates that the human immunodeficiency virus type 1 Pr55Gag I domain mediates Gag-Gag interactions. J Virol 78:1230–1242
42. Larson DR, Ma YM, Vogt VM, Webb WW (2003) Direct measurement of Gag-Gag interaction during retrovirus assembly with FRET and fluorescence correlation spectroscopy. J Cell Biol 162:1233–1244
43. Poole E, Strappe P, Mok HP, Hicks R, Lever AM (2005) HIV-1 Gag-RNA interaction occurs at a perinuclear/centrosomal site; analysis by confocal microscopy and FRET. Traffic 6:741–755
44. Magde D, Elson EL, Webb WW (1972) Thermodynamic Fluctuations in a Reacting System - Measurement by Fluorescence Correlation Spectroscopy. Phys Rev Lett 29:705–708
45. Magde D, Elson EL, Webb WW (1974) Fluorescence correlation spectroscopy. II. An experimental realization. Biopolymers 13:29–61
46. Elson EL, Magde D (1974) Fluorescence correlation spectroscopy. I. Conceptual basis and theory. Biopolymers 13:1–27
47. Chen Y, Müller JD, So PTC, Gratton E (1999) The photon counting histogram in fluorescence fluctuation spectroscopy. Biophys J 77:553–567
48. Kask P, Palo K, Ullmann D, Gall K (1999) Fluorescence-intensity distribution analysis and its application in biomolecular detection technology. Proc Natl Acad Sci USA 94:13756–13761

49. Qian H, Elson E (1990) Distribution of molecular aggregation by analysis of fluctuation moments. Proc Natl Acad Sci USA 87:5479–5483
50. Moore MD, Nikolaitchik OA, Chen J, Hammarskjöld ML, Rekosh D, Hu WS (2009) Probing the HIV-1 genomic RNA trafficking pathway and dimerization by genetic recombination and single virion analyses. PLoS Pathog 5(10):e1000627
51. Fogarty KH, Chen Y, Grigsby IF, Macdonald PJ, Smith EM, Johnson JL, Rawson JM, Mansky LM, Mueller JD (2011) Characterization of cytoplasmic gag-gag interactions by dual-color z-scan fluorescence fluctuation spectroscopy. Biophys J 100:1587–1595
52. Petersen NO, Hoddelius PL, Wiseman PW, Seger O, Magnusson KE (1993) Quantitation of membrane receptor distributions by image correlation spectroscopy: concept and application. Biophys J 65:1135–1146
53. Sergeev M, Costantino S, Wiseman PW (2006) Measurement of monomer-oligomer distributions via fluorescence moment image analysis. Biophys J 91:3884–3896
54. Wiseman PW, Petersen NO (1999) Image correlation spectroscopy. II. Optimization for ultrasensitive detection of preexisting platelet-derived growth factor-beta receptor oligomers on intact cells. Biophys J 76:963–977
55. Baumgärtel V, Ivanchenko S, Dupont A, Sergeev M, Wiseman PW, Krausslich HG, Brauchle C, Muller B, Lamb DC (2011) Live-cell visualization of dynamics of HIV budding site interactions with an ESCRT component. Nat Cell Biol 13(4):469–474
56. Seisenberger G, Ried MU, Endreß T, Büning H, Hallek M, Bräuchle C (2001) Real-time single molecule imaging of the infection pathway of an adeno-associated virus. Science 294:1929–1932
57. Arhel N, Genovesio A, Kim KA, Miko S, Perret E, Olivo-Marin JC, Shorte S, Charneau P (2006) Quantitative four-dimensional tracking of cytoplasmic and nuclear HIV-1 complexes. Nat Methods 3:817–824
58. Hübner W, McNerney G, Chen P, Dale B, Gordon R, Chuang F, Li X, Asmuth D, Huser T, Chen B (2009) Quantitative 3D video microscopy of HIV transfer across T cell virological synapses. Science 323:1743–1747
59. Miyauchi K, Kim Y, Latinovic O, Morozov V, Melikyan GB (2009) HIV enters cells via endocytosis and dynamin-dependent fusion with endosomes. Cell 137:433–444
60. Godinez WJ, Lampe M, Worz S, Muller B, Eils R, Rohr K (2009) Deterministic and probabilistic approaches for tracking virus particles in time-lapse fluorescence microscopy image sequences. Med Image Anal 13:325–342
61. Brandenburg B, Zhuang X (2007) Virus trafficking - learning from single-virus tracking. Nat Rev Microbiol 5:197–208
62. Marsh M, Helenius A (2006) Virus entry: open sesame. Cell 124:729–740
63. Rust MJ, Lakadamyali M, Zhang F, Zhuang X (2004) Assembly of endocytic machinery around individual influenza viruses during viral entry. Nat Struct Mol Biol 11:567–573
64. Endress T, Lampe M, Briggs J, Kräusslich H, Bräuchle C, Müller B, Lamb D (2008) HIV-1-cellular interactions analyzed by single virus tracing. Eur Biophys J 37:1291–1301
65. Jha NK, Latinovic O, Martin E, Novitskiy G, Marin M, Miyauchi K, Naughton J, Young JA, Melikyan GB (2011) Imaging single retrovirus entry through alternative receptor isoforms and intermediates of virus-endosome fusion. PLoS Pathog 7(1):e1001260
66. Platt EJ, Kozak SL, Durnin JP, Hope TJ, Kabat D (2010) Rapid dissociation of HIV-1 from cultured cells severely limits infectivity assays, causes the inactivation ascribed to entry inhibitors, and masks the inherently high level of infectivity of virions. J Virol 84:3106–3110
67. Daecke J, Fackler OT, Dittmar MT, Krausslich HG (2005) Involvement of clathrin-mediated endocytosis in human immunodeficiency virus type 1 entry. J Virol 79:1581–1594
68. Markosyan RM, Cohen FS, Melikyan GB (2005) Time-resolved imaging of HIV-1 env-mediated lipid and content mixing between a single virion and cell membrane. Mol Biol Cell 16:5502–5513
69. Koch P, Lampe M, Godinez WJ, Muller B, Rohr K, Krausslich HG, Lehmann MJ (2009) Visualizing fusion of pseudotyped HIV-1 particles in real time by live cell microscopy. Retrovirology 6:84

70. Pelchen-Matthews A, Kramer B, Marsh M (2003) Infectious HIV-1 assembles in late endosomes in primary macrophages. J Cell Biol 162:443–455
71. Bennett AE, Narayan K, Shi D, Hartnell LM, Gousset K, He H, Lowekamp BC, Yoo TS, Bliss D, Freed EO, Subramaniam S (2009) Ion-abrasion scanning electron microscopy reveals surface-connected tubular conduits in HIV-infected macrophages. PLoS Pathog 5(9):e1000591
72. Deneka M, Pelchen-Matthews A, Byland R, Ruiz-Mateos E, Marsh M (2007) In macrophages, HIV-1 assembles into an intracellular plasma membrane domain containing the tetraspanins CD81, CD9, and CD53. J Cell Biol 177:329–341
73. Welsch S, Keppler OT, Habermann A, Allespach I, Krijnse-Locker J, Krausslich HG (2007) HIV-1 buds predominantly at the plasma membrane of primary human macrophages. PLoS Pathog 3(3):e36
74. Jouvenet N, Neil SJ, Bess C, Johnson MC, Virgen CA, Simon SM, Bieniasz PD (2006) Plasma membrane is the site of productive HIV-1 particle assembly. PLoS Biol 4(12):e435
75. Nienhaus GU, Nienhaus K, Holzle A, Ivanchenko S, Renzi F, Oswald F, Wolff M, Schmitt F, Rocker C, Vallone B, Weidemann W, Heilker R, Nar H, Wiedenmann J (2006) Photoconvertible fluorescent protein EosFP: biophysical properties and cell biology applications. Photochem Photobiol 82:351–358
76. Hurley JH, Hanson PI (2010) Membrane budding and scission by the ESCRT machinery: it's all in the neck. Nat Rev Mol Cell Biol 11:556–566
77. Garrus JE, von Schwedler UK, Pornillos OW, Morham SG, Zavitz KH, Wang HE, Wettstein DA, Stray KM, Cote M, Rich RL, Myszka DG, Sundquist WI (2001) Tsg101 and the vacuolar protein sorting pathway are essential for HIV-1 budding. Cell 107:55–65
78. Martin-Serrano J, Yarovoy A, Perez-Caballero D, Bieniasz PD (2003) Divergent retroviral late-budding domains recruit vacuolar protein sorting factors by using alternative adaptor proteins. Proc Natl Acad Sci USA 100:12414–12419
79. Strack B, Calistri A, Craig S, Popova E, Gottlinger HG (2003) AIP1/ALIX is a binding partner for HIV-1 p6 and EIAV p9 functioning in virus budding. Cell 114:689–699
80. von Schwedler UK, Stuchell M, Muller B, Ward DM, Chung HY, Morita E, Wang HE, Davis T, He GP, Cimbora DM, Scott A, Krausslich HG, Kaplan J, Morham SG, Sundquist WI (2003) The protein network of HIV budding. Cell 114:701–713
81. Shim S, Merrill SA, Hanson PI (2008) Novel interactions of ESCRT-III with LIP5 and VPS4 and their implications for ESCRT-III disassembly. Mol Biol Cell 19:2661–2672
82. Jouvenet N, Zhadina M, Bieniasz PD, Simon SM (2011) Dynamics of ESCRT protein recruitment during retroviral assembly. Nat Cell Biol 13(4):394–401
83. Wollert T, Hurley JH (2010) Molecular mechanism of multivesicular body biogenesis by ESCRT complexes. Nature 464:864–869
84. Wollert T, Wunder C, Lippincott-Schwartz J, Hurley JH (2009) Membrane scission by the ESCRT-III complex. Nature 458:172–177
85. Sattentau Q (2008) Avoiding the void: cell-to-cell spread of human viruses. Nat Rev Microbiol 6:815–826
86. Mothes W, Sherer NM, Jin J, Zhong P (2010) Virus cell-to-cell transmission. J Virol 84:8360–8368
87. Piguet V, Sattentau Q (2004) Dangerous liaisons at the virological synapse. J Clin Invest 114:605–610
88. Sherer N, Jin J, Mothes W (2010) Directional spread of surface-associated retroviruses regulated by differential virus-cell interactions. J Virol 84:3248–3258
89. Sowinski S, Jolly C, Berninghausen O, Purbhoo MA, Chauveau A, Kohler K, Oddos S, Eissmann P, Brodsky FM, Hopkins C, Onfelt B, Sattentau Q, Davis DM (2008) Membrane nanotubes physically connect T cells over long distances presenting a novel route for HIV-1 transmission. Nat Cell Biol 10:211–219
90. Shaner N, Campbell R, Steinbach P, Giepmans B, Palmer A, Tsien R (2004) Improved monomeric red, orange and yellow fluorescent proteins derived from Discosoma sp. red fluorescent protein. Nat Biotechnol 22:1567–1572

91. Shaner NC, Lin MZ, McKeown MR, Steinbach PA, Hazelwood KL, Davidson MW, Tsien RY (2008) Improving the photostability of bright monomeric orange and red fluorescent proteins. Nat Methods 5:545–551
92. Andresen M, Stiel A, Fölling J, Wenzel D, Schönle A, Egner A, Eggeling C, Hell S, Jakobs S (2008) Photoswitchable fluorescent proteins enable monochromatic multilabel imaging and dual color fluorescence nanoscopy. Nat Biotechnol 26:1035–1040
93. Vogt A, D'Angelo C, Oswald F, Denzel A, Mazel C, Matz M, Ivanchenko S, Nienhaus G, Wiedenmann J (2008) A green fluorescent protein with photoswitchable emission from the deep sea. PLoS One 3(11):e3766
94. Miesenbock G, De Angelis DA, Rothman JE (1998) Visualizing secretion and synaptic transmission with pH-sensitive green fluorescent proteins. Nature 394:192–195
95. Nagai T, Sawano A, Park ES, Miyawaki A (2001) Circularly permuted green fluorescent proteins engineered to sense Ca2+. Proc Natl Acad Sci USA 98:3197–3202
96. Cannon MB, Remington SJ (2008) Redox-sensitive green fluorescent protein: probes for dynamic intracellular redox responses. A review. Methods Mol Biol 476:51–65
97. Stein BS, Gowda SD, Lifson JD, Penhallow RC, Bensch KG, Engleman EG (1987) pH-independent HIV entry into CD4-positive T cells via virus envelope fusion to the plasma membrane. Cell 49:659–668
98. McClure MO, Marsh M, Weiss RA (1988) Human immunodeficiency virus infection of CD4-bearing cells occurs by a pH-independent mechanism. The EMBO journal 7:513–518
99. Maddon PJ, McDougal JS, Clapham PR, Dalgleish AG, Jamal S, Weiss RA, Axel R (1988) HIV infection does not require endocytosis of its receptor, CD4. Cell 54:865–874
100. Bieniasz PD (2009) The cell biology of HIV-1 virion genesis. Cell Host Microbe 5:550–558
101. Fabrikant G, Lata S, Riches JD, Briggs JA, Weissenhorn W, Kozlov MM (2009) Computational model of membrane fission catalyzed by ESCRT-III. PLoS Comput Biol 5(11):e1000575

Index

G. Jung (ed.), *Fluorescent Proteins I*, Springer Ser Fluoresc (2012) 12: 279–282,
DOI 10.1007/4243_2011,